LABYRINTH AND PIANO KEY WEIRS –
PKW 2011

PROCEEDINGS OF THE INTERNATIONAL CONFERENCE ON LABYRINTH AND PIANO KEY WEIRS (PKW 2011), LIÈGE, BELGIUM, 9–11 FEBRUARY 2011

Labyrinth and Piano Key Weirs – PKW 2011

Editors

Sébastien Erpicum
Université de Liège, Belgium

Frédéric Laugier
Electricité de France, France

Jean-Louis Boillat
Ecole Polytechnique Fédérale de Lausanne, Switzerland

Michel Pirotton
Université de Liège, Belgium

Bernard Reverchon
Electricité de France, France

Anton Schleiss
Ecole Polytechnique Fédérale de Lausanne, Switzerland

CRC Press
Taylor & Francis Group
Boca Raton London New York Leiden

CRC Press is an imprint of the
Taylor & Francis Group, an **informa** business

A BALKEMA BOOK

First issued in paperback 2017

CRC Press/Balkema is an imprint of the Taylor & Francis Group, an informa business

© 2011 Taylor & Francis Group, London, UK

Published by: CRC Press/Balkema
                P.O. Box 447, 2300 AK Leiden, The Netherlands
                e-mail: Pub.NL@taylorandfrancis.com
                www.crcpress.com – www.taylorandfrancis.co.uk – www.balkema.nl

ISBN 13: 978-1-138-11538-5 (pbk)
ISBN 13: 978-0-415-68282-4 (hbk)

Labyrinth and Piano Key Weirs – PKW 2011 – Erpicum et al. (eds)
© 2011 Taylor & Francis Group, London, ISBN 978-0-415-68282-4

Table of Contents

Preface   IX

Acknowledgements   XI

Organization   XIII

Organizing institutions and sponsors   XV

Keynote lectures

From Labyrinth to Piano Key Weirs – A historical review   3
A.J. Schleiss

General comments on Labyrinths and Piano Key Weirs: The past and present   17
F. Lempérière, J.-P. Vigny & A. Ouamane

Hydraulic characteristics of labyrinth weirs   25
B.M. Crookston & B.P. Tullis

Physical modeling – Hydraulic capacity

PKWeir and flap gate spillway for the Gage II Dam   35
V. Dugué, F. Hachem, J.-L. Boillat, V. Nagel, J.-P. Roca & F. Laugier

Piano Key Weir design study at Raviege dam   43
S. Erpicum, V. Nagel & F. Laugier

Nine years of study of the Piano Key Weir in the university laboratory of Biskra
"lessons and reflections"   51
A. Ouamane

Influence of the Piano Key Weir height on its discharge capacity   59
O. Machiels, S. Erpicum, P. Archambeau, B. Dewals & M. Pirotton

Study of optimization of labyrinth weir   67
M. Ben Saïd & A. Ouamane

Influence of Piano Key Weir geometry on discharge   75
R.M. Anderson & B.P. Tullis

Study of piano-key morning glory to increase the spillway capacity of the Bage dam   81
G.M. Cicero, M. Barcouda, M. Luck & E. Vettori

Physical modeling – Downstream fittings

Contribution to the study of the Piano Key Weirs submerged by the downstream level   89
F. Belaabed & A. Ouamane

Flow properties and residual energy downstream of labyrinth weirs   97
R. Lopes, J. Matos & J.F. Melo

Energy dissipation on a stepped spillway downstream of a
Piano Key Weir – Experimental study 105
S. Erpicum, O. Machiels, P. Archambeau, B. Dewals, M. Pirotton & C. Daux

Coupled spillway devices and energy dissipation system at St-Marc Dam (France) 113
M. Leite Ribeiro, J.-L. Boillat, A.J. Schleiss & F. Laugier

Energy dissipation downstream of Piano Key Weirs – Case study of
Gloriettes Dam (France) 123
M. Bieri, M. Federspiel, J.-L. Boillat, B. Houdant, L. Faramond & F. Delorme

Numerical modeling

A sensitivity analysis of Piano Key Weirs geometrical parameters based on
3D numerical modeling 133
J. Pralong, F. Montarros, B. Blancher & F. Laugier

Hydraulic comparison between Piano Key Weirs and labyrinth spillways 141
B. Blancher, F. Montarros & F. Laugier

1D numerical modeling of the flow over a Piano Key Weir 151
S. Erpicum, O. Machiels, P. Archambeau, B. Dewals & M. Pirotton

Influence of structural thickness of sidewalls on PKW spillway discharge capacity 159
F. Laugier, J. Pralong & B. Blancher

Experimental study of side and scale effects on hydraulic performances
of a Piano Key Weir 167
G.M. Cicero, J.M. Menon, M. Luck & T. Pinchard

Hydraulic design

Study of optimization of the Piano Key Weir 175
A. Noui & A. Ouamane

Experimental parametric study for hydraulic design of PKWs 183
M. Leite Ribeiro, J.-L. Boillat, A.J. Schleiss, O. Le Doucen & F. Laugier

Main results of the P.K weir model tests in Vietnam (2004 to 2010) 191
M. Ho Ta Khanh, T. Chi Hien & N. Thanh Hai

Piano Key Weir preliminary design method – Application to a new dam project 199
O. Machiels, S. Erpicum, P. Archambeau, B. Dewals & M. Pirotton

Method to design a PK-Weir with a shape and hydraulic performances 207
G.M. Cicero

Planned and existing projects

Lessons learnt from design and construction of EDF first Piano Key Weirs 215
J. Vermeulen, F. Laugier, L. Faramond & C. Gille

P.K weirs under design and construction in Vietnam (2010) 225
M. Ho Ta Khanh, D. Sy Quat & D. Xuan Thuy

Spillway capacity upgrade at Malarce dam: Design of an additional
Piano Key Weir spillway 233
T. Pinchard, J.-M. Boutet & G.M. Cicero

Rehabilitation of Sawara Kuddu Hydroelectric Project – Model studies of
Piano Key Weir in India   241
G. Das Singhal & N. Sharma

A dam equipped with labyrinth spillway in the Sultanate of Oman   251
L. Bazerque, P. Agresti, C. Guilbaud, S. Al Harty & T. Strobl

Labyrinth fusegate applications on free overflow spillways – Overview of
recent projects   261
M. Le Blanc, U. Spinazzola & H. Kocahan

Nomenclature, data base, future developments

A naming convention for the Piano Key Weirs geometrical parameters   271
J. Pralong, J. Vermeulen, B. Blancher, F. Laugier, S. Erpicum, O. Machiels,
M. Pirotton, J.-L. Boillat, M. Leite Ribeiro & A.J. Schleiss

Creation of a PKW Database – Discussion   279
J.-L. Boillat, M. Leite Ribeiro, J. Pralong, S. Erpicum & P. Archambeau

Development of a new concept of Piano Key Weir spillway to increase low head
hydraulic efficiency: Fractal PKW   281
F. Laugier, J. Pralong, B. Blancher & F. Montarros

General comments on labyrinths and Piano Key Weirs – The future   289
F. Lempérière & J.-P. Vigny

Research axes and conclusions   295
S. Erpicum & J.-L. Boillat

Author index   297

Labyrinth and Piano Key Weirs – PKW 2011 – Erpicum et al. (eds)
© 2011 Taylor & Francis Group, London, ISBN 978-0-415-68282-4

Preface

Labyrinth spillways are almost as old as dam engineering. They have been installed on numerous hydraulic structures thanks to their simple structural concept and their increased specific discharge capacity. Moreover, as free flow spillways, their operation is quite reliable compared to gated ones. For instance, labyrinth weir on Beni Badhel dam (Algeria), which was built before the Second World War, has an incredibly high hydraulic performance with a discharge capacity 12 times higher than standard linear ogee crest spillway. More recently, the labyrinth weir built on Ute dam (USA) is a very large structure able to discharge a design flood of $15,600\,\mathrm{m^3/s}$ over a width of 250 m. In addition, innovative fuse plug spillways are often based on labyrinth shaped units.

Although labyrinth spillways appear as a very good technico-economical compromise, only 0.1% of large dams are equipped with such weirs. The main reason for this small rate is that traditional labyrinth weir can generally not be installed on top of concrete gravity dams because they require a large foundation surface. In 2003, Lempérière and Ouamane proposed an improved concept of traditional labyrinth spillway, with alveoli developed in overhangs from a reduced support area. Due to a plan view looking like piano keys, it was consequently called Piano Key Weir (PKW). The main advantage of this work is that it can be installed at the top of main existing concrete dams.

More recently, extreme flood issues have become more and more acute due to both increasing society demand for safety and revised hydrological calculations. The requirement for performing and reliable spillways has consequently become higher, leading to a rehabilitation need of numerous existing dams. In this context, the PKW appears as an interesting solution to increase the hydraulic capacity of spillway devices.

This innovative concept was implemented in 2006 with the construction of the first PKW on Goulours dam (France) owned by Electricité de France (EDF). Following this realization, four other PKWs were built in France between 2007 and 2010 and several other projects are under design or construction in France and other countries (India, Vietnam, Burkina Faso…).

These projects show that dam engineering is currently experiencing a strong revival of labyrinth oriented weirs. This was the motivation to organize a workshop dedicated to this specific topic. The workshop was co-organised by University of Liège, Ecole Polytechnique Fédérale de Lausanne and Electricité de France, which all together have played a significant role in recent PKW development. The conference took place at Liège University in Belgium, on 9–10 February 2011. The 2 days long workshop was followed by a technical tour in Limoges (France) where two PKWs were recently built on Saint Marc and l'Etroit dams (EDF). The workshop addressed several themes such as physical and numerical modeling. Actual design issues were discussed and a couple of planned and existing projects were presented. Finally, the development of a Database collecting all available hydraulic measurements over PKW's was discussed and is undergoing.

This book gathers the Proceedings of this first Workshop on Labyrinth and Piano Key Weirs and can be considered as Reference document on the subject.

Sébastien Erpicum, HACH – ULg
Frédéric Laugier, EDF – CIH
Jean-Louis Boillat, LCH – EPFL

Acknowledgements

The organization of a conference is only possible with motivated contributors. We are therefore grateful to practitioners and researchers who allowed taking up the challenge of proposing a workshop on the very specialized domain of Labyrinth and Piano Key Weirs. In order to warrant the high technical and scientific level of the contributions, a serious reviewing process was carried out by the International Scientific Committee. Its members merit our gratitude for the serious of their work and the respect of very short deadlines.

The reception at the University of Liège was quite simply perfect. The welcome of our Belgian colleagues contributed to develop a warm atmosphere which revealed very profitable for open discussions during and outside the meeting sessions. For this reason and for the interesting visit of the Laboratory of Engineering Hydraulics – HACH, we are indebted to the Direction and to the local staff of this institution.

The success of the workshop was also dependent of the support of professional associations of Belgium, France and Switzerland and on their technical and scientific journals who helped to announce the manifestation. A particular thank is devoted to Mrs. Alison Bartle of the International Journal of Hydropower & Dams, who fully covered the conference.

The technical tour near Limoges allowed visiting the Saint-Marc and l'Etroit dams owned by EDF, where PKW spillways were recently built. The tour was guided by competent persons in charge of management and safety of these hydraulic schemes. All our thanks go to EDF-UP Centre representatives for their pleasant availability.

The involvement of three Co-organizing institutions, University of Liège, Ecole Polytechnique Fédérale de Lausanne and Electricité de France was fundamental for the preparation of the workshop and for the edition of the proceedings. Concerning these activities, the organizers want to express their recognition to the publisher for his guidance, to Olivier Machiels and Pierre Archambeau from HACH-ULg, Julien Pralong and Julien Vermeulen from EDF-CIH, Marcelo Leite Ribeiro and Martin Bieri from LCH-EPFL for their helpful contribution.

The Scientific Secretariat of PKW 2011

Labyrinth and Piano Key Weirs – PKW 2011 – Erpicum et al. (eds)
© 2011 Taylor & Francis Group, London, ISBN 978-0-415-68282-4

Organization

MEMBERS OF THE INTERNATIONAL SCIENTIFIC COMMITTEE

Boillat, J.-L.	Ecole Polytechnique Fédérale de Lausanne, Switzerland
Cicero, G.M.	National Hydraulic and Environmental Laboratory, France
Delorme, F.	Electricité de France, France
Dewals, B.	University of Liège, Belgium
Erpicum, S.	University of Liège, Belgium
Falvey, H.T.	American Society of Civil Engineers, USA
Ho Ta Khanh, M.	VNCOLD & CFRB, France
Laugier, F.	Electricité de France, France
Lempérière, F.	Hydrocoop, France
Leite Ribeiro, M.	Ecole Polytechnique Fédérale de Lausanne, Switzerland
Matos, J.	Technical University of Lisbon, Portugal
Ouamane, A.	University Mohamed Kheider, Biskra, Algeria
Pirotton, M.	University of Liège, Belgium
Schleiss, A.J.	Ecole Polytechnique Fédérale de Lausanne, Switzerland
Sharma, N.	Indian Institute of Technology, Roorkee, India
Taquet, B.	Electricité de France, France
Tullis, B.	Utah State University, USA
Vigny, J.-P.	Hydrocoop, France

MEMBERS OF THE ORGANIZING COMMITTEE

Erpicum, S.	University of Liège, Belgium
Laugier, F.	Electricité de France, France
Boillat, J.-L.	Ecole Polytechnique Fédérale de Lausanne, Switzerland
Vermeulen, J.	Electricité de France, France
Pirotton, M.	University of Liège, Belgium
Archambeau, P.	University of Liège, Belgium

Labyrinth and Piano Key Weirs – PKW 2011 – Erpicum et al. (eds)
© 2011 Taylor & Francis Group, London, ISBN 978-0-415-68282-4

Organizing institutions and sponsors

Labyrinth and Piano Key Weirs 2011 was supported by the following institutions:

Laboratoire d'Hydraulique des Constructions – HACH Université de Liège

Laboratoire de Constructions Hydrauliques Ecole Polytechnique Fédérale de Lausanne

Centre d'Ingénierie Hydraulique Electricité de France

Journal on Hydropower & Dams

Société Hydrotechnique de France

Comité français des barrages et réservoirs

Swiss Committee on Dams

Schweizerischer Wasserwirtschaftsverband
Association suisse pour l'aménagement des eaux
Associazione svizzera di economia delle acque

Schweizerischer Wasserwirtschaftsverband

Keynote lectures

Labyrinth and Piano Key Weirs – PKW 2011 – Erpicum et al. (eds)
© *2011 Taylor & Francis Group, London, ISBN 978-0-415-68282-4*

From Labyrinth to Piano Key Weirs – A historical review

A.J. Schleiss
Director of Laboratory of Hydraulic Constructions (LCH), Ecole Polytechnique
Fédérale de Lausanne (EPFL), Switzerland

ABSTRACT: Free crest spillways are hydraulically efficient and safe in operation. Since their discharge capacity is directly proportional to the crest length several types have been developed with the purpose to increase the length of the latter. Among these types traditional labyrinth weir spillways have been studied and used for a long time. Their hydraulic performance and the effect of the involved geometrical parameters are well known. Nevertheless, their design still has to be based on experimentally derived and generalized performance curves.

The recently introduced Piano Key weirs present clear advantages regarding hydraulic performance and construction costs compared to classical labyrinth weirs. Especially its small footprint makes the PK weir an efficient and cost effective solution for the increase of the flood releasing capacity at existing concrete gravity dams. Until today only preliminary design procedures are available which cannot yet be generalized. The still ongoing research on this complex hydraulic structure is a challenge for many scientists all over the world. Despite of this, several prototypes have been installed successfully over the last years on existing dams which enhance efficiently the flood release capacity.

1   INTRODUCTION

Spillways play a major role in ensuring the flood safety of dams. Insufficient spillway capacity has been the cause of one-third of all dam failures.

The discharge capacity of free crest or ungated overfall spillways is directly proportional to the length of the ogee crest or the weir for a given upstream head. Its length can be increased by using curved, ondulated or corrugated weirs instead of straight linear weirs. Consequently the discharge for a given head also increases. Nevertheless, the channel or chute downstream of the weir should also have a sufficient evacuation capacity.

With the motivation of maximizing the crest length, the following crest geometries of weirs have so far been developed (Figure 1):

a) Duckbill spillway or bathtube spillway (in case of parallel side walls)
b) Fan spillway
c) Type Y spillway
d) Daisy-shape (marguerite), morning glory spillway

In a further step, the labyrinth spillways (Figure 2) were developed starting in the thirties of the last century. After 2000, the Piano Key weir (PK weir) was introduced as an evolution of the traditional labyrinth weirs. The paper presents a short historical review on the development and design of labyrinth and PK weirs.

2   LABYRINTH WEIRS

2.1   *Use and types of labyrinth weirs*

Labyrinth weirs can pass large flows at comparatively low heads. They are primarily used as spillways for dams where the spillway width is restricted or where the flood surcharge space is

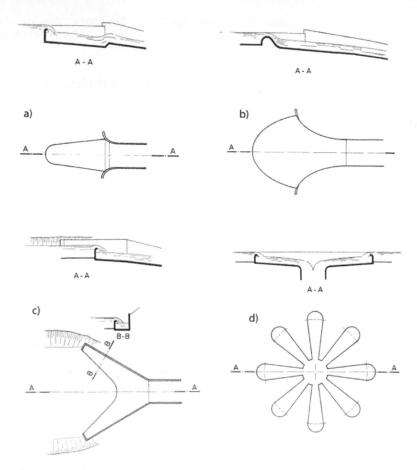

Figure 1. Curved or undulated weirs (ICOLD, 1994): a) Duckbill spillways, b) Fan spillways, c) Type Y spillways, d) Daisy-shape (marguerite), morning glory spillway.

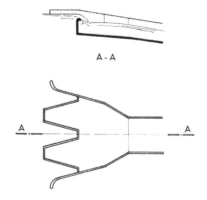

Figure 2. Layout and longitudinal section of a labyrinth weir (ICOLD, 1994).

limited (Lux & Hinchliff, 1985). Since labyrinth weirs are relatively low cost spillways compared to gated spillways, they are often used in conjunction with raising of dams for increased storage volume. Another reason for their use is often the upgrading of the flood discharge capacity at existing dams, especially at concrete gravity dams. Labyrinth weirs are also used as protection structures for canals and run-of-river hydroelectric plants (Figure 3).

Figure 3.　Labyrinth weir at Ohau river in New Zealand at the connecting canal between Ohau B and Ohau C hydroelectric power plant.

Figure 4.　Triangular labyrinth weir of the hydroelectric power plant Ohau C in New Zealand.

Furthermore, Falvey (2003) cites the use of labyrinth weirs as energy dissipaters, applied to control water quality by aering or de-aerating the flow. An overview of the characteristics of existing labyrinth installations is given by Lux & Hinchliff (1985) and Pinto Magalhães (1985).

The distinguishing characteristic of labyrinth spillways is that the plan shape is not linear but varies using a repeating planform as

- U shape (eventually rectangular),
- V or triangular shape (Figure 4) and
- trapezoidal shape (Figure 5).

5

Figure 5.   Trapezoidal labyrinth weir of Cimia dam in Italy.

Depending on the application, the crest of the weir may have different shapes as (Falvey, 2003)

- sharp or narrow crest,
- flat crest,
- quarter-round crest,
- half-round crest and
- ogee crest.

The labyrinth planform can follow a straight axis (normal case) or a curved axis as shown in Figure 6.

2.2   Labyrinth flow description

As mentioned in the introduction, the discharge passing over a labyrinth weir should increase directly proportional with crest length. However, this is only the case for labyrinth spillways operating under low heads. Qualitatively, as the upstream head increases, the flow pattern sequently passes through four basic phases namely fully aerated, partially aerated, transitional and suppressed (Lux & Hinchliff, 1985).

In the *fully aerated phase*, the flow falls freely over the entire length of the labyrinth crest. For this flow phase the thickness of the nappe and depth of tailwater has no influence on the discharge capacity of the labyrinth spillway, which has the same behavior as linear weir with the same vertical cross section.

As head increases, the flow becomes *partially aerated* due to convergence of opposing nappes and higher tailwater depths. The aeration becomes difficult because of the onset of nappe interference and results in a lowering of the discharge coefficient. A stable air pocket is formed along each side wall and downstream apex of the labyrinth.

By further increasing of upstream head and tailwater depth, the nappe becomes suppressed at various locations. The stable air pocket breaks up into smaller pockets that intermittently move upstream along the side walls causing instability in the nappe. This condition is the beginning of the *transitional phase*, which is difficult to observe precisely in the laboratory. Nevertheless, the

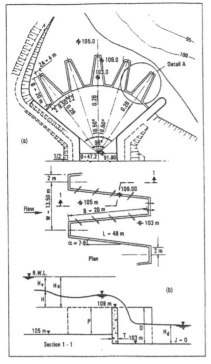

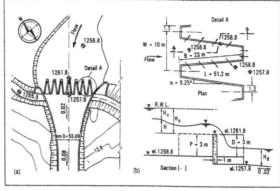

Figure 6. Curved and straight planform axis of Kizilcapinar (left) and Sarioglan (right) labyrinth spillway in Turkey (from Yildiz & Üzücek, 1996).

transitional region can be easily identified as a slope change in the discharge coefficient curve (Figure 9).

Finally, if the flow forms a solid and non-aerated nappe, it is in the *suppressed phase* and no air is drawn under the nappe. Complete submergence will occur if the head above the crest is greater than the height of labyrinth weir. Its efficiency decreases rapidly approaching that of a linear crest with a length equal to the canal or chute width but with a rather low discharge coefficient. This phase should obviously be avoided for the design flood of the spillway, since the upstream head would increase more rapidly with increasing discharge.

2.3 *Determination of labyrinth discharge capacity*

Flow over labyrinth weirs is very complex as it is three-dimensional and influenced by many parameters. Therefore, a large number of laboratory tests are needed to determine empirically the influence of the various parameters on the discharge capacity.

The first reported study of labyrinth weirs was conducted by Gentilini (1941) in Italy. In a laboratory flume several configurations of corrugated, triangular sharp crested weirs were tested and compared with linear oblique or perpendicular weirs (Figure 7). He compared the ratio of the Bazin/Rehbock discharge coefficient for linear weirs and those configurations with the ratio head over weir height (Figure 8).

It lasted almost 30 years before the first study was started by Taylor (1968) with the purpose to produce design-oriented labyrinth weir data (Hay and Taylor, 1970).

Darvas (1971) utilized the results from hydraulic models studies (Avon and Woronora Dam) to expand the theory and developed a family of curves to evaluate spillway performance.

Then the USBR performed also some labyrinth flume studies because the spillway design dimensions of Ute Dam exceeded the range of application of Taylor's study (Houston, 1982). It has to be mentioned that Taylor's analysis used piezometric head over the weir h rather than the total head H_t

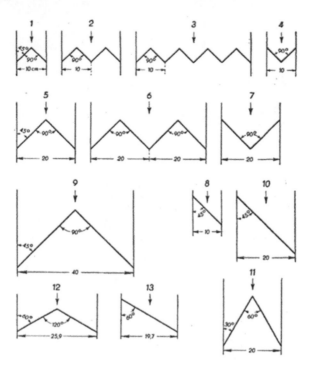

Figure 7. Oblique and corrugated triangular weir configurations tested by Gentilini (1941).

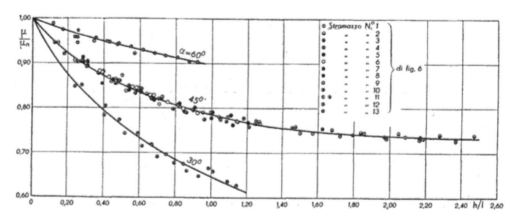

Figure 8. Discharge coefficient ratio μ/μ_n as a function of total head h over weir height l ($\mu = q/\sqrt{2g}h^{3/2}$; μ_n for linear weirs according Bazin/Rehbock) (Gentilini, 1941). Tested configurations No 1 to 13 according Figure 7. Original notations have been conserved.

(or H_o) as the USBR tests. If comparing test results from different sources, the data has to be converted since the velocity head in the flume cannot be neglected ($H_t = h + V^2/2g$). An overview of the tests performed by USBR is given in Lux & Hinchliff (1985). The derived discharge coefficients were defined according to

$$Q = C_w \left(\frac{\dfrac{W}{P}}{\dfrac{W}{P} + k} \right) W H_0 \sqrt{g H_0} \tag{1}$$

8

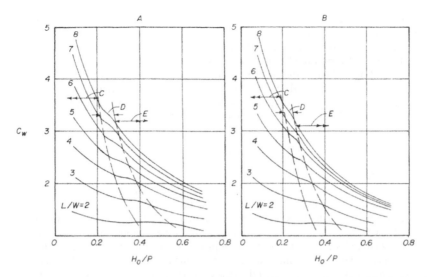

Figure 9. Design curves for quarter-round-crested labyrinth weirs: (A) Triangular labyrinth weir (left); (B) Trapezoidal labyrinth weir (right); (C) Aerated nappe region; (D) Transitional and (E) Suppressed nappe region (from Lux & Hinchliff, 1985).

where Q is the discharge per cycle, W is the cycle width, P is the upstream wall height and H_o is the total upstream head. The fitted parameter k is a constant having values of 0.18 and 0.10 for triangular and trapezoidal ($A/W = 0.0765$) planforms respectively. The crest length L per cycle width W becomes

$$L = 4A + 2B \qquad (2)$$

with apex half length A and wall length B. Design curves for quarter-round crests labyrinth weirs were developed (Figure 9). The discharge coefficient C_w uses the subscript w since it is based on the vertical aspect ratio correlations. The total discharge of the labyrinth weir Q_t can be calculated by multiplying Q by the number of cycles n. Alternatively the total spillway width W_t instead of the cycle width W can be used in Eq. (1).

Pinto Magalhães (1985) modified Darvas (1971) design chart by using a dimensionless discharge coefficient and new experimental results. The latter confirmed that the measured discharges were systematically lower that those indicated by Darvas (1971), in particular for high values of H_o/P.

Based on extensive experimental work at Utah Water Research Laboratory (UWRL), Tullis et al. (1995) developed a simplified design method using the standard weir equation:

$$Q = \frac{2}{3} C_d L_c \sqrt{2g} H_T^{3/2} \qquad (3)$$

where Q is the discharge of a labyrinth weir, C_d is a dimensional discharge coefficient and H_T the total upstream head. They replaced the total weir length in Eq. (3) with an effective weir length L_e. This is a more physical based approach to take into account for apex influences on discharge efficiency than the "black box mixing" of all influences in the discharge coefficient. In his detailed compilation and discussion of the published information on labyrinth weirs, Falvey (2003) favored also this design method. Tullis et al. (2005) conducted further tests to optimize the performance of low head 7° and 8° labyrinth weirs.

The effect of tailwater submergence on a dimensionless submerged head discharge relationship was studied by Tullis et al. (2007). Lopez et al. (2008) evaluated characteristics tailwater depth and energy dissipation downstream of a labyrinth weir. More recently Khode et al. (2010) performed experiments on wall angles of 21°, 26° and 30° in order to validate the linear interpolation of the C_d curves for 25° and 35° by Tullis et al. (1995).

Crookston (2010) recently tested 32 new hydraulic labyrinth weir models at UWRL. The discharge coefficient data obtained for quarter-round labyrinth weirs with side wall angles $6° \leq \alpha \leq 35°$ are compared with those of Tullis et al. (1995) and Willmore (2004). The improved C_d data is recommended for use in determining labyrinth weir discharge relationships (Crookston & Tullis, 2011) (see these Proceedings). Finally, the cycle efficiency is introduced as new parameter which allows a more rapid evaluation of relative discharge efficiencies of various labyrinth weir geometries. It has to be noted that the discharge coefficient was calculated as function of the weir centerline length and not of the effective length as used in Tullis et al. (1995).

2.4 *Hydraulic and structural design consideration*

Due to its geometry and large discharge capacity, a labyrinth spillway is particularly sensitive to *reservoir approach flow conditions* (Lux & Hinchliff, 1985). The direction of the approach flow should be perpendicular to the alignment axis for the labyrinth.

The design of the upstream apexes aims an optimum compromise between using a larger trapezoidal apex for ease of construction as well as to minimize nappe interference and an almost triangular planform to limit entrance loss. The approach condition of the end cycle towards the lateral canal wall is also very important. Most efficient is the use of a vertical wall laid out on a radius adjacent to each end cycle of the labyrinth (Houston, 1982) (see Figure 6 right). If the labyrinth weir is placed as far upstream into the reservoir as possible, the entrance losses are reduced since the flow does not have to be channelized before passing over the labyrinth.

The *flow out of the labyrinth* in the chute downstream should be supercritical. If this is not possible the labyrinth should be designed such that maximum head to crest ratio H_o/P remains in the aerated region.

The *number of labyrinth cycles* is directly related to the nappe interference. Therefore, for normal operating conditions, the vertical aspect ratio W/P, should be higher than 2.0 and 2.5 for trapezoidal and triangular shapes respectively (Hay & Taylor, 1970).

At *low flow conditions* a non-aerated clinging nappe can occur, which produces nappe oscillation and noise. This should be avoided to prevent structural problems caused by vibrations and resonance. A common solution against a clinging nappe is the installation of splitter piers. Air may also be supplied through vents in labyrinth walls if thick enough.

Further recommendations on structural analysis and construction can be found in Lux & Hinchliff (1995).

3 PK WEIRS

3.1 *Use and types of PK weirs*

The recent developed Piano Key weir (PK weir) spillway is a variation of traditional labyrinth weirs. The planform has a rectangular shape (Figure 10). Contrary to a labyrinth weir the apex is not vertical but inclined by turns both, in upstream and in downstream direction. This arrangement explains the name Piano Key weir. According to the chosen slopes of the inlet and outlet keys they have a certain upstream and downstream overhang. This result in a smaller footprint of the structure compared to a rectangular labyrinth weir with vertical walls. Therefore, besides an improved hydraulic performance, the PK weir has the advantage that it may be easily installed even at very limited foundation space as for example on gravity dam crests. This is also the reason why PK weir spillways are a efficient and economical solution for the increase of the flood releasing capacity at existing gravity dams. The first PK weir was installed in 2006 at Golours dam in France (Laugier, 2007). Since then PK weirs have been used to increase the flood discharge capacity of the three other EDF dams, namely St. Marc (2008), Etroit (2009) and Gloriettes (2010). PK weir spillways can be easily combined with stepped chutes which lead to a pronounced downstream energy dissipation (Bieri et al., 2009). Lessons learnt from the design of these four PK weir spillways can be found in Laugier et al. (2009) and Vermenten et al. (2011) (in these Proceedings).

Other PK weir developments are presently conducted in Vietnam (Chi Hien et al., 2006), India (Sharma & Singhal, 2008) and France (Gage, Malarce and La Raviège Dam).

Figure 10.   View of a PK weir spillway of Gloriettes dam in France during construction (Photo: EDF).

Initially, two main types of PK weir have been identified (Lempérière & Ouamane, 2003):

Configuration A: The chutes (apex) are overhanging on both the upstream and downstream sides (Figure 11). This self-balanced structure favors the use of precast concrete elements and may be used for specific discharges up to $20\,\text{m}^3/(\text{sm})$. The relatively small footprint of the foundation allows placing this configuration on existing gravity dam crests in order to increase the flood release capacity as already mentioned.

Configuration B: The chutes (apex) have only an upstream overhang (Figure 12). Structural loads are less for high specific discharges which makes this configuration attractive for new dam projects. Specific discharges up to $100\,\text{m}^3/(\text{sm})$ can be allowed.

Many optional features have been developed compared to the basic configurations as:

a) *Parapet Walls*: Vertical parapet walls placed on the crest of the PK weir transform its upper part to a rectangular labyrinth weir. The parapet wall on the outlet key increases the discharge capacity since it improves the stream line pattern of the approaching flow and increases the outlet key volume (Vermeulen et al., 2011).

b) *Width of inlet and outlet keys*: choosing a higher width of the inlet key than that of the outlet key results in a better hydraulic performance (Le Doucen et al., 2009).

c) *Sidewall angle*: a sidewall angle narrowing the inlet key and widening the outlet key are likely to improve the discharge capacity.

3.2   *PK weir flow description*

Similar flow features may be observed as for traditional labyrinth weirs. Nevertheless, the influence of the downstream water level is of low relevance.

For low heads the transition from a partially clinging nappe to a leaping and then to a springing nappe can be observed on the different parts of the PK weir crest (Machiels et al., 2009b). For high heads the downstream crest of the inlet key is more supplied than the lateral crest. On the other

11

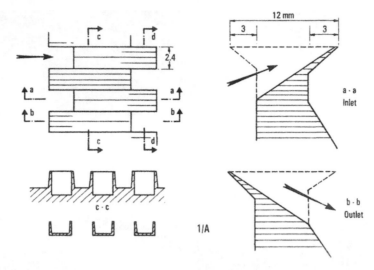

Figure 11.   Configuration A of a PK weir (from Lempérière & Ouamane, 2003).

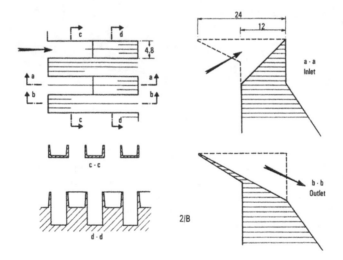

Figure 12.   Configuration B of a PK weir (from Lempérière & Ouamane, 2003).

hand the upstream crest of the outlet key is similarly supplied for high and low upstream head. For increasing heads a critical section appears from the downstream along the inlet key.

3.3   *Determination of PK weir discharge capacity*

In the case of a PK weir even more parameters influence the three-dimensional flow than compared to a labyrinth weir. Therefore, a large number of systematic laboratory experiments on physical models of PK weirs are required in order to establish general applicable design rules.

Preliminary design criteria were presented by Lempérière & Ouamane (2003) and Ouamane & Lempérière (2006) which were based on experiments performed at Biskra University in Algeria and at Roorkee University in India (Lempérière et al., 2003). Since then more systematic laboratory experiments have been performed at EDF National Hydraulic Laboratory (EDF-LNHE Chatou) in France (Cicéro et al., 2010), at HACH-Hydraulic Laboratory of University of Liège, Belgium (Machiels et al., 2009a) and at the Laboratory of Hydraulic Constructions (LCH-EPFL) in Lausanne, Switzerland (Le Doucen et al., 2009). Further systematic studies are still under way at these laboratories.

As for the traditional labyrinth weirs two different approaches may be used to describe the hydraulic performance of PK weirs.

The *first approach* uses the standard weir equation and considers all specific PK weir parameter in a modified discharge coefficient:

$$Q = C_w L_u \sqrt{2g} H^{3/2} \tag{4}$$

with Q as discharge, H as total head, L_u as developed weir length and C_w as global discharge coefficient. In such a way Machiels et al. (2009a,b) have presented C_w curves as a function of the ratio head over weir height H/P for certain PK weir designs.

The *second approach* uses the concept of effective crest length (Leite Ribeiro et al., 2007, 2009). A discharge enhancement ratio r between PK weir discharge Q_{PKW} and a sharp crested weir discharge Q_w has been defined:

$$r = \frac{Q_{PKW}}{Q_w} = \frac{C_d L_{eff} \sqrt{2g} H^{3/2}}{C_d W \sqrt{2g} H^{3/2}} = \frac{L_{eff}}{W} \tag{5}$$

Here, W corresponds to the total width of the PK weir. The discharge coefficient C_d of the sharp crested standard weir can be assumed as almost constant with $C_d = 0.42$ (Hager & Schleiss, 2009).

Based on the different configurations tested in laboratory for Saint-Marc Dam and Gloriettes Dam in France, the following equation could be proposed (Leite Ribeiro et al., 2009)

$$r = \frac{L_{eff}}{W} = \frac{1}{\left(\frac{H}{W} + \frac{1}{n\sqrt{\frac{L_u}{W}-1}} \right)^n} \tag{6}$$

For the two mentioned prototype spillways the exponents n varied between 7.5 and 14.5 depending on the geometrical configuration.

Further analysis of systematic physical experiments are still needed in order to derive general applicable design rules taking into account all complex geometric features of PK weirs.

The first more classical approach by using a modified discharge coefficient is less physically based but probably easier to assess. The approach using the effective PK weir length describes better the flow physic involved but is probably more difficult to put into operation. A combination of these two approaches may perhaps be promising.

3.4 *Comparison of labyrinth with PK weirs*

Comparing a PK weir with a corresponding rectangular labyrinth weir reveals that the PK weir has a significantly better discharge efficiency if comparing the global discharge coefficient (Anderson & Tullis, 2011). This is also the case if considering trapezoidal labyrinth weirs for comparison (Blancher et al., 2011). The gain of efficiency with a PK weir can reach 20%.

Finally, the use of PK weirs can also result in considerable construction cost savings compared to traditional labyrinth weirs.

3.5 *Further developments*

The PK weir offers many geometrical options as already presented in 3.1, whose influence has to be tested more systematically in future. Especially the use of side wall angle is promising, which would result in a converging inlet key.

Another not yet considered feature would be to incline slightly the sidewall which enlarges the width of the inlet key. Furthermore a very early springing of the nappe as well as better aeration could be ensured.

4 CONCLUSIONS

The hydraulic performance of traditional labyrinth weirs is well known since they have been studied for a long time. Nevertheless, analytical design equations considering all the involved parameters are not yet available. The design has to be based on experimentally derived and generalized performance curves.

In the case of the recent developed PK weirs even such generalized performance curves are not yet available. More systematic laboratory studies on physical models are still needed as well as alternative parametric approaches. The complexity of flow and geometry of PK weir is a fascinating challenge for future experimental and numerical research.

REFERENCES

Anderson, R. M. and Tullis B. P. 2011. Influence of Piano Key Weir Geometry on Discharge. *Proc. International Workshop on Labyrinth and Piano Key Weirs, Liège, Belgium.*

Bieri, M., Leite Ribeiro, M., Boillat, J.-L., and Schleiss, A.J. 2009. Réhabilitation de la capacité d'évacuation des crues – intégration de PK-Weir sur des barrages existants. *Proc., Colloque CFBR-SHF, Dimensionnement et fonctionnement des évacuateurs de crues (CD-ROM), Paris France.*

Blancher, B., Montarros, F. and Laugier, F. 2011. Hydraulic comparison between piano-keys weir and labyrinth spillways. *Proc. International Workshop on Labyrinth and Piano Key Weirs, Liège, Belgium.*

Chi Hien, T., Thanh Son, H. and Ho Ta Khanh, M. 2006. Results of some "Piano Keys" weir hydraulic model tests in Vietnam. *Proc., 22nd Congress of Large Dams, Question 87, Response 39. Barcelona, Spain. Volume IV,* 581–595.

Cicéro, G.M., Guene, C., Luck, M., Pinchard, T., Lochu, A., and Brousse, P.H. 2010. Experimental optimization of a piano key weir to increase the spillway capacity of the Malarce dam. *1st IAHR European Congress, Edinbourgh, 4–6 mai 2010.*

Crookston, B. M. 2010. Labyrinth weirs. Ph.D. Dissertation. Utah State University, Logan, Utah.

Crookston, B. M. and Tullis B. P. 2011. Hydraulic characteristics of labyrinth weirs. *Proc. International Workshop on Labyrinth and Piano Key Weirs, Liege, Belgium.*

Darvas, L. A. 1971. Discussion of Performance and Design of Labyrinth Weirs. *Journal of the Hydraulics Division,* 97(8), 1246–1251.

Falvey, H.T. 2003. *Hydraulic Design of Labyrinth Weirs.* ASCE Press, Reston, VA, United States.

Gentilini, B. 1941. Stramazzi con cresta a pianta obliqua e a zig-zag. Memorie e studi dell'Instituto di Idraulica e Costruzioni Idrauliche del Regio Politecnico di Milano. No. 48. (in Italian).

Hager, W., and Schleiss, A.J. 2009. Traité de Génie Civil, Volume 15 – *Constructions Hydrauliques – Ecoulements Stationnaires.* Presses polytechniques et universitaires romandes, Switzerland.

Hay, N., and Taylor, G. 1970. Performance and design of labyrinth weirs. *Journal of the Hydraulics Division,* 96(11), 2337–2357.

Houston, K. 1982. Hydraulic model study of ute dam labyrinth spillway Report No. GR-82-7. U.S. Bureau of Reclamation, Denver, Colorado.

ICOLD, 1994. *Technical Dictionary on Dams.*

Khode, B. V., Tembhurkar, A. R., Porey, D. and Ingle, R. N. 2010. Improving Discharge Capacity over Spillway by Labyrinth Weir. *World Applied Sciences Journal* 10(6): 709–714.

Lempérière, F., and Ouamane, A. 2003. The Piano Keys weir: a new cost-effective solution for spillways. *Hydropower & Dams,* 7(5): 144–149.

Lempérière, F., Sharma, N., Mourya, R. N., Shukla R. and Gupta, U. P. 2003. Experimental Study on Labyrinth Spillways. *Proc. International Conference on Engineering of Dams and Appurtenant Works Including Power Houses & Transmission Systems (CD-ROM), New Delhi, India.*

Leite Ribeiro, M., Boillat, J.-L., Schleiss, A., Laugier, F. and Albalat, C. 2007. Rehabilitation of St-Marc Dam – Experimental Optimization of a Piano Key Weir. *Proc., 32nd Congress of IAHR (CD-ROM), Venice, Italy.*

Laugier, F. 2007. Design and construction of the first Piano Key Weir (PKW) spillway at the Goulours dam. *Hydropower & Dams,* 14(5): 94–101.

Laugier, F., Lochu, A., Gille, C., Leite Ribeiro, M. and Boillat, J-L. 2009. Design and construction of a labyrinth PKW spillway at St-Marc dam, France. *Hydropower & Dams,* 16(5): 100–107.

Le Doucen, O., Ribeiro, M.L., Boillat, J.-L., Schleiss, A. J. and Laugier, F. 2009. Etude paramétrique de la capacité des PK-Weirs. Modèles physiques hydrauliques – outils indispensables du XXIe siècle, SHF, Lyon.

Leite Ribeiro, M., Bieri, M., Boillat, J-L., Schleiss, A., Delorme, F. and Laugier, F. 2009. Hydraulic capacity improvement of existing spillways – Design of piano key weirs. *Proc., 23rd Congress of Large Dams. Question 90, Response 43 (CD-ROM), Brasilia, Brazil.*

Lopez, R., Matos, J. and Melo, J. 2008. Characteristic depths and energy dissipation downstream of a labyrinth weir. *2nd Int. Junior Researcher and Engineer Workshop on Hydraulic Structures, Pisa, Italia.*

Lux, F. and Hinchliff, D. 1985. Design and construction of labyrinth spillways. *15th Congress ICOLD, Vol. IV, Q59-R15, Lausanne, Switzerland,* 249–274.

Machiels, O., Erpicum, S., Archambeau, P., Dewals, B.J. and Pirotton, M. 2009a. Analyse expérimentale du fonctionnement hydraulique des déversoirs en touches de piano. *Colloque CFBR-SHF: "Dimensionnement et fonctionnement des évacuateurs de crues", Paris, France.*

Machiels, O., Erpicum, S., Archambeau, P., Dewals, B. J. and Pirotton, M. 2009b. Large scale experimental study of piano key weirs. *Proc., 33rd Congress of IAHR (CD-ROM), Vancouver, Canada.*

Magalhães Pinto, A. 1985. Labyrinth-weir spillways. *15th Congress ICOLD, Vol. IV, Q59-R24, Lausanne, Switzerland,* 395–407.

Ouamane, A. and Lempérière, F. 2006. Design of a new economic shape of weir. *Proc., International Symposium on Dams in the Societies of the 21st Century, Barcelona, Spain.* 463–470.

Sharma, N., and Singhal, G. 2008. A dam safety solution by Piano Key Weir for enhanced spillway capacity. *Proc., International Conference on Hydrovision, Sacramento, United States.*

Taylor, G. 1968. The performance of labyrinths weirs. PhD thesis. University of Nottingham, Nottingham, United Kingdom.

Tullis, J.P., Amanian, N., and Waldron, D. 1995. Design of Labyrinth Spillways. *Journal of Hydraulic Engineering,* 121(3), 247–255.

Tullis, B.P., Willmore, C.M. and Wolfhope, J.S. 2005. Improving performance of low head labyrinth weir. *J. Hydr. Eng. ASCE,* 173: 418–426.

Tullis, B., Young, J., and Chandler, M. 2007. Head-discharge relationships for submerged labyrinth weirs. American Society of Civil Engineering, *Journal of Hydraulic Engineering* 133(3), 248–254.

Vermeulen, J., Laugier, F., Faramond, L. and Gille, C. 2011. Lessons learnt from design and construction of EDF first Piano Key Weirs. *Proc. International Workshop on Labyrinth and Piano Key Weirs, Liège, Belgium.*

Willmore, C. 2004. Hydraulic characteristics of labyrinth weirs. M.S. Report, Utah State University, Logan, Utah.

Yildiz, D. and Üzücek, E. 1996. Modelling the performance of labyrinth spillways. *Hydropower & Dams,* 3 (3): 71–76.

15

Labyrinth and Piano Key Weirs – PKW 2011 – Erpicum et al. (eds)
© *2011 Taylor & Francis Group, London, ISBN 978-0-415-68282-4*

General comments on Labyrinths and Piano Key Weirs: The past and present

F. Lempérière & J.-P. Vigny
Hydrocoop, France

A. Ouamane
Biskra University, Algeria

ABSTRACT: Labyrinth weirs are the first sills based on the concept of an increase in developed crest length for a given weir width. This article begins with a comparison between data of existing weirs. After these considerations, the studies performed by Hydrocoop and Biskra University on Piano Ney Weirs are presented. Data concerning PKW under construction are then given, and the article ends with a proposition of reference designs for PKW.

1 INTRODUCTION

Hydrocoop is a non profit making Association promoting international technical cooperation in dam engineering with special focus on flood control, spillways and sedimentation. It has accordingly, over the past ten years, studied two innovative devices for existing or new free-flow spillways: Piano Keys Weirs and low cost Concrete Fuse Plugs. Hydrocoop has promoted and coordinated relevant hydraulic model tests in five countries: France, Algeria, India, China, Vietnam.

Since ten years the University of Biskra (Algeria) has been deeply associated with Hydrocoop as well for theoretical studies as for hydraulic model tests (Mr Ouamane). The University has built a specific hydraulic flume mainly devoted to the studies of PKW.

2 EXISTING LABYRINTH WEIRS AND ASSIMILATED

2.1 *Typical design*

Most existing traditional labyrinths have similar data: vertical walls on a horizontal bottom slab with trapezoidal or triangular layout with same shape upstream and downstream. The most usual properties are gathered in the following table.

These designs require approx. 5 to 10 m^3 of reinforced concrete per m^2 of nape saved.

Table 1.   Typical designs.

Properties	Unit	Value
Developed length/Spillway width	–	3 to 5
Wall height P	m	2 to 4
Nape depth	m	1.5 to 2 ($0.5 \times P$)
Depth saving (vs Creager weir)	m	~1 ($0.3 \times P$)
Specific discharge capacity at design head	m^3/s/m	~10
Specific discharge capacity increase (vs Creager weir)	m^3/s/m	~5

Table 2. Ute labyrinth weir main properties.

Properties	Unit	Value
Wall height P	m	9
Weir width W	m	250
Specific discharge capacity at design head	m³/s/m	60
Depth saving (vs Creager weir)	m	3.5
Area of walls and basis as compared with depth saving	m³/m²	>15

Figure 1. Beni Bahdel labyrinth spillway upstream view (left) and top view (right).

2.2 *Very large labyrinth weirs: the example of Ute (U.S.)*

Three very large labyrinth weirs have been constructed. The best example is the spillway built at Ute Dam, U.S.

15 000 m³ of concrete and 3 000 t of reinforcing steel have been used for the construction of Ute labyrinth weir. The nape depth saving is about 900 m², hence over 15 m³ of reinforced concrete are required per m² of depth saving.

2.3 *Drawbacks of traditional labyrinth spillways*

The main advantage of labyrinth weirs is the easy construction of vertical walls. However, there are three main drawbacks. First, the reinforced concrete quantities are high. Then, the hydraulic efficiency is much reduced for high discharge. Moreover, labyrinth weir implementation requires a massive basis which prevents their construction upon usual spillways footprint. Most existing labyrinth weirs have been placed on a flat bank of the river.

2.4 *Special labyrinths with inclined walls*

The Beni Bahdel labyrinth weir has been constructed in 1938 in Algeria, and was designed with upstream overhangs (Fig. 1). Beni Badhel dam is a multiple arch 55 m high dam. The spillway, which can be assimilated to a labyrinth weir, has a rectangular layout with very long upstream overhangs, developing a crest length 15 times higher than the weir width. Its discharge capacity is about 1 200 m³/s, hence approx. 10 m³/s/m with an upstream head of 0.5 m, instead of 2.8 m for a Creager weir. However, the area of wall per m² of nape depth saved is more than 30 m². This spillway is rather similar with a PKW having only upstream overhangs.

The Bakhada dam, a 45 m high rockfill dam, has been built in 1936 in Algeria and raised with labyrinth in 1960. The spillway discharge has a 2 000 m³/s discharge capacity and can be assimilated to a labyrinth weir. Its specific discharge capacity is about 15 m³/s/m, and its shape appears more cost-effective than Beni Badhel spillway. The layout is rather similar to a PKW having only upstream overhangs, with a trapezoidal shape on a curved basis (Fig. 2).

Figure 2.   Bakhada labyrinth spillway dowstream view (left) and upstream view (right).

Figure 3.   Hydroplus fusegates.

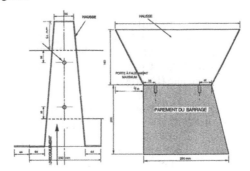

Figure 4.   First design – trapezoidal shape with overhangs.

2.5   *Fuse gates (Hydroplus)*

Since 1991, Hydroplus has built labyrinth fusegates for about 40 dams with labyrinth height between 1 and 7 m. They are made of steel or reinforced concrete. For tilting purpose, the shape usually includes a single downstream overhang. The developed length is usually close to 3 times the spillway width with a trapezoidal layout. Such fusegates may be placed upon existing spillways. Tilting has already happened for 10 dams, each time in accordance with design level.

3   STUDIES OF PINAO KEY WEIRS (PKW) SINCE 1998 BY HYDROCOOP AND BISKRA UNIVERSITY

3.1   *A brief historic*

The first model tests were performed for Hydrocoop in 1998 at the LNH in Chatou (France). As shown on the scheme here below, the final rectangular shape was not yet adopted, but the main principle of overhangs was included.

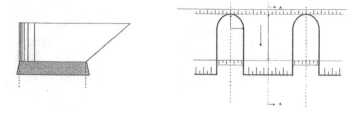

Figure 5.   Later design – rectangular shape with overhangs.

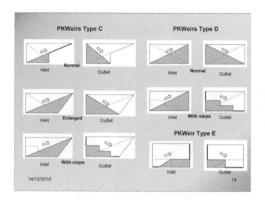
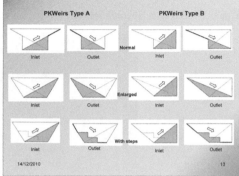

Figure 6.   Catalogue of the configurations tested.

Improvements were then suggested by P. Blanc and F. Lempérière in Hydropower and Dams (2001 – Issue 4) for traditional labyrinths and for associating a rectangular layout with overhang. Tests in Biskra University (Algeria) were performed in 2002 on the here below model. Many tests have followed in 2002 and 2003, in Biskra, Chatou and Roorke (India) leading to a basic paper on PKW by F. Lempérière and A. Ouamane (Hydropower & Dams – Issue Five, 2003).

Since 2003, hundreds of tests were promoted by Hydrocoop and performed in Marolles (Hydroplus, France), in India (Roorkee University), in China (IWHR Laboratory in Beijing) and in Vietnam (Ho Chi Minh University). Many model tests for theoretical studies and researches on various types of PKW have been performed at Biskra.

All these studies and tests have the same goals. On the one hand, they aim in finding a compromise between the hydraulic efficiency, the structural and relevant construction problems, and the economical considerations. On the other hand, they aim in elaborating some standard models of PKW to facilitate preliminary designs, cost estimation and comparison with other solutions.

Many models have been tested including various widths and fillings of the inlet or outlet keys, various noses shapes, study of the consequences of floating debris, etc. The following plots show the existing "catalogue" of various configurations envisaged and tested.

3.2   *Main results*

It seems that PKW Type B (only upstream overhang) appears as the most hydraulic efficient solution, with probably same width for inlet and outlet keys. However, floating debris may require special care. This solution is maybe the best solution for new dams.

The PKW Type A (upstream and downstream overhangs) has a good hydraulic efficiency, with inlet keys larger than outlet keys (ratio about 1.2). This is maybe the best solution for most existing dams.

The PKW Type C (only downstream overhang) doest not appear more interesting except for fusegates or for huge floating debris management.

PKW Type D and E (without overhang) are improved classic labyrinths. They may be a solution when there is space enough in banks, in river or if there is lack of skilled workers.

Table 3. Bambakari dam PKW main properties.

Properties	Unit	Value
Wall height P	m	3.2
Height P_m	m	2.13
Weir width W	m	200
PKW longitudinal length	m	8.52
Outlet key width	m	1.55
Inlet key width	m	1.2
Developed length ratio	–	6
Number of PKW units	–	58
Specific discharge capacity at design head	$m^3/s/m$	5

In each case, the following conclusions of PKW shapes have been shown.

Profiling the nose under the outlet key is efficient.

Developed length ratio L/W around 4 or 5 seems a reasonable compromise. Ratio of 6 or 7 may be used in special cases.

Filling the outlet keys with steps (to combine PKW with a stepped spillway) does not reduce much the discharge capacity providing that the height of the steps remains under certain limitation.

The hydraulic efficiency remains good even when the PKW is submerged with a high downstream level.

The best dimensions for an optimal design vary with the discharge. For instance, the optimal ratio between the widths of the keys (W_i/W_o) is not the same for a small or a high discharge through the PKW.

The trapezoidal layout is not necessary when associated with inclined walls because the adjustment of the wet section to the flow is made by the walls slope (PKW). The trapezoidal layout appears logical hydraulically when associated with vertical walls and flat bottom. It keeps a same speed along the inlet. However even in this case the rectangular lay out may be more cost efficient than the trapezoidal layout because the average width of the inlet and outlet may be reduced as well as the length (and the area and cost) of the structure. PKW model D and E are thus proposed with vertical walls and rectangular layout.

4 CURRENT PKW PROJECTS

4.1 Overview

EDF, which was the first Owner to build PKW, presents in separate papers its projects and its improvements on various points of the design and construction methods for PKW already built or under design.

Hydrocoop was asked for giving advices for various PKW projects. Among those, the following ones are under construction: Bambakari dam in Burkina Faso, Lhasi dam in India, Van Phong dam in Vietnam and Dakmi 2 dam in Vietnam.

The first two are presented here while the last ones are presented by M. Ho Ta Khanh in separate papers as well as other projects under study in Vietnam.

4.2 Bambakari dam (Burkina Faso)

The PKW spillway of Bambakari dam is designed in order to spill 1 000 m^3/s for a design upstream head of 0.86 m. The preliminary design was presented in March 2009 at Liège PKW workshop. The final design (PKW type A) is now under completion. Its main properties are gathered in the following table.

This cross section has been selected to be used as a test for first construction of a PKW type A "normal" in Burkina Faso. Because of the local conditions (PKW set on the ground) another solution such as type A "enlarged" or type E would have been probably also convenient.

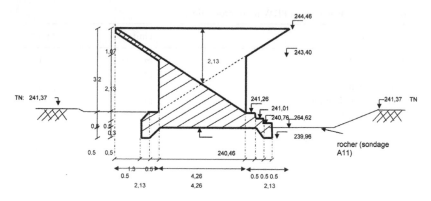

Figure 7.   Bambakari dam PKW – cross section.

Table 4.   Lhasi dam PKW main properties.

Properties	Unit	Value
Wall height P	m	6.5
Height P_m	m	4
Weir width W	m	115
PKW longitudinal length	m	16
Outlet key width	m	2.4
Inlet key width	m	3
Developed length ratio	–	6
Number of PKW units	–	18
Specific discharge capacity at design head	$m^3/s/m$	9.7

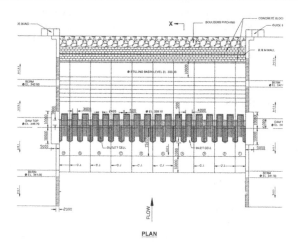

Figure 8.   Lhasi dam PKW – plan view.

4.3   *Lhasi dam (India)*

The PKW spillway of Lhasi dam is designed in order to spill $1\,115\ m^3/s$ for a design upstream head of 0.86 m. Following Hydrocoop advices, the preliminary PKW design presented in October 2009 at Lyon PKW workshop, was slightly changed. The total length of the weir W has been reduced from 128 to 115 m, the number of elements N_u from 20 to 18, and a PKW type A "enlarged" was preferred to a "normal" one.

Plan view and cross section as presently proposed by the local designer as per the following drawings (Fig. 8 and 9).

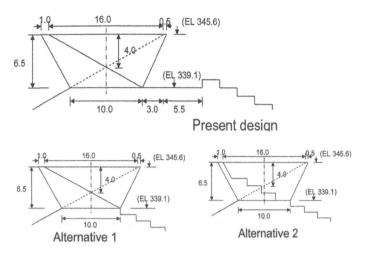

Figure 9.   Lhasi dam PKW – cross sections.

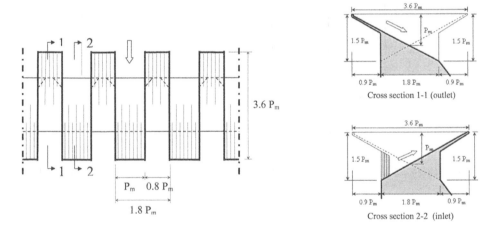

Figure 10.   Proposed reference design pour PKW model A – plan view (left) and cross sections (right).

5   CONCLUSION – REFERENCE DESIGNS

5.1   *Reference designs*

PKW are cost efficient but it is not easy to optimize the shape for each scheme. It appeared useful, using results of many hydraulic model tests, to propose some reference designs and relevant hydraulic performances in order to help owners and consultants, either for using directly these models, or for using them for preliminary designs to be optimized by specific tests.

The principle of these models is to have the best ratio of cost per saving in nape depth or per specific flow increase. They may not represent the maximum saving.

They are possibly not the optimum shape but are probably most often rather close to it. They may be easily tested in existing laboratories.

Hereafter are presented the reference configurations for PKW model A (upstream and downstream overhangs), model B (only upstream overhangs) and model E (without overhangs). PKW model C (one downstream overhang) is not presented here but should be studied.

PKW model A and B are presented in Appendix to ICOLD Bulletin 144: Cost Savings in Dams.

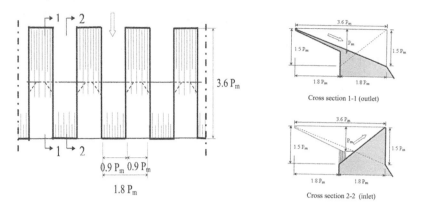

Figure 11.   Proposed reference design pour PKW model B – plan view (left) and cross sections (right).

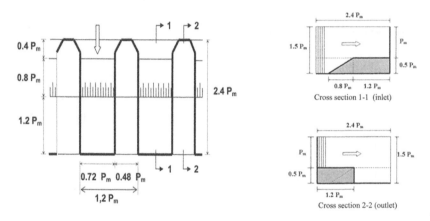

Figure 12.   Proposed reference design pour PKW model E – plan view (left) and cross sections (right).

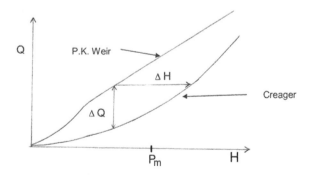

Figure 13.   Hydraulic performance of reference designs.

5.2   *Hydraulic performance of reference designs*

The hydraulic performance is measured by comparison with a Creager Weir. The key economic parameters are the saving ΔH in nape depth H or the increase ΔQ in specific flow Q as compared with a Creager Weir (Q (m^2/s/m) $= 2{,}15\,H^{1,5}$ (H in m)).

For reference designs A and B, the saving ΔH in depth is close to $0{,}5 \times P_m$ (maximum wall height of the PKW) for usual nape depths (H between $0{,}4\,P_m$ and 1,5 m). The discharge increase ΔQ is close to the discharge of a Creager weir overtopped by a nape depth equal to P_m. The volume of reinforced concrete is about half the volume of traditional labyrinths for a same saving.

24

Labyrinth and Piano Key Weirs – PKW 2011 – Erpicum et al. (eds)
© *2011 Taylor & Francis Group, London, ISBN 978-0-415-68282-4*

Hydraulic characteristics of labyrinth weirs

B.M. Crookston & B.P. Tullis
Utah Water Research Laboratory, Department of Civil and Environmental Engineering,
Utah State University, Logan, Utah, USA

ABSTRACT: The experimental results of 32 physical models were used to develop a hydraulic design and analysis method for labyrinth weirs. Discharge coefficient data for quarter-round labyrinth weirs are presented for $6° \leq$ sidewall angles $\leq 35°$. The influence of cycle geometry, nappe flow regimes, artificial aeration (vents, nappe breakers), and nappe stability on hydraulic performance are discussed. The validity of this method is presented by juxtaposing discharge coefficient data from this study and previously published labyrinth weir studies.

1 INTRODUCTION

1.1 *Labyrinth weirs*

A labyrinth weir is a series of linear weirs 'folded' in plan-view for the purpose of increasing weir length for a given channel or spillway width. Because weir discharge is proportional to weir length, at a given upstream head, the discharge capacity of a spillway channel of fixed width can be increased using a labyrinth weir over a standard linear weir. This characteristic has made labyrinth weirs popular in recent years, particularly with existing spillway retrofits, because of increased discharge requirements associated with more extreme hydrologic events. Figure 1 shows a schematic, identifies common geometric parameters, and illustrates common crest shapes associated with labyrinth weirs.

The significant design flexibility of labyrinth weirs, associated with the numerous geometric parameters, also presents a challenge when trying to develop a hydraulically optimal labyrinth weir design. For example, the sidewall angle (α), crest length (L_c), crest shape, number of cycles (N), the configuration of the labyrinth cycles, and the orientation and placement of a labyrinth weir must all be determined. As the driving head increases, flow efficiency, as measured by the discharge coefficient (C_d), declines. Submergence effects, caused either by a high tailwater or by narrow outlet cycles where the inflow passing over the weir is more efficient than the flow exiting the upstream cycle, a term referred to as "local" submergence can also affect discharge efficiency. Changing nappe aeration conditions and nappe stability affect both discharge capacity and the static and dynamic forces experienced by the weir wall. Historically, physical models have been useful tools to design and optimize labyrinth weirs. Numerical modeling of labyrinth weirs, often

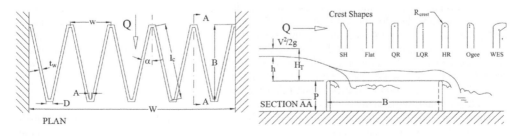

Figure 1. Labyrinth weir schematic.

Table 1. Labyrinth weir design methods.

Design Methods			
( )	Authors	Labyrinth Cycle Type*	Crest Shape**
1	Taylor (1968), Hay & Taylor (1970)	Tri, Trap, Rect	SH, HR
2	Darvas (1971)	Trap	LQR
3	Hinchliff & Houston (1984)	Tri, Trap	SH, QR
4	Lux & Hinchliff (1985)	Tri, Trap	QR
5	Magalhães & Lorena (1989)	Trap	WES
6	Tullis et al. (1995)	Trap	QR
7	Melo et al. (2002)	Trap	LQR
8	Tullis et al. (2007)	Trap	HR
9	Lopes et al. (2008)	Trap	LQR

*Tri = Triangular, Trap = Trapezoidal, Rect = Rectangular
**See Figure 1

in conjunction with physical models (composite modeling), has also proven to be a useful design tool in recent years.

1.2 Labyrinth Weirs

Initial insights into labyrinth weirs come from Gentilini (1940) and Kozák and Sváb (1961), but the first study to produce design-oriented labyrinth weir data came from Taylor (1968) and Hay & Taylor (1970). This study and subsequent notable research studies that have provided hydraulic design guidance for labyrinth weirs is presented in Table 1. The US Bureau of Reclamation (USBR) noted discrepancies between their Ute Dam model experimental data (Houston 1982) and the Hay & Taylor (1970) predictive method. Subsequent USBR labyrinth weir research produced two publications (based upon total head, H_T) for labyrinth weir design (Hinchliff & Houston 1984, Lux & Hinchliff 1985).

Based upon model studies for Avon and Woronora Dam, Darvas (1971) simplified labyrinth weir design by introducing an empirical discharge equation and a discharge coefficient. Magalhães & Lorena (1989) expanded this approach by presenting a dimensionless discharge coefficient and new experimental results and design curves, which were found to be systematically lower than Darvas' (1971).

Tullis et al. (1995) developed a design method based on the standard weir equation (Equation 1) and laboratory-scale model testing at the Utah Water Research Laboratory (UWRL) by Waldron (1994), Tullis (1993), and Amanian (1987).

$$Q = \frac{2}{3} C_d L_c \sqrt{2g} H_T^{3/2}$$ (1)

In Equation 1, Q is the discharge of a labyrinth weir, C_d is a dimensionless discharge coefficient, g is the acceleration constant of gravity, and H_T is defined as $H_T = V^2/2g + h$ (V is the average cross-sectional velocity at the gauging location, h is the piezometric head upstream of the weir).

The supporting data for the Tullis et al. (1995) method (quarter-round crest shape) was limited to C_d data for $6° \leq \alpha \leq 18°$ and linearly interpolated C_d curves for $\alpha = 25°$ and $35°$. In an effort to account for apex influences on discharge efficiency, Tullis et al. (1995) replaced the total weir length (L_c) in Equation 1 with an effective weir length, L_e. They also provided their labyrinth weir design method as a spreadsheet-based computer program; Falvey (2003) favored this design method. Willmore (2004), however, found that the $\alpha = 8°$ C_d data was over-predictive and also corrected a minor mathematical error in the design program.

Melo et al. (2002) expanded the work of Magalhães & Lorena (1989) by adding an adjustment parameter for labyrinth weirs located in a channel with converging channel sidewalls. Tullis et al. (2007) developed a dimensionless submerged head-discharge relationship (tailwater submergence) for labyrinth weirs. Lopez et al. (2008) evaluated downstream characteristic depths and energy dissipation of labyrinth weirs. Finally, Falvey (2003) compiled a significant part of the published information on labyrinth weirs.

The objectives of this study are to improve labyrinth weir design and analysis and provide additional hydraulic characteristic information regarding cycle efficiency, nappe aeration conditions, and nappe instability. Data from this study (quarter-round crest shape) are compared to data sets from Tullis et al. (1995) and Willmore (2004). For this study, Equation 1 was used to quantify the labyrinth weir head-discharge relationships. Unlike the Tullis et al. (1995) method, the crest centerline length (total weir length), L_c, was used as the characteristic length.

2 EXPERIMENTAL METHOD

32 laboratory-scale physical labyrinth weir models were tested at the Utah Water Research Laboratory (UWRL), located at Utah State University in Logan, Utah, USA. Labyrinth weirs were fabricated from High Density Polyethylene (HDPE) sheeting and were tested in a laboratory flume (1.2 m × 14.6 m × 1.0 m). The models were installed on an elevated horizontal platform (level to ±0.4-mm). Details of the physical model test program are summarized in Crookston (2010).

Experimental data were collected under steady-state conditions. A significant number of head-discharge data points were collected for all tested weir geometries; approximately 10% of the data were repeated to ensure accuracy and determine measurement repeatability. Flow rates were measured using calibrated orifice meters in the flume supply piping. The upstream flow depth relative to the weir crest, h, was determined a point gauge (±.15 mm) installed in stilling well that was hydraulically connected to the flume.

Weirs were tested with and without a nappe aeration apparatus consisting of an aeration tube located near the downstream side for each labyrinth weir sidewall. The test program also evaluated the performance of wedge-shaped nappe breakers in a variety of locations (upstream apex, weir sidewall, downstream apex). A dye wand was used to observe the unique and complex local flow patterns associated with labyrinth weir flow. Observations noted nappe aeration conditions and behavior, nappe stability, nappe separation point, nappe interference, and areas of local submergence.

3 EXPERIMENTAL RESULTS

3.1 *Discharge coefficients*

The quarter-round crest, non-vented nappe, labyrinth weir C_d vs. H_T/P data presented in Figure 2 include data from the current study, Tullis et al. (1995), and Willmore (2004); all of the C_d data were determined using Equation 1 with L_c as the characteristic weir length (the Tullis et al. 1995 data were changed from L_e to L_c to accommodate a direct comparison). Note that, in general, the C_d values decrease with decreasing sidewall angle and that, with the exception of the smaller H_T/P values, C_d for a given sidewall angle decreases (discharge efficiency decreases) as H_T/P increases.

At the larger H_T/P values, there is relatively good agreement between data sets for $\alpha = 6°$, $12°$, and $15°$ data sets. The current study $\alpha = 8°$ C_d data are consistent with the Willmore (2004) data and, as suggested by Willmore (2004), the Tullis et al. (1995) $\alpha = 8°$ data is higher than expected. For $\alpha > 15°$, the correlation between the current and the Tullis et al. (1995) data sets decreases with increasing α. Contrary to the general trend, the $\alpha = 20°$ data from the current study, though similar to the Tullis et al. (1995) $\alpha = 18°$ data, actually dips below the $\alpha = 18°$ data at higher H_T/P values. The Tullis et al. (1995) $\alpha = 35°$ data, which was interpolated using the $\alpha = 18°$ and $90°$ (linear weir with a quarter-round crest shape), correlates relatively poorly with the $\alpha = 35°$ experimental data from the current study.

Though the reasons for the limited correlation between the data from the current study and Tullis et al. (1995) study are not necessarily clear, there were some notable differences in experimental method, which could have had an influence. The Tullis et al. (1995) data set represents a compilation of data from three separate investigations, with a different researcher responsible for each (Waldron 1994, Tullis 1993, Amanian 1987); the data collection and quality control processes in the current study were closely managed by a single project manager. Size scale effects could also be a potential

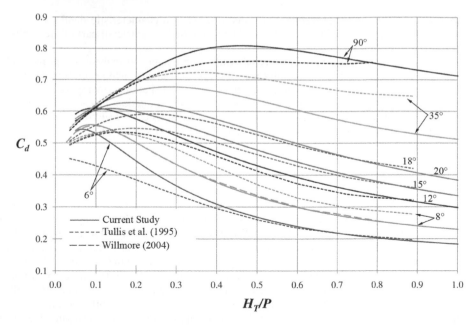

Figure 2. Comparison of C_d vs. H_T/P data from the current study, Tullis et al. (1995), and Willmore (2004) for non-vented, quarter-round labyrinth weirs with L_c as the characteristic weir length.

factor as the weir heights in the current study were two-to-three times taller than the Tullis et al. (1995) weirs and were approximately twice the thickness. The Tullis et al. (1995) weirs included partial outside cycles (the weirs were designed such that the apron length (B) was maintained constant rather than maintaining complete cycles. Based on good agreement between the C_d data from current study and Willmore (2004) study for $\alpha = 8°$ ($\alpha = 10°$ and 12° also showed good correlation-data not present here), and the fact that the data for the larger α values in the Tullis et al. (1995) study were interpolated rather that determined experimentally, the C_d data from the current study is recommended for use in determining labyrinth weir head-discharge relationships for labyrinth weirs with a quarter-round crest shape

3.2 Nappe aeration conditions and nappe instability

Nappe aeration behavior and nappe stability influence the hydraulic performance of labyrinth weirs. Four different aeration conditions were observed during labyrinth weir testing: clinging, aerated, partial-aerated, and drowned. Clinging refers to a nappe that is attached to the downstream side of the weir wall. The partially aerated nappe (see Fig. 3) refers to a condition where nappe aeration condition along a sidewall leg is non-uniform (i.e. the nappe over parts of the weir is aerated and other parts not). The drowned nappe is one where there the nappe does spring away from the weir but the underside of the nappe is devoid of any air (water filled).

Aeration conditions are influenced by the crest shape, H_T, the momentum and trajectory of the nappe or jet passing over the crest, the depth and turbulent nature of flow behind the nappe (and in the outlet cycle in general), and the pressure behind the nappe (sub-atmospheric for non-vented or atmospheric for vented nappes). As H_T increases, the labyrinth weir nappe typically transitions from clinging to aerated, to partially aerated, and finally to a drowned condition. Table 2 presents the H_T/P ranges, segregated by α, for each of the nappe aeration conditions.

Nappe instability is characterized by a nappe whose trajectory changes without any net changes in weir discharge. Nappe instability is a low frequency phenomenon that often causes the formation and removal of the air cavity behind the nappe, which produces an audible, strong flushing noise. Nappe instabilities were observed for the larger sidewall angled labyrinth weir (e.g. $\alpha = 12°$, 15°, 20°, and 35°) over certain H_T/P ranges. There are little to no discernable indicators for predicting

Figure 3. Partially aerated nappe aeration condition observed for trapezoidal labyrinth weir, $\alpha = 12°$, $H_T/P = 0.3$.

Table 2. Nappe aeration flow conditions for labyrinth weirs.

α (°)	(H_T/P) Clinging	Aerated	Semi-Aerated	Drowned
6	<0.05	0.05–0.26	0.26–0.32	>0.32
8	<0.06	0.06–0.29	0.29–0.36	>0.36
10	<0.06	0.06–0.29	0.29–0.48	>0.48
12	<0.06	0.06–0.28	0.28–0.51	>0.51
15	<0.05	0.05–0.26	0.26–0.51	>0.51
20	<0.05	0.05–0.24	0.24–0.52	>0.52
35	<0.06	0.06–0.23	0.23–0.52	>0.52

Table 3. Unstable nappe H_T/P regions for quarter-round crest labyrinth weirs.

α (°)	H_T/P Range
6	–
8	–
10	–
12	0.30–0.35
15	0.27–0.47
20	0.22–0.53
35	0.22–0.70

specifically when the nappe instability will occur, but the phenomena intensity decreases and eventually disappears as H_T increases. It is suggested that these ranges be avoided, as vibrations, pressure fluctuations, and noise may reach sufficient levels as to be undesirable or structurally harmful. Artificial aeration by vent or nappe breakers minimizes but does not prevent nappe instability. The nappe instability H_T/P ranges are summarized in Table 3 as a function of α.

3.3 *Aeration vents and nappe breakers*

Artificial aeration was found to greatly improve nappe instability and decrease noise; however, the phenomenon was still observed (to a lesser degree) for $\alpha \geq 20°$. Aeration vents were found to have

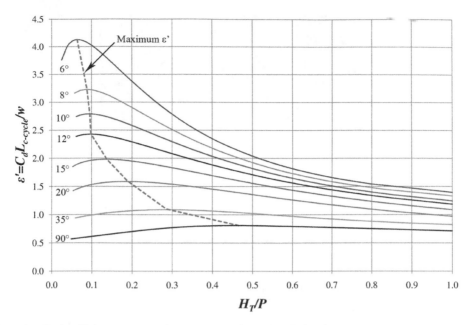

Figure 4.   Cycle efficiency vs H_T/P for quarter-round crest labyrinth weirs.

little to no effect on the discharge capacity of quarter-round labyrinth weirs, relative to non-vented conditions. Because the colliding nappes at the upstream apex did not allow air to be passed from one sidewall to the adjacent sidewall under most flow conditions, aeration vents should be provided for each labyrinth weir sidewall. Aeration vents placed near the downstream apex were found to be less effective.

Nappe breakers with a triangular cross-section (plan view) placed on the downstream apexes with the point oriented into the flow were very effective at facilitating nappe aeration. The leading edge should be protected to minimize the potential damage from debris impact. The triangular nappe breaker oriented as described produced no measurable reduction in the labyrinth weir discharge capacity relative to the aeration vent performance. The same nappe breakers were also evaluated at the mid-point of each weir side leg and at the upstream apexes. Nappe breakers at the upstream apexes proved to be effective at facilitating nappe aeration and breakers located on the side wall caused a greater disruption in the flow pattern and reduced the discharge efficiency of the weir.

3.4   Cycle efficiency

Optimizing a labyrinth weir geometric design can be difficult. As shown in Figure 2, C_d decreases with decreasing α. Since C_d is representative of a weir's discharge per unit length, according to the data in Figure 2, the linear weir ($\alpha = 90°$) weir is the most hydraulically efficient per unit length. Although the discharge per unit length decreases as α decreases, the increase in weir length for a give channel width more than compensates for the decrease in C_d, giving the labyrinth weirs a higher overall discharge capacity relative to the linear weir.

In an effort to develop a better understanding of the relationship between C_d, weir length, both of which are α dependent, and weir discharge capacity, cycle efficiency [ε' ($\varepsilon' = C_d L_{c\text{-}cycle}/w$)] was developed, which is representative of the labyrinth weir's discharge per cycle. The ε' vs. H_T/P data for the quarter-round crest labyrinth weirs are presented in Figure 4. Maximum ε' values occur at low H_T/P (as delineated by the dashed line); ε' increases as α decreases; and the benefits of smaller α angles decrease with increasing H_T/P. The data in Figure 4 therefore suggests that, based solely on discharge capacity considerations, the smaller sidewall angle geometries are more efficient than the larger sidewall angles. Though no labyrinth weirs with sidewall angles smaller than 6° were tested, it is very likely that there is a limiting value of α, below which ε' begins to decrease. The economics associated with construction costs should also be considered as part of the labyrinth weir design process. Even though the $\alpha = 6°$ had that highest cycle efficiency (largest discharge per cycle), relative to the other labyrinth weir geometries tested, it was also the longest weir and produced the lowest discharge per unit length.

4 CONCLUSIONS

In an effort to improve the hydraulic design and analysis of labyrinth weirs and provide additional flow characteristics insights, 32 laboratory-scale labyrinth weirs were tested (Crookston 2010). Of the models tested, seven two-cycle labyrinth weirs with a quarter-round crest shape replace the C_d data presented in the Tullis et al. (1995) labyrinth weir design method. All C_d in this study were calculated as a function of the weir centerline length, rather than the effective length, which was used in the Tullis et al. (1995) method.

The C_d data are presented in Figure 2 as a function of H_T/P for labyrinth weirs with $6° \le \alpha \le 35°$ and for a linear weir with a quarter-round crest shape ($\alpha = 90°$). C_d data from this study are proposed to replace the data from the Tullis et al. (1995) design method, based upon the good agreement with the Willmore (2004) C_d data, inconsistencies identified in the Tullis et al. (1995) data, and the fact that the Tullis et al. (1995) method presents interpolated curves for $\alpha > 18°$.

C_d decreases and, for a given channel width, the weir length increases with decreasing α. To aid in evaluating the relative discharge efficiencies of various labyrinth weir geometries, a new parameter, Cycle efficiency (ε'), was introduced. ε' ($\varepsilon' = C_d L_{c\text{-}cycle}/w$) represents the discharge per labyrinth weir cycle. The ε' vs. H_T/P data presented in Figure 3 show that ε' increases with decreasing α (relative to the range of α values tested). The $\alpha = 6°$ produced the highest discharge per cycle for a given head. As H_T/P increases, however, the head-discharge benefits of the small α weirs diminish. As α gets smaller, the weir length becomes longer; a labyrinth weir design should also include an economic analysis based on the associated construction costs when evaluating feasibility.

Nappe aeration conditions and nappe stability should also be considered in the hydraulic and structural design of labyrinth weirs. Tables 2 and 3 give the ranges of H_T/P for each aeration condition and for nappe instability. Regions where the nappe is unstable should be avoided as the fluctuating pressures at the weir wall, noise, and vibrations may be undesirable. Nappe ventilation by means of aeration vents or nappe breakers diminished but did not eliminate nappe instability. From the experimental results, it is recommended that one vent be placed per sidewall and one breaker be centered on each downstream apex.

Despite the additional information presented in the current study relative to labyrinth weir hydraulic characteristics, it is generally recommended that a labyrinth weir design be verified with a physical and/or numerical model studies in order to account for site-specific conditions that may be outside the scope of this study.

ACKNOWLEDGEMENTS

This study was funded by the State of Utah and the Utah Water Research Laboratory.

NOTATIONS

A Inside apex width
α Sidewall angle
B Length of labyrinth weir (Apron) in flow direction
C_d Discharge coefficient, data from current study
D Outside apex width
ε' Cycle efficiency
g acceleration constant of gravity

h	depth of flow over the weir crest
H_T	Unsubmerged total upstream head on weir
H_T/P	Headwater ratio
L_c	Total centerline length of labyrinth weir
l_c	Centerline length of weir side wall
$L_{c\text{-}cycle}$	Centerline length for a single labyrinth weir cycle
L_e	Total effective length of labyrinth weir
M	Magnification ratio, $L_{c\text{-}cycle}/w$
N	Number of labyrinth weir cycles
P	Weir height
Q	Discharge over weir
R_{crest}	Radius of crest shape
t_w	Thickness of weir wall
V	Average cross-sectional flow velocity upstream of weir
W	Width of channel
w	Width of a single labyrinth weir cycle

REFERENCES

Amanian, N. 1987. Performance and design of labyrinth spillways. M.S. thesis, Utah State University, Logan, Utah.

Crookston, B.M. 2010. Labyrinth weirs. Ph.D. Dissertation. Utah State University, Logan, Utah.

Darvas, L. 1971. Discussion of performance and design of labyrinth weirs, by Hay and Taylor. *American Society of Civil Engineering, Journal of Hydraulic Engineering* 97(80): 1246–1251.

Falvey, H. 2003. *Hydraulic design of labyrinth weirs*. Reston, Va: ASCE.

Gentilini, B. 1940. Stramazzi con cresta a piñata oblique e a zig-zag. Memorie e studi dell Instituto di Idraulica e Costruzioni Idrauliche del Regio Politecnico di Milano. No. 48. (in Italian).

Hay, N. & Taylor, G. 1970. Performance and design of labyrinth weirs. *American Society of Civil engineering, Journal of Hydraulic Engineering*, 96(11), 2337–2357.

Hinchliff, D. & Houston, K. 1984. Hydraulic design and application of labyrinth spillways. *Proceedings of 4th Annual USCOLD Lecture.*

Houston, K. 1982. Hydraulic model study of Ute Dam labyrinth spillway *Report No. GR-82-7.* U.S. Bureau of Reclamation, Denver, Colo.

Kozák, M. & Sváb, J. 1961. Tort alaprojzú bukók laboratóriumi vizsgálata. Hidrológiai Közlöny, No. 5. (in Hungarian).

Lopes, R. Matos, J. & Melo, J. 2008. Characteristic depths and energy dissipation downstream of a labyrinth weir. *Proc. of the Int. Junior Researcher and Engineer Workshop on Hydraulic Structures (IJREWHS '08)*, Pisa, Italy.

Lux III, F. & Hinchliff, D. 1985. Design and construction of labyrinth spillways. *15th Congress ICOLD, Vol. IV, Q59-R15*, Lausanne, Switzerland, 249–274.

Magalhães, A. & Lorena, M. 1989. Hydraulic design of labyrinth weirs. *Report No. 736*, National Laboratory of Civil Engineering, Lisbon, Portugal.

Melo, J. Ramos, C. & Magalhães, A. 2002. Descarregadores com soleira em labirinto de um ciclo em canais convergentes. Determinação da Capacidad de Vazão. *Proc. 6° Congresso da Água*, Porto, Portugal. (in Portuguese).

Taylor, G. 1968. The performance of labyrinth weirs. Ph.D. thesis, University of Nottingham, Nottingham, England.

Tullis, B. Young, J. & Chandler, M. 2007. Head-discharge relationships for submerged labyrinth weirs. *American Society of Civil Engineering, Journal of Hydraulic Engineering* 133(3): 248–254.

Tullis, P. Amanian, N. & Waldron, D. 1995. Design of labyrinth weir spillways. *American Society of Civil Engineering, Journal of Hydraulic Engineering* 121(3): 247–255.

Waldron, D. 1994. Design of labyrinth spillways. M.S. thesis, Utah State University, Logan, Utah.

Willmore, C. 2004. Hydraulic characteristics of labyrinth weirs. M.S. Report, Utah State University, Logan, Utah.

Physical modeling –
Hydraulic capacity

Labyrinth and Piano Key Weirs – PKW 2011 – Erpicum et al. (eds)
© 2011 Taylor & Francis Group, London, ISBN 978-0-415-68282-4

PKWeir and flap gate spillway for the Gage II Dam

V. Dugué, F. Hachem & J.-L. Boillat
Laboratory of Hydraulic Constructions (LCH), Ecole Polytechnique Fédérale de Lausanne (EPFL), Switzerland

V. Nagel, J.-P. Roca & F. Laugier
Electricité de France (EDF), Centre d'Ingénierie Hydraulique, Savoie Technolac, Le Bourget-du-Lac, France

ABSTRACT: The Gage II concrete arch dam shows some limiting constraints in its structural behavior when submitted to particular hydrostatic and thermal load conditions. Its sensitivity to low temperature has led to determine two different maximum water levels during winter and summer, respectively. Therefore, a new complementary spillway, composed by a PKWeir and a flap gate weir has been designed by EDF and optimized using a physical scaled model at LCH. The experimental tests have been carried out for different configurations of the spillway. The output flow capacities of the PKWeir and the flap gate weir including their interaction through the restitution channel have been also investigated. The geometric optimization process has been conducted on several spillway parameters: number of the PKWeir units, its position relative to the flap gate weir, and the height, width and bottom slope of the restitution channel. The tested configurations have been compared based on their flow output capacity.

1 PROJECT DESCRIPTION

1.1 *Context*

The Gage II dam, operated by Electricité de France, is located on the Gage River in the Ardèche department in France (Fig. 1a). This concrete arch dam, of 40 m high, was built in 1967. Flood events are actually evacuated by a free spillway over the dam crest, with a capacity of 230 m³/s (Fig. 1b).

The particular characteristics of the dam (high slenderness ratio and high developed crest length over height ratio) lead to important high tensile stresses at the upstream foot of the arch. This phenomenon is particularly marked in winter when the contraction of the arch concrete promotes a downstream tipping displacement of the dam. Therefore, it has been decided to lower the water operation level by 5 m during cold periods, in order to decrease the water pressure loads on the

Figure 1. a) Aerial view of the Gage II dam and reservoir. b) Photo of the downstream face of the dam and the existing crest spillway.

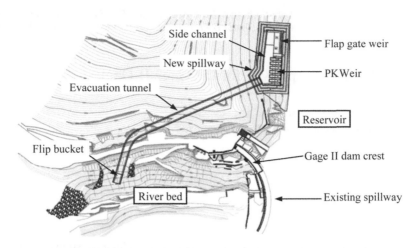

Figure 2.   Plan view of the dam with the new spillway on its right bank side.

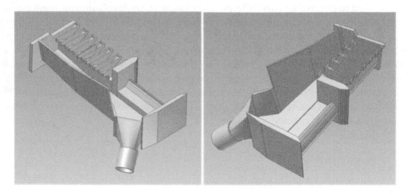

Figure 3.   3D views of the new spillway with the PKWeir and flap gate weir combination.

dam. The solution promoting the construction of an additional new spillway on the right bank side of the dam has been adopted.

The design of the new spillway was done according to the results of a recent hydrologic study of the Gage II watershed. The estimated new extreme floods are higher than the original design values of the existing spillway. For example, the new design peak flow for a 1000 years return period is evaluated to 675 m³/s and is approximately three times higher than the previous one.

1.2   The concept of the new spillway

The design of the new additional spillway is an axial arrangement of a PKWeir and a flap gate weir overtopping in a side channel. The latter is connected to an underground evacuation tunnel of approximately 200 m length (Fig. 2). The PKWeir spillway is an efficient measure to reduce the total spillway width and the rock excavation volume needed for the side channel construction (Falvey 2003, Leite et al. 2009).

The new spillway (Fig. 3) will be operated as follows: (i) During hot season, the flap gate will be closed (high position) and the main floods discharge will be evacuated by the existing and new spillways, (ii) during cold season, the flap gate will be opened (low position) and the reservoir operation level will be lowered down to el. 1005 m, 5 m below the hot season operation level (el. 1010 m). Floods discharge will then be evacuated only through the flap gate weir.

The new evacuation system (weirs, side channel and tunnel) is designed and checked for 1000 years return periods floods and its sensitivity to more extreme floods (5000 years return periods floods) is also considered. The water level and discharge constraints for hot and cold seasons are listed in Table 1.

Table 1. Design objectives for the new additional spillway.

Flood return period	Season	Maximum Water Level (MWL) [m]	Peak flood discharge to be evacuated by the new spillway [m³/s]
1000	Hot	1011.50	459 (PKWeir + flap gate)
5000	Hot	1011.77	557 (PKWeir + flap gate)
1000	Cold	1009.00	468 (Flap gate only)
5000	Cold	1009.64	610 (Flap gate only)

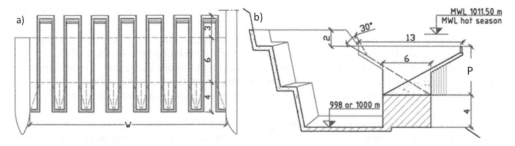

Figure 4. PKWeir geometry: a) Plan view; b) Transversal cross-section at an outlet unit.

Table 2. Geometric parameters of the PKWeir units adopted for the new spillway.

Nomenclature*		[m]
Upstream-downstream length of elements	B	13
Upstream overhang crest length	B_o	4
Downstream overhang crest length	B_i	3
Base length	B_b	6
Wall height	P	6
Width of a unit	W_u	3.7
Inlet key width	W_i	1.6
Outlet key width	W_o	1.3
Crests & wall thickness	$T_s = T_i = T_o$	0.4
Total developed length of a unit	L_u	28.9
Developed length ratio of a unit	n_u	7.8
Aspect ratio	$2W_u/P$	1.23
Parapet walls height	$P_{pi,o,s}$	0.8
Parapet walls crest thickness	T_{px}	0.2
Nose length (triangular profile)	B_n	2

(*) See Pralong et al. 2011

1.3 Preliminary design

The preliminary design of the new spillway showed that a flap gate weir around 25 m wide is sufficient to evacuate the 1000 years return period flood (United States Department of the Interior 1987, Hager & Schleiss 2009). A flap gate width of 20 m has been adopted as a minimum start value for the optimization process to be carried out on the physical model of the spillway.

For the PKWeir, the design procedure was aiming to obtain the best economical and technical configuration by minimizing the width W and maximizing the height P shown on Figure 4. Different PKWeir geometries have been previously tested numerically using the Flow-3D model (Flow-3D 2010). The results reveal that a PKWeir of 34 m width and 6 m height could evacuate the design flood discharge under a water level head below the authorized maximum value. They show that the expected specific discharge referred to the width W is around 15 m³/s/m for a water head of 1.5 m. The other fundamental geometry parameters for the PKWeir keys or units are depicted in Table 2.

It should be emphasized here that PKWeirs of 6 m height have never been built before in France. A height of 4 m is generally used.

2 PHYSICAL MODELLING

2.1 *Description of the physical scaled model*

An experimental set-up of the new additional spillway of Gage II dam, has been constructed at the Laboratory of Hydraulic Constructions (LCH) of the Ecole Polytechnique Fédérale de Lausanne (EPFL) using a geometrical scale factor of 1:40. The model was operated respecting the Froude similarity that preserves the ratio of inertial to gravitational forces in the model like in prototype. It covers an area of 160×160 m^2 of the reservoir including the bank topography where the new spillway is located.

The flap gate, the designed PKWeir and the restitution side channel have been constructed with grey PVC material. The topography of the reservoir was reproduced by a series of PVC vertical profiles filled with sand and covered by a thin layer of cement. Figure 5a presents the new spillway set-up with a movable flap gate that can have two different positions corresponding to the hot and cold seasons. Figure 5b shows the models of the spillway and part of the water reservoir after their assembly.

The tailrace tunnel has been modeled using transparent PVC material rolled around a special mould to produce the horse shoe cross-section of the tunnel. A limited part of the topographic area of the valley downstream from the dam has been also reproduced to examine the jet trajectory issued from the deflector of the tunnel.

The model is supplied by a variable speed pump and the input water discharge is measured by an electromagnetic flowmeter, with 1% accuracy. The water level in the tank was measured by two ultrasonic sensors with ± 1 mm accuracy while the water depth in the restitution channel was estimated using two limnimeters fixed against the side walls of the channel near the PKWeir and the flap gate weir.

2.2 *PKWeir capacity*

The spillway capacity has been evaluated based on the physical measurements performed for three different geometries. The number of units N_u and the total width W of the PKWeir are listed in Table 3 for the three tested configurations. The developed length, L, of the PKWeir along its

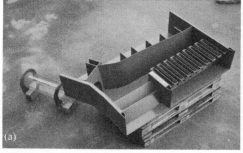

Figure 5. Physical scaled model: a) New spillway (PKWeir, flap gate, and side channel); b) The model with the reservoir topography and the water supply system.

Table 3. Global geometric parameters of the tested geometries of the new spillway.

Geometry	N_u [unit]	W [m]	L [m]	Flap gate width [m]
1	9	34.0	264	20.0
2	7	26.6	201	20.0
3	7	26.6	201	27.4

overflowing crest axis is also shown in Table 3. The difference between Geometries 1 and 2 lies only in the number of units of the PKWeir. The spillway in Geometries 2 and 3 has the same PKWeir configuration but with different flap gate width. For each Geometry, different flood scenarios during hot and cold periods have been tested.

The experimental results regarding the output discharge of the spillway and the water level in the reservoir are presented in Figure 6 for both hot and cold periods. During the hot period (Fig. 6, left), the main discharge is evacuated by the PKWeir. As expected, the spillway capacity is greater for Geometry 1 than for the two others (26% more than Geometry 2 and 19% more than Geometry 3), as the total discharge is proportional to the number of PKWeir units. The comparison of Geometries 2 and 3 shows that the width of the flap gate has a small impact on the capacity. However, the discharge capacity is 6% higher in Geometry 3.

The spillway capacity relative to the cold period scenario is shown in Figure 6 (right). During this period of the year, the floods are evacuated only through the flap gate weir. During the tests, the maximum operation level has been modified by EDF from el. 1010 to 1009 m. The two Geometries 1 and 2 had already been tested for the maximum operation level 1010 m. The tests done with Geometry 3 have been adapted to the new maximum operation level. This explains why the discharge points are shifted down for this latter spillway configuration.

The results show that Geometry 3 with a PKWeir of 7 units and a flap gate width of 27.4 m satisfies the flood evacuation capacity and the reservoir level constraints for 1000 years return period floods, during hot and cold periods. This geometry, called Optimized Preliminary Configuration (OPC), was adopted as start configuration for the following physical tests dedicated to the flow interaction inside the channel.

Figure 7 shows an interesting comparison of the PKWeir capacity versus water head for different projects that have been designed or built in France (Leite et al. 2007, Bieri et al. 2010). It shows

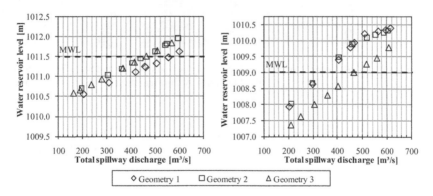

Figure 6.   Spillway capacity for the three different geometries during hot (left) and cold periods (right).

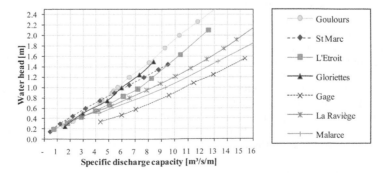

Figure 7.   PKWeir unit discharge capacity versus water head for different projects including the Gage II spillway.

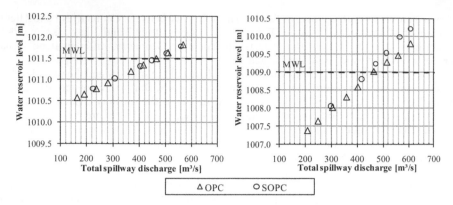

Figure 8. Spillway capacity for the two configurations OPC and SOPC during hot (left) and cold (right) periods.

Figure 9. Flow in the side channel for the 5000 years return period flood: a) and b) during hot period; c) and d) during cold period; a) and c) for the Optimized Preliminary Configuration (OPC); b) and d) for the Switch Optimized Preliminary Configuration (SOPC).

that the capacity of Gage II PKWeir spillway is significantly improved. This is mainly due to the additional experience gained by designers in the geometrical optimization of PKWeirs (Vermeulen et al. 2011).

2.3 *Flow interaction between the flap gate and the PKWeir*

By spilling into the same side channel, the flap gate and the PKWeir overtopping jets interact together and with the flow inside the channel.

In the preliminary design, the PKWeir was placed in the upstream part of the side channel and the flap gate in its downstream part in front of the entrance of the tailrace tunnel (Figure 5a). The width of the side channel is variable and grows from 7.6 m in the upstream side to around 13.5 m in the downstream side near the tunnel entrance. The interaction of both the PKWeir and the flap gate weir has been studied for two different configurations: (i) the Optimized Preliminary Configuration (OPC), and (ii) the (SOPC) configuration obtained from the OPC by switching the relative positions of the PKWeir and the flap gate weir.

Figure 8 (left) shows the experimental results for these two configurations, during hot period. The new spillway conserves its capacity for both configurations. The PKWeir, placed in front of the tunnel entrance, creates less turbulent flow in the side channel compared to the original design (Figs. 9a, 9b).

During cold period, the water flow evacuated by the opened flap gate is strongly decelerated by the turbulent flow generated in the side channel before entering the tunnel. Therefore, the water level rises in the upstream side of the channel, approaching the crest level of the flap gate. This

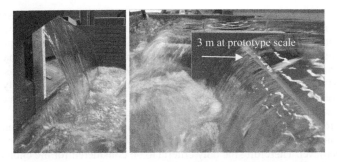

Figure 10. Downstream view of the flap gate jet in the original axial position (left) and in the adapted axial configuration (right).

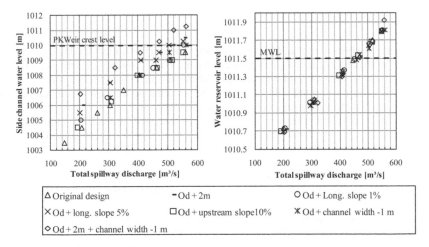

Figure 11. Water level at the upstream end of the side channel (left) and spillway capacity (right) for the different tested channel geometries.

induces a significant drop of the spillway capacity especially near and above the maximum reservoir level (Figs. 8 (right), 9c and 9d).

The first conclusion coming out from the above observations penalizes the SOPC configuration. In spite of the presence of relatively high turbulences in the channel flow of the OPC configuration, this latter has been retained for the following test series.

Another interaction phenomenon has been observed between the PKWeir output flow near the flap gate and the jet of the latter. As shown in Figure 10 (left), the flap gate jet does not adhere to the central spillway pier. The flow evacuated by the near PKWeir unit infiltrates under the flap gate jet and creates instable flow oscillations. In order to protect the output flap gate jet from this underneath flow, it was decided to move the longitudinal axis of the flap gate 3 meters towards the reservoir. The flap gate jet will be then protected by the central spillway pier (Fig. 10 right).

2.4 Influence of the water level in the side channel on the PKWeir capacity

The final optimization step concerns the side channel dimensions and slope. Different geometries have been tested and compared relative to their flow capacity and general behavior (Fig. 11). The modifications of the side channel geometry have been done by changing its width, height and the slope of its bottom.

A first series of tests has been carried out using two different channel depths. The so-called "Original design" (Od) has a horizontal bottom level at el. 998 m. This channel depth was used in all the previous tests discussed above. The configuration called "Od + 2 m", which corresponds to the preliminary design, has the horizontal bottom level at el. 1000 m. To accelerate the flow, two longitudinal channel bottom slopes of 1 and 5% have been simultaneously tested. This defines two new configurations called "Od + longitudinal slope 1%" and "Od + longitudinal slope 5%", respectively. Another configuration for the channel bottom with a longitudinal slope of 10% applied on the first 15 m upstream has been also tested ("Od + upstream slope 10%). Finally, the width of the channel has been reduced by 1 m for the "Od" and "Od + 2 m" configurations.

For all the tests carried out on each of the defined configurations, the capacity of the spillway and the water level at the upstream end of the side channel have been measured, in order to evaluate the downstream water level effect on the PKWeir output capacity (Fig. 11).

The obtained results show that the different tested configurations of the side channel have a very low impact on the discharge capacity of the new spillway, even so the PKWeir crest lies below the channel water level. Therefore, the only criteria to be considered for the choice of the optimal channel configuration were the flow capacity of the entrance of the tailrace tunnel and the rock excavation volume. The side channel configuration "Od + 2 m" revealed to be the best one. It offers also a sufficient capacity for the 5000 years return period floods.

3 CONCLUSIONS

The experimental study presented in this paper has been a powerful support for designers to optimize the new spillway structure of the Gage II dam. Such complicated spillway, composed of a PKWeir, a flap gate weir and a side evacuation channel, is the master piece for reducing the water pressure loads on the arch dam during the cold season and for evacuating the new estimated flood events that almost triples the old flow peak used for the original design.

The preliminary spillway configuration that has been designed by the EDF Engineers could be optimized through a series of physical tests carried out in the Laboratory of Hydraulic Constructions (LCH-EPFL) in Lausanne. The optimized configuration is the combination of a 7 units PKWeir and a 27.4 m wide flap gate weir whose longitudinal axis was shifted upstream by 3 m relative to the PKWeir axis. The flap gate weir reveals to be more efficient when it is located in front of the entrance of the tailrace tunnel. The side channel depth and width were maintained as in the preliminary design.

In this ongoing project, additional laboratory tests will be carried out to study the flow in the tailrace tunnel and the behavior of the jet issued from a flip bucket. These tests are focused on the optimization of the geometry of the flip bucket with the aim to mitigate the jet impact on the river bed downstream.

REFERENCES

Bieri, M., Federspiel, M., Boillat J-L., Houdant B. and Delorme, F. 2010. Spillway discharge capacity at Gloriettes dam. *Hydropower and Dams*. Issue Five, 88–93.

Falvey, T. H. 2003. *Hydraulic design of labyrinth weirs*. ASCE Press.

Flow-3D 2010, Computational fluid dynamics software. Flow science Inc (www.flow3d.com)

Hager, W. H. & Schleiss, A.J. 2009. *Traité de Génie Civil Volume 15 - Constructions Hydrauliques, Ecoulements stationnaires*. Switzerland: PPUR – Presses Polytechniques Romandes.

Leite Ribeiro, M., Boillat, J-L., Schleiss, A., Laugier, F. and Albalat, C. 2007. Rehabilitation of St-Marc dam – Experimental optimization of a Piano Key Weir. *32nd Congress of IAHR, The International Association of Hydraulic Engineering and Research*. Venice, Italy.

Leite Ribeiro, M., Bieri, M., Boillat, J-L., Schleiss, A., Delorme, F. and Laugier, F. 2009. Hydraulic capacity improvement of existing spillways – Design of piano key weirs. *23rd Congress of Large Dams*. Question 90, Response 43. 25–29 May 2009. Brasilia, Brazil.

Pralong, J., Blancher, B., Laugier, F., Machiels, O., Erpicum, S., Pirotton, M., Leite Ribeiro, M., Boillat. J-L., and Schleiss, A.J. 2011. Proposal of a naming convention for the Piano Key Weir geometrical parameters. *International Workshop on Labyrinth and Piano Key weirs*, Liège, Belgium.

United States Department of the Interior 1987. *Design of Small Dams*. A water resources technical publication, Third edition.

Vermeulen, J. & Faramond, L. & Gille, C. 2011. Lessons learnt from design and construction of EDF first Piano Key Weirs. *Workshop on Labyrinth and Piano Key Weir*, Liège. Belgium.

Labyrinth and Piano Key Weirs – PKW 2011 – Erpicum et al. (eds)
© 2011 Taylor & Francis Group, London, ISBN 978-0-415-68282-4

Piano Key Weir design study at Raviege dam

S. Erpicum
Laboratory of Engineering Hydraulics (HACH), ArGEnCo Department –
MS²F – University of Liège (ULg), Liège, Belgium

V. Nagel & F. Laugier
Electricité de France, Centre d'Ingénierie Hydraulique, Savoie Technolac,
Le Bourget du Lac, France

ABSTRACT: The 40-m high Raviege dam, built in 1957 and operated by Electricité de France (EDF), is located in South West of France, on the Agout River. The discharge capacity of the two existing gated spillways located in the center of the dam is not sufficient regarding the updated design flood. One of the solutions studied is a project combining the use of the reservoir storage capacity and a new Piano Key Weir spillway on top of the dam, beside the existing gates, aims to increase the dam safety towards floods. Scale model tests have been carried out at the Laboratory of Engineering Hydraulics of the University of Liège to optimize the PKW design and precisely characterize its discharge capacity. The scale model study also focused on the design of energy dissipation structures downstream of the PKW. This rehabilitation project is depicted in the paper, with focus on the PKW design.

1   PROJECT OVERVIEW

1.1   *Context*

The Raviege dam, operated by Electricité de France (EDF), is located in South West of France, near the town of Toulouse, on the Agout River. This concrete buttress 40-m high dam (Fig. 1) was

Figure 1.   Raviege Dam at the end of the construction.

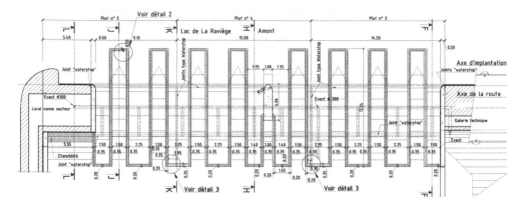

Figure 2. Plan view of the PKW.

built in 1957 for electricity production. The hydropower plant is located at the dam toe, on the right side of the river.

The discharge capacity of the two existing gated spillways, located in the center of the dam, is 1,000 m³/s under the Maximum Water Level. Hydrology calculations of the dam catchment area have been recently updated. The new extreme flood is far more significant than the one considered at the dam design stage. The design flood peak discharge has indeed been raised from 1,000 m³/s to 1,720 m³/s.

The dam release capacity is thus not sufficient regarding extreme floods and a project aiming at increasing the dam safety has been studied. One of the solutions combines the use of the reservoir storage capacity and a new PKW spillway on top of the dam, beside the existing gates, to face safely the new extreme floods.

1.2 *General features*

The rehabilitation project may consist in building a new spillway on the available crest of the dam, on the left side of the existing spillways. As the two existing spillways are gated, it has been decided to design the new structure as a free overflow spillway, which is considered to be more secure. Indeed, it doesn't need human action nor necessitate a mechanical process to operate.

In this framework, the PKW spillway appears to be a good solution to minimize the new spillway width on the dam crest while reducing the volume of the secondary structures to be built. Indeed, more than the spillway itself, the project includes demolishing the upper part of the dam and building a bridge upon the new spillway to maintain the existing road and an energy dissipation structure.

1.3 *Focus on the PKW design*

The width available on the dam crest to build the PKW is about 40 m. Due to operation and stability constraints, the maximum head available above the normal reservoir level should be limited to 1 to 1.5 m. Indeed, in the existing configuration, the maximum reservoir level is 1 m higher than the normal reservoir level.

Considering the reservoir flood routing and the existing spillways capacity, the PKW has to be able to release around 400 m³/s under the maximum head. On the basis of these entry data, the specific discharge required for the PKW is approximately 10 m³/s/m.

The PKW has been designed on the basis of the "Model B" from Ho Ta Khanh (Truong Chi et al., 2006). It is 39.60 m wide, 4.67 m high and counts for a rather important number of keys (10 inlets and 8 outlets – Fig. 2) compared to the similar structures already built in France (Goulours Dam, St Marc Dam, l'Etroit Dam and Gloriettes Dam) (Laugier, 2007, Laugier et al., 2009).

A particularity of the PKW design is the presence of a 1 m wide central pier to support the road bridge. 1 m high and 0.2 m wide parapet walls with a semi circular crest have been placed along

Table 1. Main dimensions of the PKW.

Description	Notation	[m]
Upstream-downstream length of the PKW	B	13.24
Upstream overhang crest length	B_o	4
Downstream overhang crest length	B_i	3.24
Base length	B_b	6
Total height	P	4.67
Total width of the PKW	W	39.6
Width of a unit	W_u	4.45
Inlet key width	W_i	2.25
Outlet key width	W_o	1.5
Crests & wall thickness	$T_{[i,o,s]}$	0.35
Developed length of a unit	L_u	30.13
Developed length ratio of a unit	n_u	6.77
Parapet walls height	$P_{p[i,o,s]}$	1
Parapet walls crest thickness	$T_{p[I,o,s]}$	0.2
Nose length (triangular profile)	B_n	2

the whole weir crest. Triangular noses have been designed on the upstream face of the outlet keys. Table 1 summarizes the main geometrical parameters of the structure.

2 SCALE MODEL STUDY

2.1 Numerical-experimental interaction

The Raviege dam is located in a bend of the Agout River, 2 km upstream of the Ponviel dam. In order to determine the physical model boundaries in the upstream reservoir as well as the flow conditions in the natural river downstream of the Raviege dam, numerical modeling of the flow in the Agout River has been performed using the software WOLF, developed by the HACH team.

A 2D approach, considering turbulence effects by means of a depth averaged k-ε mathematical model, has been used to model the flow in the reservoir and to match the scale model surface with the space available in the laboratory. In a first step, the bathymetry of a 1-km long river reach upstream of the dam has been generated on a 1 to 1 m regular grid by digitalizing contour lines provided by EDF. In a second step, the solver WOLF2D has been applied on a multiblock grid covering the whole reach to determine the flow fields for two extreme floods peak discharge, with dam equipment conditions corresponding to the existing situation (two gated spillways) or to the projected one (additional spillway on the left side of existing ones). These operating configurations have been chosen to maximize the flow velocities in the reservoir, and thus inertia forces, in order to help in defining a scale model reservoir surface consistent with the main approach currents to the spillways. In parallel, modeling of the flow in the scale model reservoir has been performed for the same dam operating conditions and discharges. The comparison of the full reservoir flow patterns with the scale model ones enables to optimize the scale model dimensions and to define the position and geometry of the discharge repartition wall upstream of the model, while preserving the equivalence between approach currents to the spillways.

The 2-km long natural river reach between the Raviege dam and the Ponviel dam has been modeled in one dimension with the solver WOLF1D. The reach slope is weak, and the Ponviel dam spillway controls the water depth upstream. The discharge in the river reach is the one spilled at the Raviege dam. Modeling of the back water curves along the whole reach for various discharges and considering different values of the roughness coefficients provides a true prediction of the water levels at the downstream extremity of the scale model. These free surface elevation data have been used to define the downstream boundary condition and the maximum level of the topography to be reproduced in the scale model. The numerical model also helps in studying the impact of the supercritical flow coming out of the spillways on the Agout river regime near the Raviege dam.

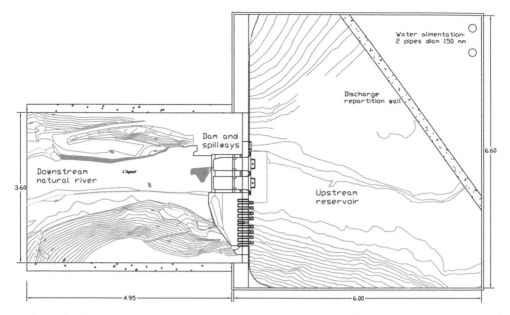

Figure 3. Plan view of the scale model – Dimensions in model m.

In particular, it has been showed that no boundary condition needs to be imposed at the downstream extremity of the scale model for most of the discharge configurations tested with an additional PKW spillway as the flow remains supercritical and the hydraulic jump occurs far downstream from the dam toe.

2.2 Physical model characteristics

The scale model study has been realized considering the Froude similitude, with a 1/35 geometric scale factor. The scale model surface in the laboratory is around $70\,m^2$ (Fig. 3). It represents the whole dam structures, a 150-m long reach of the natural Agout River downstream and a $230 \times 210\,m^2$ area in the reservoir.

The bathymetry and the topography have been modeled with a 5 to 10 cm thick mortar layer posed on concrete blocks inside two waterproof basins (an upstream and a downstream one) made of steel panels or concrete walls and linked by the dam structure. The dam body, existing and projected spillways components, bottom outlets and hydroelectric power plant walls have been represented using PVC, aluminum and synthetic resin.

The water alimentation system is a closed circuit with auto-regulated pumps delivering up to 300 l/s in the model through two 150-mm pipes. Electromagnetic flow-meters, limnimeters, ultrasonic sensors and electromagnetic probes have been used to accurately measure the discharges, the water levels and depths as well as the flow velocities in various points on the scale model.

2.3 PKW design optimization and discharge capacity

The initial design of the PKW considers a 20.22° constant inlet keys slope with vertical 1 m high parapet walls. The influence on the weir discharge capacity of two geometrical variants (Fig. 4) has been assessed during the scale model tests. In variant a, the inlet keys slope is increased to 30° at their downstream extremity to suppress the vertical face of the parapet wall. In variant b, the inlet keys slopes is increased to 25° along the whole keys length, while suppressing the parapet wall at the downstream extremity.

In order to determine the discharge capacity of the PKW in each geometrical configuration, 12 points of the head/discharge curve of the weir have been measured on the scale model for upstream heads between 0.3 and 2.3 m, i.e. considering a PKW level 2 m under the maximum reservoir level

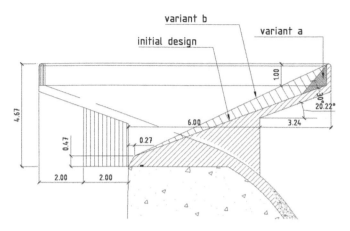

Figure 4.   Initial design and variants in the inlet keys slope.

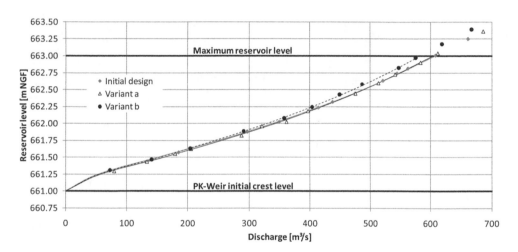

Figure 5.   Head/discharge relation of the PKW. Experimental results for initial design, variant a and variant b of the inlet keys slope.

(Fig. 5). These tests show that the increased constant slope in the inlet keys (variant b) is slightly less efficient than variant a or initial design. Initial design of the inlet keys slope has thus been considered to continue the study.

Experimental results also enable to fit the parameters of an analytical relation linking the discharge on the PKW to the upstream head. Using a similar relation previously determined for both gated spillways, the PKW optimal crest level has been defined on the basis of analytical modeling of the flood storage in the reservoir (Fig. 6).

Additional tests show very limited influence of the existing and projected spillways on their respective discharge capacity, as well as negligible effect of the central pier.

The final design of the new Raviege dam flood control structures enables to maintain the normal reservoir level at its current elevation while releasing the updated design flood with a reservoir elevation of 1.22 m above this normal level. In that case, peak discharge on the PKW is 400 m³/s and 1,050 m³/s through the gates fully opened (Fig. 6).

The difference between the peak discharge entering the reservoir and the global peak discharge downstream of the dam is thus of more than 250 m³/s for the design flood.

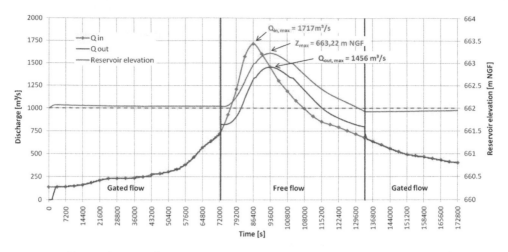

Figure 6. Effect on the design flood hydrogram of the reservoir storage and the spillways operation.

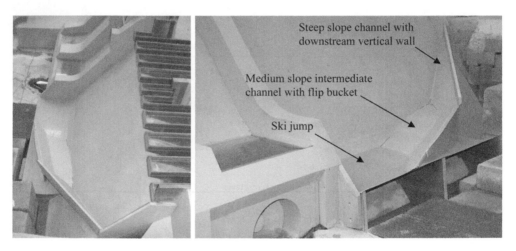

Figure 7. Final design of the energy dissipation structure downstream of the PKW.

2.4 *Energy dissipation system*

Special care has been devoted to the study and design of the energy dissipation system downstream of the PKW. Indeed, as the new weir is located on the left side of the dam, water released on the PKW has to be deviated to the right, towards the main river bed, to avoid damage to the left bank of the valley.

The main difficulty of the design lied in the relative small energy of the flow coming out of the PKW combined to the reduced length of the spillway channel. This made hard the concentration of the flow and its deviation using a flip bucket or convergent walls.

On the basis of a preliminary analysis of the flow patterns downstream of the PKW and considering the satisfactory design of a ski jump obtained during the previous study of a fuse gates solution instead of the PKW, the energy dissipation structure has been designed as a converging smooth channel with varied slopes and downstream deflectors.

The final structure is made of a steep slope channel with a vertical downstream wall on the left side of the PKW, a medium slope channel with a downstream flip bucket in the middle and a ski jump on the right side (Fig. 7). It enables to concentrate the strongly aerated flows coming out of the PKW on around half the width of the PKW and to turn them towards the main river bed, on a place where the flows released through the gated spillways induce a thick water cushion (Fig. 8).

Figure 8. Flow over the existing and projected spillways at Raviege dam – Discharge of 1240 m³/s.

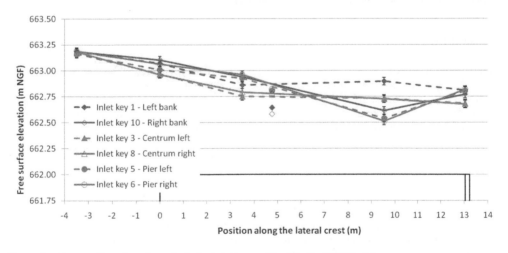

Figure 9. Free surface elevations along the inlet keys – Head on the PKW is 1.2 m.

2.5 *Other measurements*

Water depths and flow velocities have been measured in the reservoir upstream and inside the inlet keys of the PKW.

Flow velocities are rather uniformly distributed along the weir width, as well as the free surface elevation values in the reservoir near the PKW.

Detailed measurements of the water depths in the inlet keys (Fig. 9) show that the PKW behavior is symmetrical. Free surface lines in the central inlet keys 3 and 8 are almost identical, as well as in inlet keys 5 and 6 from side to side of the central pier. This symmetry doesn't exist for inlet keys 1 and 10 as their developed length is different.

Figure 10.   Free surface for a head of 1.2 m on the PKW.

These measurements also show the variation of the flow features in inlet keys of variable width. For high head, the free surface elevation in the central inlet keys 3 and 8 decreases continuously, while in the extremity inlet keys 1, 5, 6 and 10 it shows variations of opposite sign. From upstream to downstream, it begins to decrease, and then it increases, before passing over the parapet wall. This sequence is slightly different in the inlet key 1 with a smaller developed length. The back water curves observed for high heads in the inlet keys 5, 6 and 10 suggest their saturation and a decrease in their developed length because of a critical section located inside the inlet, while extremity inlet keys continue to work with a full developed length (Fig. 10).

3   CONCLUSIONS

In the global framework of a rehabilitation project to significantly increase the discharge capacity at the Raviege dam, a PKW solution has been designed and its geometry has been optimized by means of a scale model study.

Thanks to preliminary numerical modeling, the geometric scale factor of the model has been increased up to 1/35 by drastically limiting the model surface in the reservoir and the downstream river reach.

The experimental test shows that, at constant height of the PKW, a continuous 20.22° slope of the inlet key bottom with vertical parapet wall downstream is more efficient in terms of discharge capacity than an increased continuous slope without the parapet wall. The tests also show the very limited influence of a nearby existing gated spillway on the PKW hydraulic behavior, as well as the limited effect of pier in the central inlet key.

An energy dissipation structure has been designed to deviate the flow coming out of the weir to the right, towards the main river bed. The geometry of this structure is complex because of the small length of the spillway and the low energy of the flow downstream of the PKW.

Free surface measurements show the symmetrical behavior of the PKW and suggest the saturation of the less wide inlet keys for high heads.

REFERENCES

Laugier, F. 2007. Design and construction of the first Piano Key Weir spillway at Goulours dam. *Hydropower & Dams* 14(5): 94–100.
Laugier, F., Lochu, A., Gille, C., Leite Ribeiro, M. & Boillat, J-L. 2009. Design and construction of a labyrinth PKW spillway at Saint-Marc dam, France. *Hydropower & Dams* 16(5): 100–107.
Truong Chi, H., Huynh Thanh, S. & Ho Ta Khanh, M. 2006. Results of some piano keys weir hydraulic model tests in Vietnam. *22ème Congrès des grands barrages – Barcelona*: 581–596. CIGB-ICOLD: Paris.

Labyrinth and Piano Key Weirs – PKW 2011 – Erpicum et al. (eds)
© 2011 Taylor & Francis Group, London, ISBN 978-0-415-68282-4

Nine years of study of the Piano Key Weir in the university laboratory of Biskra "lessons and reflections"

A. Ouamane

Laboratory of Hydraulic Planning and Environment, University Mohamed Kheider, Biskra, Algeria

ABSTRACT: Since more than nine years studies and tests were carried out in association with Hydrocoop-France for developing a new shape of labyrinth weir. This new design was baptized in 2003 Piano Key Weir (PK-Weir). Studies and general tests which were realized during these nine years allowed a better understanding of this type of weir behavior and their geometrical shape optimization, so two geometrical forms were adopted (Model A and B).

This paper presents a synthesis of the results of these studies and recommendations on the geometry of the PK-Weir and the adopted experimental procedure. Two examples of weirs similar to the PK-Weir constructed in Algeria are exposed.

1 GENERALITIES

The labyrinth weirs are characterized by a crest with broken axis in plan, so that the length of the crest is longer than the width of the weir and occupies the same side space as a linear weir. The objective of this design is to increase the discharge capacity by unit of width for a given hydraulic head. The labyrinth weir is the ideal structure to discharge strong floods with low heads, in situations where the maximum head is limited. This type of weir represents an adequate alternative when the width of the weir is limited by the topography and/or the design flood is important.

The first analysis of the hydraulic performance of a labyrinth weir is attributed to Hay and Taylor (1970). It is considered as a base of the labyrinth weir design and was issued concomitantly with criterion and procedure to determine the discharge on a labyrinth weir (Darvas, 1971), (Magalhaes, 1985), (Houston et al. 1982).

2 EXPERIMENTAL STUDY OF THE LABYRINTH WEIR

Since 1995, various research works within the field of spillways were committed to the laboratory of hydraulics of Biskra University (Algeria). These works concerned theoretical and experimental studies of morning glory weirs, stepped weirs, labyrinth weirs and PK-Weirs.

The experimental study of a labyrinth weir showed that the flow over the weir is characterized by two opposite napes overtopping the lateral walls of the weir. When increasing the hydraulic head, the two opposite flows meet and a central jet appears with a disturbed flow around the corners, whose length evolves proportionally with the length of disturbance.

The connection between the length of disturbance ld and the relative head H^*/P was defined to be a linear relation of the shape: $ld/lc = a \times H^*/P + b$, wherea and b are a constant for every corner. The relevant parameters are defined on Figures 1 and 3.

The connection between the relative length of disturbance ld/lc and the relative length of the central jet ljc/lc corresponds to a unique straight line whatever the corner (Fig. 2).

$$ljc/lc = 0.6 \times ld/lc \qquad (1)$$

This relationship shows that the length of the central jet ljc is closely linked to the evolution of the disturbance that affects the flow over a labyrinth weir.

The effect of the height of the labyrinth weir upstream (P) and downstream (D) was studied for the three configurations (P = D, D/P = 1), (P < D, D/P > 1), (P > D, D/P < 1) showed on Figure 3.

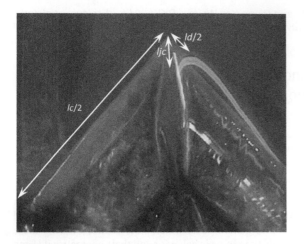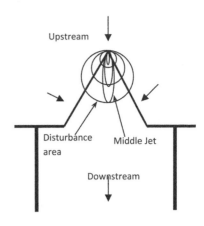

Figure 1. Disturbance flow on the labyrinth weir.

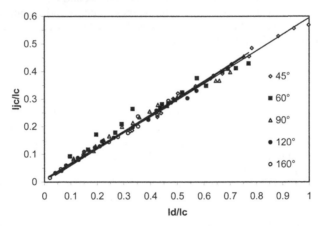

Figure 2. Relative length of the central jet according to the length disturbance at the corner.

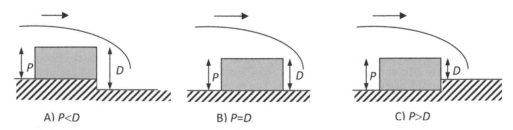

Figure 3. Schematic diagram of labyrinth weir with various cases of slab levels upstream and downstream.

The results revealed that no improvement of the performance of the labyrinth weir can be obtained by raising the downstream slab higher than the upstream one (D/P < 1). However, when the downstream height of the weir is superior to the upstream one (D/P > 1) the performance is improved.

Studies conducted on four models of labyrinth weirs, of triangular, trapezoidal, rectangular and trapezoidal upstream rounded shapes (Fig. 4), showed that the most recommended shapes, as far as hydraulics and economics are concerned, correspond to the trapezoidal upstream rounded and the trapezoidal shapes.

To a certain extent, it is more convenient to increase the number of cycles for a given width of the labyrinth weir. This reduces the longitudinal length of the weir and therefore the concrete volume of the slab between the sidewalls of the weir.

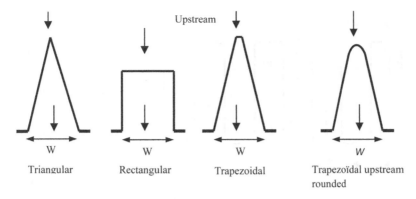

Upstream

Triangular   Rectangular   Trapezoidal   Trapezoïdal upstream rounded

Figure 4.   Schematic plan view of the experimented models of labyrinth weirs.

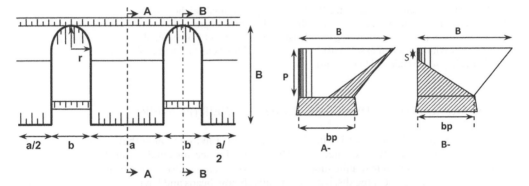

Figure 5.   Schematic model of labyrinth weir with downstream overhang and inclined slab.

Figure 6.   Flow on the labyrinth weir with downstream overhang and inclined slab.

Generally, the labyrinth weirs are designed with vertical walls; but, it is possible to design a portion of the crest weir in overhang by inclining the downstream frontal wall (Figs. 5 and 6).

This layout represents an economic shape which allows reducing the base of the weir and consequently, the construction of this type of weir on a concrete dam becomes possible. The advantage of this type of weir lies in its base length which is reduced, according to the layout with overhang, and the aeration of the flow downstream.

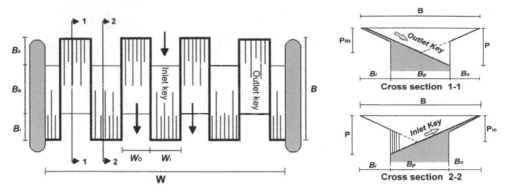

Figure 7.   Schematic diagram of PK-Weir.
Where:
B   : Upstream-downstream length of the PKW P_m: Maximal height of walls
B_i : Upstream (inlet key) overhang crest length B_b : Base length
B_o : Downstream (outlet key) overhang crest length   W : Total width of the PKW
W_i : Inlet key width (sidewall to sidewall)   P   : Total height of the PKW
W_o: Outlet key width (sidewall to sidewall)

3   EXPERIMENTAL STUDY OF THE PIANO KEY WEIR (PK-WEIR)

For ten years, the Laboratory of Hydraulic Planning and Environment at theUniversity of Biskra (Algeria) has been deeply associated with Hydrocoop France for conducting theoretical studies and experimental tests on hydraulic models in order to develop a form of labyrinth weir presenting the advantage to evacuate large discharges at relatively low heads and having low construction and maintenance costs relatively to gated spillways.

Consequently, a shape of weir with free surface flow, baptized Piano Key Weir (PK-Weir), was developed; this shape presents the advantage to be realized either on the top of concrete dams, or in channels laterally to earth dams.

The geometry of the PK-Weir is defined by a rectangular arrangement of the inlet and outlet keys, with tilted slabs and a reduced width due to the rectangular shape (Fig. 7).

For a better understanding of the functioning, of the flow behavior, as well as the effect of various geometrical parameters on the hydraulic performance, tests were conducted on several models, all taking into account the construction cost. More precisely, following themes were tackled:

(i) the efficiency of the PK-Weir according to the total head on the weir, (ii) the total height of PK-Weir (P), (iii) the width of the inlet (W_i) and the outlet (W_o), (iv) the length of overhangs (B_i) and (B_o), (v) the ratio L/W, (vi) the location of the weir (on the wall of a dam or in a canal), (vii) the difference of height between the inlet and outlet slabs, (viii) the effect of the slope of outlet, (ix) the shape of the nose under upstream overhangs, (x) the shape of the part of entrance above the sill of the weir, and (xi) the effect of floating debris.

The tests were conducted at first on models of type A (with upstream and downstream overhangs) and in a second time on models of type B (with only a downstream overhangs).

The results obtained from these tests constituted a base for designing the PK-Weir and better understanding its functioning for various conditions.

4   LESSONS AND REFLEXIONS ON PK-WEIRS

The flow on a PK-Weir is completely different from that on the labyrinth weir. For moderate heads, the flow is characterized by two layers: the first as a jet coming from the bottom, which flows along the inlet slab and the second more or less thin according to the head over the weir. The second layer, which is superficial, favors the aeration of the weir. When the head on the weir becomes large, the two layers converge and form an integral single nappe with a disturbed flow downstream.

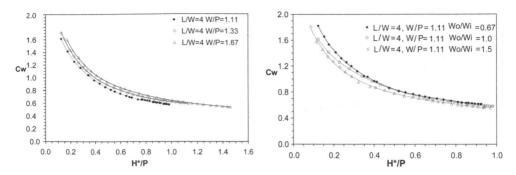

Figure 8. Discharge coefficients with respect to the ratio of the vertical aspect (left) and to the ratio of the inlet and outlet widths (right).

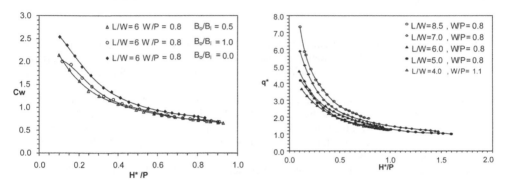

Figure 9. Discharge coefficient according to the overhangs lengths ratio (left) and the performance according to the ratio L/W (right).

The flow on a weir is generally influenced by the geometrical parameters, the configuration of PK-Weir is defined by several geometrical parameters, and consequently, the discharge can be affected by an important number of these parameters.

Indeed, the various experimental studies have highlighted that the weir height has an effect on the PK-Weir discharge (Fig. 8, left), which is more important at relative small heads (an increase of 25% of the height of the PK-Weir increases the discharge by 6%). It has to be noted that the interest of increasing the height of the weir is limited to a given value of the relative head. In other words, beyond a certain head on the PK-Weir the effect of a height increase becomes insignificant. The choice of the height must also take into account the site conditions and the construction cost.

The rectangular shape of both inlet and outlet allows to distinguish between three configurations in plan: (i) the inlet is wider than the outlet $W_i > W_o$, (ii) the inlet and outlet have a same width $W_i = W_o$ and (iii) the inlet is smaller than the outlet $W_i < W_o$. Tests showed that the most advantageous arrangement corresponds to an inlet width greater than the outlet's one (Fig. 8, right). The optimal width of the inlet lies between 1,2 and 1,5 times the width of the outlet.

The geometry of the PK-Weir, which is characterized by a large part of the crest in overhang, can have two different configurations: the first with two overhangs, one upstream and one downstream (model A), symmetric or not symmetric, and the second configuration with only an upstream overhang (model B). According to the experiments (Fig. 9, left), the configuration with only an upstream overhang is characterized by a greater efficiency than the one with overhangs upstream and downstream, especially for high discharges. The configuration without downstream overhang can be recommended for high discharges provided that it is structurally stable during the periods when the level in reservoir is lower than the level of the sill.

The solution with similar overhangs favors the use of precast concrete elements which can be used for specific discharges of up to about $20 \, m^3/s/m$. This solution can be more interesting to improve the discharge capacity of several existing weirs.

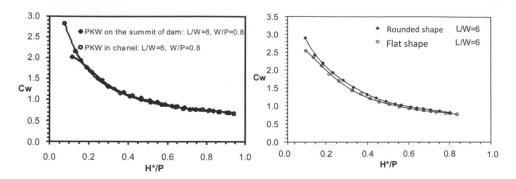

Figure 10. Discharge coefficient according to the location of the PK-Weir (left) and to the shape entrance under the overhangs (right).

The solution without downstream overhang offers a significant saving (about 10%) with regard to the solution with symmetric overhangs and structural efforts are less important for large specific discharges. So, this could be the most attractive choice for several future large dams.

Generally, the elongation of the crest, which results from the non-linear shape of PK-Weir, is expressed by L/W which is the ratio between the total length of the crest and the width of the weir. The discharge increases with the ratio L/W, and is more significant for L/W values greater than 6 (Fig. 9, right). The results show that even when L/W is greater than 8, the PK-Weir continues to have a increasing efficiency.

However, the increase of efficiency of PK-Weir decreases with the head increase; this implies a reduction of the performance which is more significant on the weirs with large values of L/W. The large values of elongation of the crest are profitable only for lower relative heads (about 0.3), and the low values L/W do not favor a big improvement of the capacity of the PK-Weir. So, the most optimal ratio lies between L/W = 5 and 6.

One advantage of the PK Weir is its ability to be designed on the top of concrete dams or inserted into the channels of earth dams or through the natural waterways. Studies have shown that the hydraulic performance is the same for various cases of location (Fig. 10, left). Therefore, the PK Weir can be a cost effective solution for both concrete and earth dams.

The design of the slab of the outlet as a ski jump does not affect the capacity of a PK-Weir. This design of slab allows taking the discharged water jet away from the downstream face of the dam.

The entrance shape under the upstream overhangs affects the hydraulic performance of the PK-Weir. The rounded shape is significantly better than a flat one (Fig. 10, right). Consequently, it is recommended to use a profiled shape under the overhangs upstream.

The plugging of the weir by floating debris can have serious consequences on the safety of the dam. Therefore, it was important to verify the operation of the PK-Weir in such conditions. The results indicate that for hydraulic heads lower than P/3, the weir operates normally. Beyond this head the floating debris begin to pile up the entrance of the weir thereby causing a partial blockage and consequently, a reduction of the PK-Weir capacity (Fig. 11, left). As the head increases gradually, the floating debris are evacuated downstream, without any permanent blocking in the inlets and outlets (Fig. 11, right) and finally the weir retrieve its initial capacity.

The variation of the shape of the PK-Weir is possible; its geometrical configuration can undergo some modifications of secondary importance according to the type of the dam and the local conditions of the site. These modifications can concern, (i) the shape of overhangs and their arrangement, (ii) the shape of the outlet apron (inclined, horizontal or in steps). For that purpose, three forms of PK-Weir were suggested and studied in the laboratory to verify the type of flow and the hydraulic performance of these geometries (Fig. 12).

To verify the hydraulic performance of these three forms of PK-Weir several tests were realized recently with free surface flow conditions and in the presence of floating debris.

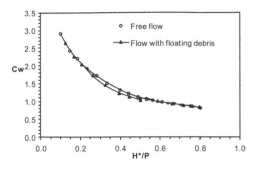

Figure 11.   Discharge coefficient according to the presence of floating debris (left) and flow with floating debris on the PK-Weir (right).

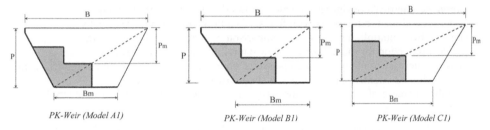

Figure 12.   Various shapes of PK-Weir.

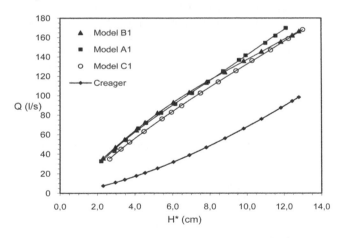

Figure 13.   Rating curves of the three different models A1, B1, C1 and comparison with a linear Creager.

The effect of overhangs was studied for three cases:

1-PK-Weir with upstream and downstream overhangs (Model A1)
2-PK-Weir with upstream overhang and vertical wall downstream (Model B1)
3-PK-Weir with vertical wall upstream and downstream overhang (Model C1)

The experiments realized on the PK-Weir for verifying the effect of overhangs on the hydraulic performance showed that the profiled shape under overhangs is recommended. It is possible that the overhangs will be tilted from the top to the base of the PK-Weir. With such geometry, the overhang can have a profiled shape over all its height, in order to improve the hydraulics and avoid the jamming of floating debris during floods. The results obtained on the three models (Fig. 13) show that the hydraulic performance of weirs with upstream overhang is better than without upstream overhang.

57

The results show that, in comparison to the linear Creager weir, the discharge capacity at low heads (H* < 0,2P) is four times higher for arrangements with upstream overhangs (model A1 and B1) and 3,5 times without upstream overhang (model C1). For greater heads (H* > 0,45P) the performance is double.

5 CONCLUSIONS

The experimental studies, performed on more than eighty models of spillways over ten years, were mainly devoted to the investigation of two types of labyrinth weirs: the first and the second piano key. The results provide a basis for the design of these types of weirs.

The studies showed that the piano keys weir (PK-Weir) represents an effective solution for increasing the storage capacity and/or the flood evacuation capacity on the majority of existing dams. It can also be an economic solution for new dams. The PK-Weir is characterized by:

- a simple geometrical configuration which allows the use of prefabricated units,
- an operation similar to the weirs with free surface flow but much more effective,
- a specific flow multiplied by 2 to 4 in comparison with a standard linear weir,
- a big cost reduction for the majority of new dams and a guarantee of safety,
- a possible increase of many existing reservoirs storage at a cost in the range of 0,05 $ per m³ in most developing countries, under 0.5 $/m³ in industrialized countries,
- a possible increase of the capacity for many existing dams with 0,5 m³ of reinforced concrete by additional m³/s.

The PK-Weir with overhangs from the top to the base represents an effective solution for the reshaping of existing weirs and on new dams. The shape is easy to build. It can be recommended for dams on rivers characterized by important floating debris, and can be an adequate solution for RCC dams by settling the PK-Weir on the crest of the dam and by fitting out steps inside the outlet.

Nevertheless, a lot of work remains to be done to better understand the flow characteristics over these weirs and define standard forms of PK-Weir according to site conditions and type of dam.

As quoted by J. Gruat and C. Thirriot (1983), "the practice of the model is a slow education which can't be transmitted only at the Socratic way of the daily encounters".

REFERENCES

Darvas, L.A. 1971. Performance and design of labyrinth weirs. *Journal of Hydraulic Division*, N° 08, pp. 1246–1251.
Hay, N., Taylor, G. 1970. Performance and design of labyrinth weirs. *Journal of the hydraulics division Proceedings of ASCE*, Vol 96, N° 11, pp. 2337–2357.
Houston, K. L. and DeAngelis C. S. 1982. A Site of Specific Study of a Labyrinth Spillway. *Proceeding of the conference Applying Research to Hydraulic Practice, Hydraulics Division of ASCE*, pp. 86–95.
Lempérière, F. & Ouamane, A. 2003. The Piano Keys Weir: a new cost-effective solution for spillways. *The International Journal on Hydropower & Dams*, Issue Four.
Magalhães, A. P. 1985. Labyrinth-Weir Spillways. *15th Congress on Large Dams, ICOLD*, Vol. IV, Q59, R 24, pp. 395–407, Lausanne.
Ouamane, A. & Lempérière, F. 2006. Design of a new economic shape of weir. *International Symposium on Dams in the Societies of the XXI Century*, Barcelona, Spain.
Ouamane, A. 2006. Hydraulic and Costs data for various Labyrinth Weirs. *22th Congress on Large Dams, ICOLD*, Q84, Vol IV, Barcelona, Spain.
Ouamane, A. & Lempérière, F. 2007. Increase of the safety of existing dams – Rehabilitation of weir. *Symposium on Dam Safety Management. Role of State, Private Companies and Public in Designing, Constructing and Operating of Large Dams, ICOLD 75th Annual Meeting* Saint Petersburg, Russia.
Ouamane, A. & Lempérière, F. 2008. The Piano Key Weir is the solution to increase the capacity of the existing spillways. *Hydro 2008 Conference*, Ljubljana, Slovenia.
Ouamane, A. & Lempérière, F. 2010. Study of various alternatives of shape of piano key weirs. *HYDRO 2010 – Meeting Demands in a Changing World*, Congress Centre, and Lisbon, Portugal.

Labyrinth and Piano Key Weirs – PKW 2011 – Erpicum et al. (eds)
© 2011 Taylor & Francis Group, London, ISBN 978-0-415-68282-4

Influence of the Piano Key Weir height on its discharge capacity

O. Machiels*, S. Erpicum, P. Archambeau, B. Dewals & M. Pirotton
Laboratory of Hydrology, Applied Hydrodynamics and Hydraulic Constructions (HACH), ArGEnCo Department, Liège University, Liège, Belgium
*Fund for education to Industrial and Agricultural Research, F.R.I.A.

ABSTRACT: Within the framework of Piano Key Weir (PKW) development, the hydraulic optimization of its geometry seems essential. First systematic studies of all the geometrical parameters of the PKW highlight a more important influence of some parameters compared to other ones. A former study led, at the University of Liège, on a large scale model of PKW highlights the main influence of the inlet cross section on the global discharge capacity of the weir. Based on this finding, the geometrical parameters influencing the inlet cross section seem to be the most influential parameters of the weir efficiency. This paper presents results of an experimental study using models of PKW with varying height. The mean non-dimensional parameters characterizing the geometry of the models are maintained constant, only varying the slope of the alveoli. The results of the 7 models studied, with a wide range of slopes (height over length) from 0.25 to 1.5, are presented. Based on these results, an analytical formulation of the unit discharge over a PKW as a function of its height is developed.

1   INTRODUCTION

The Piano Key Weir (PKW) is a particular form of labyrinth weir, developed by Lempérière (Blanc & Lempérière 2001, Lempérière & Ouamane 2003), using up- and/or downstream overhangs to limit its basis length and permit its use directly on the dam crest. The PKW is so a cost effective solution for rehabilitation but also for new dam projects with limited spillway apron space or reservoir segment available to release a large design discharge. The first scale model studies showed that this new type of weir can be four times more efficient than a traditional ogee-crested weir at constant head and crest length on the dam (Ouamane & Lempérière 2006a).

The first prototype size PKWs were built by "Electricité de France (EDF)" in France since 2006 (Laugier 2007, Bieri, et al. 2009, Laugier, et al. 2009, Leite Ribeiro, et al. 2009), the definition of the optimal geometry of the structure, however, has been still poorly approached. Until now, the hydraulic design of a PKW is mainly performed on the basis of experimental knowledge and scale model studies, modifying step by step an initial geometry following the ideas of the project engineers (Leite Ribeiro, et al. 2007, Cicero, et al. 2010).

The geometric specificities of the PKW involve a large set of parameters increasing the difficulty of a systematic optimization of the geometry. The "PKW-unit" can be defined as the basic structure of a PKW, composed of two transversal walls, an inlet and two half-outlets. The main geometric parameters of a PKW are the weir height P, the weir width W, the number of PKW-units N_u, the basis length B_b, the inlet and the outlet widths W_i and W_o, the up- and downstream overhang lengths B_o and B_i, and the wall thickness T (Figure 1).

The first parametrical studies, carried out to characterize the influence of a number of these geometrical parameters, have shown the interest of increasing the inlet/outlet widths ratio W_i/W_o, the upstream overhang length B_o and the weir height P, to increase the release capacity of the PKW (Ouamane & Lempérière 2006b, Le Doucen, et al. 2009, Machiels, et al. 2010a). A profiling of the shape of the upstream overhang also contributes to increase the discharge capacity (Ouamane & Lempérière 2006b). Some studies have finally shown the interest of the PKW in terms of aeration capacity and floating debris response (Lempérière & Ouamane 2003, Hien, et al. 2006). Through all

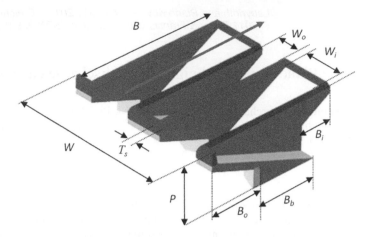

Figure 1.   3D sketch of two PKW-units and main geometric parameters (Pralong, et al. 2011).

these parametrical studies, only one optimization curve, concerning the influence of the inlet/outlet widths ratio, has been proposed (Machiels, et al. 2010a).

In order to improve the understanding of the flow in PKW, a physical approach of the flow behavior along the structure has been initiated at the University of Liège (Machiels, et al. 2009). The results of this study highlight the apparition of a critical section in the inlet when velocities become significant. This critical section apparition limits the efficient length of the crest and so permits to explain the observed decreasing of the PKW efficiency for increasing head (Machiels, et al. 2010b). It is thus interesting to increase the inlet cross section, limiting high velocities occurrence, in order to increase the release capacity of the PKW. That explains the former observations concerning the inlet/outlet widths ratio, the position of the overhangs and the weir height, which are the three mean geometric parameters acting on the inlet cross section.

In order to follow the physical approach of the design optimization of the PKW started for the flow behavior study, an experimental study of the influence of the three main geometric parameters highlighted has been initiated. The first results of this study, presented in this paper, show the influence of the bottom slope of the alveoli on the release capacity, by the study of seven models of variable height.

2   EXPERIMENTAL SETUP

A specific experimental facility made of a 7.2 m long, 1.2 m wide and 1.2 m high channel has been built to perform the scale model tests (Figure 2). The channel is fed by two pumps delivering up to 300 l/s in an upstream stilling basin. The upstream entry of the channel is equipped with a metal grid and a synthetic membrane ensuring uniform flow conditions. Two Plexiglas plates on both channel sides allow the observation of the flow patterns on the whole channel height at the location of the PKW models. Specific convergent structures reduce the channel width to the one of the tested model. Finally, a 0.2 m high support simulates the dam under the tested models.

Seven models of PKW, with only varied height (other geometrical parameters are kept constant) providing varied bottom slopes of the inlet and outlet ($S_o = S_i = 0.25$; 0.375; 0.5; 0.6; 0.75; 1; 1.5), have been tested (Figure 2). The inlet/outlet widths ratio W_i/W_o of all models is 1.5, the ratio between the total length of the crest and the width of the weir L/W is 5, and the two overhangs are symmetric and equal to the third of the lateral crest length B.

Discharge measurements have been performed with an electromagnetic flowmeter with an accuracy of ± 1 l/s. Nine electronic limnimeters have been used to measure the water free surface level with an accuracy of ± 0.5 mm. Two of them have been placed at the entrance of the channel to allow the water height measurements for the drawing of the head/discharge curves. Four have been placed along the center of the inlet and three along the outlet center to provide the free surface profiles (Figure 2).

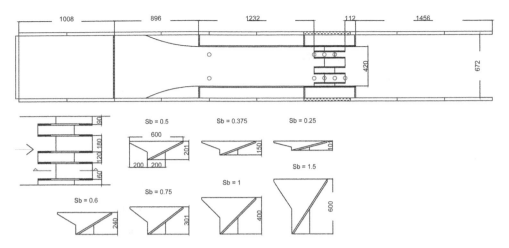

Figure 2. Experimental layout (Dimensions in mm) and position of limnimeters (circles).

3 RESULTS

The study of the effect of bottom slopes has been carried out in three ways to isolate different expected effects. In parallel to the study of the effect of bottom slopes of the alveoli on the PKW release, a study of the effect of bottom slope variations on the release capacity of a sharp crested weir has been done to characterize the influence on the release capacity of the downstream part of the PKW crest. A second part of the study on the PKW models aims also to verify that hysteresis effects will not occur comparing rising or decreasing discharge approach.

3.1 Effect of the bottom slope on the downstream crest capacity

To highlight the influence of the approach bottom slope on release capacity of the downstream crest, five sharp-crested weir models with varied bottom slopes ($S_b = 0.27$; 0.36; 0.47; 0.7; 1.19) have been studied. All the models have been placed normal to the experimental channel with the same crest level $P = 0.5$ m over the bottom of the channel.

The Figure 3 shows the variation of the unit discharge q with the upstream head H for the five tested models. The head/discharge curves of the five weirs are similar compared with the measurements accuracy. It seems thus that there is no significant effect of the variation of the bottom slope on the release capacity of the sharp-crested weir. However, the curves don't correspond with the SIA (1926) equation results for sharp-crested weirs:

$$q_{SIA} = 0.41\left(1 + \frac{1}{1000H + 1.6}\right)\left(1 + 0.5\left(\frac{H}{H + P}\right)^2\right)\sqrt{2gH^3} \qquad (1)$$

Boussinesq (1877) proposed a correction of the discharge equation of sharp-crested weirs to take into account the weir inclination, in function of the angle i, in degrees, with the vertical:

$$q = q_{Vertical}\left(1 + 0.3902\frac{i}{180}\right) \qquad (2)$$

The results have shown that this correction is applicable until the angle i is equal to 40°, corresponding with $S_b = 1.2$ (Figure 3). For higher values of the inclination, corresponding with smaller bottom slopes, there is no more effect of the angle variation observed in the results.

For usual PKW geometries, with bottom slopes varying from 0.3 to 0.6, there is thus no effect of the slope variation on the downstream crest capacity. The discharge of the downstream part of the inlet crest could then be calculated combining equations (1) and (2) with $i = 40°$.

61

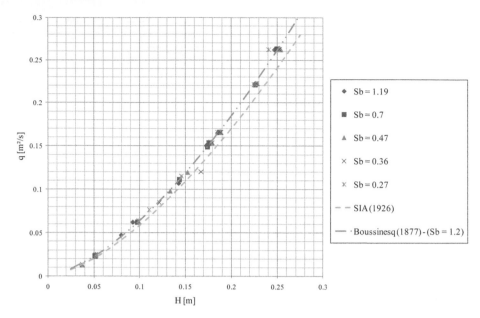

Figure 3. Head/discharge curves of the inclined sharp-crested models and comparison with SIA and Boussinesq formulations.

Figure 4. Variation of the air entrainment for high heads ($S_i = 0.25$, left – $q = 0.167\,\mathrm{m^2/s}$; $H/P = 1.55$, right – $q = 0.399\,\mathrm{m^2/s}$; $H/P = 3.17$).

3.2 *Hysteresis effect*

With variations of the upstream head, variations of the air entrainment can be observed on the tested models of PKW. A former study on a larger scale model of PKW had already highlighted the variation of the nappe shape, from a leaping or an adherent nappe to a springing one, for very low heads ($H/T < 0.4$) (Machiels, et al. 2009). For high heads, the transition from fully aerated nappes to leaping nappes can also be observed (Figure 4). This transition is observed, on the four smallest tested models, for the same range of the ratio between the upstream head and the crest height from the bottom of the channel H/P_T near 0.55.

The air entrainment may influence the discharge capacity of the weir. The transition heads highlighted may thus change in function of the rising or decreasing approach of the head. Hysteresis effects may thus occur. However, the comparison of the head/discharge curves of the tested PKW considering rising or decreasing heads shows that there is no hysteresis effect despite variation of the nappe aeration (Figure 5). This result, confirmed on all the tested geometries, allows the study of a particular head whatever the approach of it.

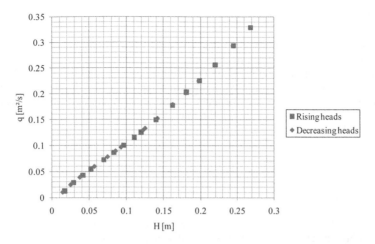

Figure 5. Head/discharge curves of the PKW ($S_i = 0.25$) considering rising or decreasing heads.

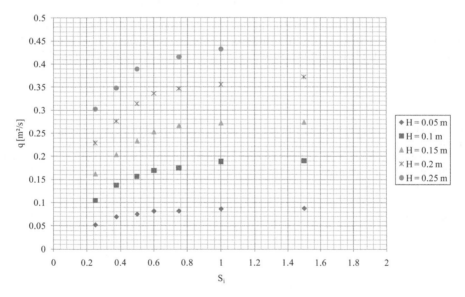

Figure 6. Evolution of the discharge with the bottom slope for different heads.

3.3 *Effect of the bottom slope on the PKW capacity*

The results of the study of the flow along a large scale model of PKW (Machiels, et al. 2010b) show that a control section appears in the inlet when the upstream head becomes significant. This control section moves upstream along the inlet with increasing head, decreasing the effective length of the crest and so the release capacity of the weir. Regarding this assumption, there is an interest of increasing the inlet cross section to limit high velocities, involving this critical section apparition.

Figure 6 shows the variation of the unit discharge q on the tested PKW in function of the bottom slope of the inlet S_i for different values of the upstream head H. As expected, increasing the bottom slope of the inlet increases the global release capacity of the weir. However, there is a limit over which increasing this slope no longer changes the release capacity of the PKW. This limit increases with increasing head corresponding with the minimal slope necessary to avoid critical section apparition in the inlet.

4 ANALYTICAL APPROACH

Combining the results of the studies of the sharp crested inclined weirs and of the PKW with varied bottom slopes, with the conclusions of the study on a large scale model of PKW, an analytical approach of the effect of the inlet slope on the PKW release capacity has been developed. The discharge over a PKW geometry can be divided in three parts: the discharge evacuated on the upstream crest of the outlet, the one evacuated on the downstream part of the inlet and the one evacuated on the lateral crests in-between the inlet and the outlet. The unit discharge of a PKW geometry could thus be calculated as:

$$q = q_{upstream} \frac{W_o}{W_u} + q_{downstream} \frac{W_i}{W_u} + q_{lateral} \frac{2B}{W_u} \tag{3}$$

where W_u is the PKW-unit width.

Regarding the results of the inclined sharp-crested weirs study, the discharge on the downstream crest is only a function of the ratio between the water head and the weir height H/P and can be calculated combining equations (1) and (2) for and an angle i of $40°$:

$$q_{downstream} = 0.445 \left(1 + \frac{1}{1000H + 1.6}\right) \left(1 + 0.5 \left(\frac{H}{H+P}\right)^2\right) \sqrt{2gH^3} \tag{4}$$

Assuming the same behavior for the upstream crest, the discharge could be calculated using the same equations with an angle i of $-40°$. However, for the upstream crest, the crest height P_T must be considered from the bottom of the channel adding the weir and dam heights P and P_d:

$$q_{upstream} = 0.374 \left(1 + \frac{1}{1000H + 1.6}\right) \left(1 + 0.5 \left(\frac{H}{H+P_T}\right)^2\right) \sqrt{2gH^3} \tag{5}$$

For the lateral crest, the SIA equation (1) can be used considering the water height as the head and correcting the discharge to take in account of the flow inertia in the inlet direction (Hager 1987). Considering a variation of the water height from the upstream head to the critical height respectively at the upstream and downstream limits of the lateral crest, the discharge on the lateral crest could be calculated as:

$$q_{lateral} = 0.41 \left(1 + \frac{1}{833H + 1.6}\right) \left(1 + 0.5 \left(\frac{0.833H}{0.833H + P_e}\right)^2\right) \left(\frac{P_e^\alpha + \beta}{(0.833H + P_e)^\alpha + \beta}\right) \sqrt{2gH^3} \tag{6}$$

where α and β are parameters depending on the weir geometry and P_e is the mean lateral crest height:

$$P_e = P_T \frac{B_o}{B} + \frac{P}{2} \left(1 - \frac{B_o}{B}\right) \tag{7}$$

The corrective coefficient assures that the lateral discharge corresponds to the discharge of a sharp-crested weir when the head tends to 0, corresponding with negligible longitudinal velocity. It also assures that the lateral discharge became negligible when head tends to infinity, corresponding with very high longitudinal velocity.

A least square approach of the experimental results permits to define the values of the two parameters α and β function of the inlet slope:

$$\alpha = \frac{0.7}{S_i^2} - \frac{3.58}{S_i} + 7.55 \tag{8}$$

$$\beta = 0.029 e^{\frac{1.446}{S_i}} \tag{9}$$

Figure 7 shows the comparison between the unit discharges computed with equations (3) to (9) and the experimental results. It highlights that the experimental results are approached with accuracy within 10% except for very low heads where effects of the nappe change are not taking in account.

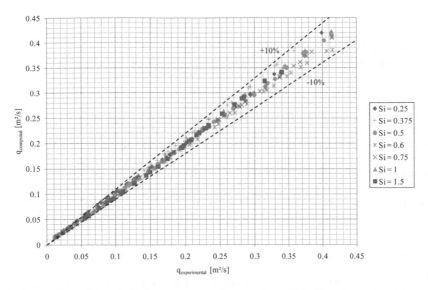

Figure 7. Comparison of the unit discharges computed by equations (3) to (9) and experimental results.

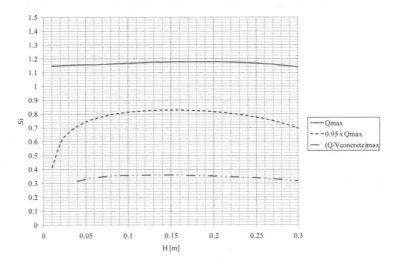

Figure 8. Comparison of the variation of the optimal slope with the head for various design criteria.

Regarding the equations (3) to (9), the slope providing the maximal discharge capacity can be computed for each value of the upstream head (Figure 8). The optimal slope, in terms of discharge capacity, is between 1.1 and 1.2, for the tested values of the steady non-dimensional ratios ($L/W = 5$, $W_i/W_o = 1.5$, $B_o/B_i = 1$, $B_o/B = 0.33$). However, 95% of the maximum discharge capacity is already provided with slopes between 0.4 and 0.8 depending on the upstream head. The technical-economical optimum may thus differ from the hydraulic optimum. For example, the optimal value of the ratio between the discharge capacity and the volume of the PKW-unit, which can directly be related to the cost of the structure building, is obtained with bottom slopes between 0.3 and 0.4 (Figure 8).

5 CONCLUSION

In order to design PKW, an experimental study of the influence on the release capacity of three geometrical parameters (W_i/W_o, B_o/B_i and P), highlighted as main influent ones by former experiments, has been undertaken. The results of experiments led on models with varied inlet slope, induced by variation of the weir height, highlight the interest on the hydraulics standpoint of increasing the inlet slope to increase the release capacity of the weir. However, over a limit slope near 1.2, the increase of the slope doesn't more change the release capacity of the PKW. This limit corresponds to the slope assuring sufficiently low velocity in the inlet, avoiding critical section apparition.

Regarding the results of the experimental study an analytical formulation of the unit discharge on a PKW has been developed. This formulation assures that the discharge of the PKW corresponds to the discharge of a sharp-crested weir with the total developed crest length L for very low heads or when the longitudinal velocity in the inlet became negligible. The discharge for infinitely high head corresponds with the discharge of a sharp-crested weir with a crest length equal to the weir width W. This formulation has an accuracy of 10% for the tested values of the steady non-dimensional ratios. For different values of these ratios, the proposed equation must be considered with different values of the two empirical parameters α and β, which can be determined by exploitation of a unique scale model. The formulation enables then to define the optimal slope in function of varied design criteria.

REFERENCES

Bieri, M., Leite Ribeiro, M., Boillat, J.-L., Schleiss, A., Laugier, F., Delorme, F. & Villard, J.-F. 2009. *Réhabilitation de la capacité d'évacuation des crues: Intégration de «PK-Weirs» sur des barrages existants.* Colloque CFBR-SHF: "Dimensionnement et fonctionnement des évacuateurs de crues". Paris, France.

Blanc, P. & Lempérière, F. 2001. Labyrinth spillways have a promising future. *Int. J. Hydro. Dams* 8 (4): 129–131.

Boussinesq, J. 1877. Essai sur lar théorie des eaux courantes. *Mem. Présentés Acad. Sci* 23: 46.

Cicero, G.-M., Guene, C., Luck, M., Pinchard, T., Lochu, A. & Brousse, P.-H. 2010. *Experimental optimization of a Piano Key Weir to increase the spillway capacity of the Malarce dam.* 1st IAHR European Congress, Edinburgh.

Hager, W.H. 1987. Lateral Outflow Over Side Weirs. *J. Hydr. Eng.* 113 (4): 491–504.

Hien, T.C., Son, H.T. & Khanh, M.H.T. 2006. *Results of some piano keys weir hydraulic model tests in Vietnam.* 22nd ICOLD congress, Barcelona: CIGB/ICOLD.

Laugier, F. 2007. Design and construction of the first Piano Key Weir (PKW) spillway at the Goulours dam. *Int. J. Hydro. Dams* 14 (5): 94–101.

Laugier, F., Lochu, A., Gille, C., Leite Ribeiro, M. & Boillat, J.-L. 2009. Design and construction of a labyrinth PKW spillway at Saint-Marc dam, France. *Int. J. Hydro. Dams* 16 (5): 100–107.

Le Doucen, O., Leite Ribeiro, M., Boillat, J.-L., Schleiss, A. & Laugier, F. 2009. *Etude paramétrique de la capacité des PK-Weirs.* Modèles physiques hydrauliques – outils indispensables du XXIe siècle, Lyon: SHF.

Leite Ribeiro, M., Albalat, C., Boillat, J.-L., Schleiss, A.J. & Laugier, F. 2007. *Rehabilitation of St-Marc dam.* Experimental optimization of a piano key weir. 32th IAHR Congress, Venice, Italy.

Leite Ribeiro, M., Bieri, M., Boillat, J.-L., Schleiss, A.J., Delorme, F. & Laugier, F. 2009. *Hydraulic capacity improvement of existing spillways – Design of Piano Key Weirs.* 23rd congress of CIGB/ICOLD, Brasilia.

Lempérière, F. & Ouamane, A. 2003. The piano keys weir: a new cost-effective solution for spillways. *Int. J. Hydro. Dams* 10 (5): 144–149.

Machiels, O., Erpicum, S., Archambeau, P., Dewals, B.J. & Pirotton, M. 2009. *Large scale experimental study of piano key weirs.* 33rd IAHR Congress. Vancouver, Canada: ASCE.

Machiels, O., Erpicum, S., Archambeau, P., Dewals, B. & Pirotton, M. 2010a. Analyse expérimentale de l'influence des largeurs d'alvéoles sur la débitance des déversoirs en touches de piano. *La Houille Blanche* (2): 22–28.

Machiels, O., Erpicum, S., Archambeau, P., Dewals, B.J. & Pirotton, M. 2010b. *Piano Key Weirs, experimental study of an efficient solution for rehabilitation.* FRIAR, Milano: Wessex Institute of Technology, 95–106.

Ouamane, A. & Lempérière, F. 2006a. *Design of a new economic shape of weir.* International Symposium on Dams in the Societies of the 21st Century, Barcelona, Spain: 463–470.

Ouamane, A. & Lempérière, F. 2006b. *Nouvelle conception de déversoir pour l'accroissement de la capacité des retenues des barrages.* Colloque international sur la protection et la préservation des ressources en eau, Bilda, Algérie.

Pralong, J., Blancher, B., Laugier, F., Machiels, O., Erpicum, S., Pirotton, M., Leite Ribeiro, M., Boillat. J-L., & Schleiss, A.J. 2011. Proposal of a naming convention for the Piano Key Weir geometrical parameters. *International Workshop on Labyrinth and Piano Key weirs*, Liège, Belgium.

SIA 1926. *Contribution à l'étude des méthodes de jaugeages.* Bulletin 18, Bern: Swiss Bur. of Water Res.

Labyrinth and Piano Key Weirs – PKW 2011 – Erpicum et al. (eds)
© *2011 Taylor & Francis Group, London, ISBN 978-0-415-68282-4*

Study of optimization of labyrinth weir

M. Ben Saïd
Hydraulics and environment laboratory facilities, University Mohamed Khider, Biskra, Algeria
Centre for Scientific and Technical Research on Arid Regions CRSTRA, Biskra, Algeria

A. Ouamane
Hydraulics and environment laboratory facilities, University Mohamed Khider, Biskra, Algeria

ABSTRACT: The spillway represents a fundamental importance for the safety of dams; however, its cost represents a significant part of the global cost of dam. The labyrinth weir represents an effective alternative for the control of floods with a low realization cost. It is characterized by length crest longer than that of a rectilinear weir for a same width of influence; this allows to increase the discharge significantly. The labyrinth weir is conceived using several geometrical simple and repetitive forms. The two concepts, simplicity and repetition, makes the conception and construction easy and economic. However, the variety of forms of the labyrinth and the complexity of the flow did not allow the determination of an optimal shape for this type of weir. This paper is interested in the study of optimization of the labyrinth weir by an experimental way. The study showed that the flow on the labyrinth weir is dependent of various geometrical parameters which characterize this particular type of weir.

1   INTRODUCTION

Dams are of great importance for country socio-economic development. However, these works represent a major risk for persons and material security. Consequently, it is worthy understanding problems related to dams, especially the submersion risk during exceptional floods. It is essential to make a conception allowing the flood evacuation without prejudice to dams and downstream territories.

Therefore the labyrinth weir plays a key role for floods to be passed with low upstream head and to increase the reservoir capacity in a significant manner. The labyrinth weir conception also helps economically thanks to the reduction of structural expense.

This kind of weir is often used in the case of limited width or reduction of maximal upstream head conditions. Generally, it is characterized by no linear plan shapes, represented by a repetition of trapezoidal, triangular or rectangular shapes. Of course, this disposition increases the weir length and consequently the discharge on the labyrinth will be increased by comparison to a linear creager weir of same width and upstream head.

The variables which will be taken into consideration in the labyrinth weir conception include (Fig. 1) the length L, width W, height P, the angle α, the cycle number n and additional parameters of secondary importance as the thickness of walls, crest shape and the flow approach condition.

The improvement of weir hydraulic performance generally requires physical model experimentation; due to the important number of dimensional parameters that govern the flow over this weir type.

The objective of the present research is to assess the impact of various geometrical parameters of the labyrinth weir by an experimental way. The assessment is linked to three dimensionless factors;

– The ratio between the widths of upstream and downstream alveoli a/b.
– The increasing length ratio L/W.
– The ratio between the width of the intake channel to the width of the spillway W_c/Wt.

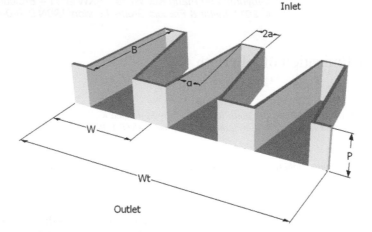

Figure 1. Labyrinth weir with trapezoidal shape.

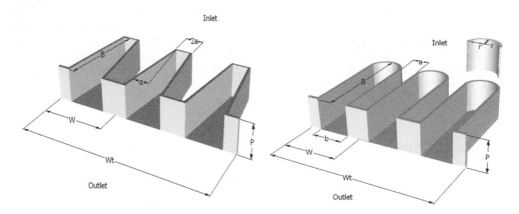

Figure 2. Labyrinth weirs, with round shape (left) and trapezoidal shape (right).

2 EXPERIMENTAL PROGRAM

2.1 *Definition of the experimental models*

The examination of the Impact of these dimensionless parameters on the design capacity requires the design of six models of labyrinth weir. The models were made with 2 mm thickness walls, allowing to consider the crest of the various models as thin-wall. The characteristics of each model are listed in Table 1.

2.2 *Description of the test facility*

The experimental work was conducted in the laboratory facility comprising a supply channel with 0.75×0.75 m cross section and 4.30 m length (Fig. 3). This channel is connected to the simulation basin, with a square shape 3×3 m and 1.1 m height. The labyrinth model is inserted at the outlet of the simulation basin. The so-called return channel is 2 m long and 1 m wide, connected to outlet. The labyrinth models were constructed with 2 mm thickness steel walls.

Table 1. Geometrical characteristic of the experimental models

Models	n°	n	L cm	Wt cm	P cm	B cm	W cm	a cm	b cm	r cm	L/Wt	W/P	a/b	B/P
Round	01	6	353	90.3	15	25	15	6	9	5	3.91	1	0.66	1.66
Round	02	6	355	90.8	15	25	15	9	6	5	3.91	1	1.5	1.66
Round	03	6	351	90	15	25	15	7.5	7.5	5	3.91	1	1	1.66
Trapézoïdal	04	6	358	90.5	15	27	15	–	–	–	3.96	1	–	1.8
Round	05	6	538.5	90.2	15	39	15	9	6	3	5.97	1	1.5	2.6
Trapezoidal	06	6	606.6	102.3	15	48	15	–	–	–	5.93	1	–	3.2

a: half frontal wall length.   B: length of lateral wall.
P: height of labyrinth weir.   Wt: width of labyrinth weir.
L: length of crest development labyrinth weir.
r: radius of round shape.
a & b: width of the upstream and downstream alveoli for round shape.
α: angle between the side wall of labyrinth weir and the flow direction for trapezoidal shape .

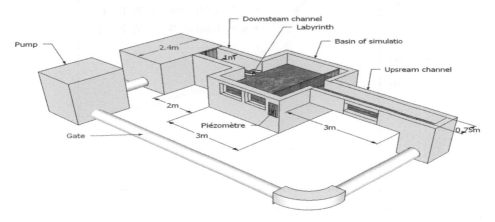

Figure 3.   The experimental facility.

3   EXPERIMENTAL RESULTS

3.1   Introduction

This chapter proposes to present the tests results of different models. These results are expressed by the relation between the discharge coefficient and the dimensionless upstream head

$$CW = f(h^*/P).\qquad(1)$$

where C_W is the discharge coefficient and h^* the upstream head

3.2   Influence of the ratio a/b

In order to verify the ratio a/b impact on the labyrinth weir performance, three models of rectangular shape with rounded upstream, with ratios $a/b = 0.67$, 1 and 1.5 and $L/W = 4$ were performed.

The tests clearly illustrate the effect of varying the width of the input and output alveoli. The increase of the ratio a/b let automatically grow the labyrinth weir performance. The discharge coefficient for ratio $a/b = 1.5$ is clearly superior to the one with a ratio $a/b = 1$ and the latter is higher than that of ratio $a/b = 0.67$. This result does not warrant that the optimum will be reached with a ratio $a/b = 1.5$ but this one is the best among the tested value.

Figure 4. Labyrinth weir during the test, low upstream head (left) and high upstream head (right).

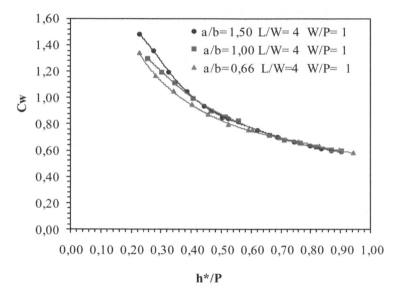

Figure 5. Variation of the discharge coefficient Cw for different a/b ratios and $L/W = 4$.

The difference between the various configurations is reduced with increasing upstream head. This can be explained by the fact that for low and medium upstream head the downstream alveoli are not entirely filled with water (Fig. 4 left).

3.3 *Influence of the ratio L/W*

To check the influence of the ratio L/W on labyrinth performance, four models have been tested: two models of rectangular shape with rounded upstream with ratio $L/W = 4$ and 6 and two models of classic trapezoidal shape with ratio $L/W = 4$ and 6.

The graphs of Figure 6 illustrate clearly that, in the case of the labyrinth weir with rectangular shape and rounded upstream, the increase of L/W from 4 to 6 can improve the performance by 10% for a upstream head of $h^*/P = 0.3$ and 5% for an upstream head on $h^*/P = 0.7$, although the length of the walls was increased by 50 %. An increase of ratio $L/W = 4$ to 6 for the labyrinth weir with trapezoidal shape produces an efficiency gain of 30% for upstream head for $h^*/P = 0.3$ and 17% for $h^*/P = 0.7$ (Fig. 7).

This shows that the transition of ratio $L/W = 4$ to 6 permit a higher efficiency increase of the labyrinth weir with trapezoidal shape than rectangular rounded shape.

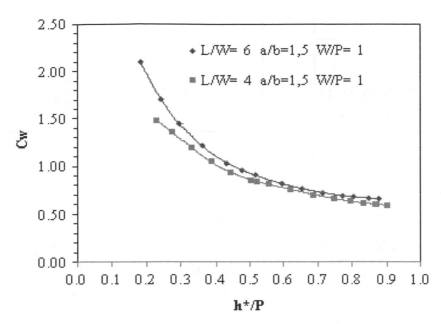

Figure 6.   Variation of discharge coefficient *Cw* with *L/W* for rectangular shape rounded upstream.

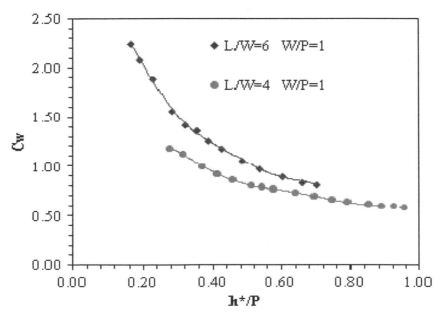

Figure 7.   Variation of discharge coefficient *Cw* with *L/W* for trapezoidal shape.

3.4   *Impact of lateral contraction Wc/Wt*

To test the effect of the lateral contraction on the hydraulic performance of labyrinth weir two types of implementation were experienced (Fig. 8). The first arrangement corresponds to a width of inlet channel equal to the width of the labyrinth weir *Wc = Wt*. The second disposition corresponds to a width of inlet channel equal four times the width of the spillway *Wc = 4Wt*.

The graphs of Figure 9 clearly show the effect of the lateral contraction on the discharge coefficient. Comparing the discharge coefficient of both installations shows that the discharge coefficient values corresponding to the weir without contraction are much higher than the values obtained for

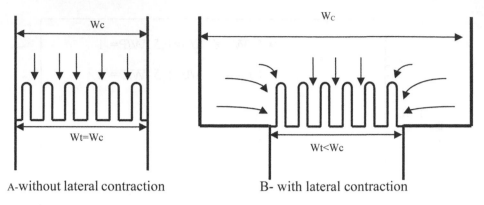

A-without lateral contraction   B- with lateral contraction

Figure 8.   Labyrinth weir with and without lateral contraction.

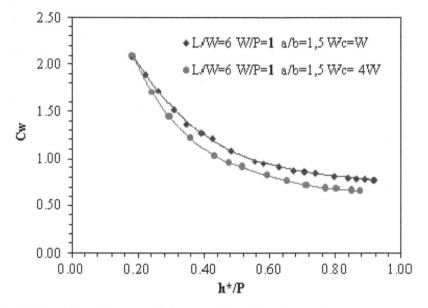

Figure 9.   Variation of the discharge coefficient Cw with model implementation.

the spillway with contraction. The difference between the values of the two dispositions increases with h^*. For $h^*/P = 0.3$ the difference is 11% for $h^*/P = 0.5$, the difference increases to 17% and for $h^*/P = 0.7$ the difference between the two curves is 22%.

3.5   *Impact of filling alveoli*

Partial filling of the labyrinth alveoli can reduce the construction cost of the work (Fig. 10). However, this solution may have effects on the hydraulic performance. To determine the degree of filling under which the hydraulic performance will not be affected, several height and length filling ratios have been tested.

A- Filling 1/4 the length upstream and downstream, with filling heights 1/3, 1/2 and 2/3P.
B- Filling half the length upstream and downstream, with filling heights 1/3, 1/2 and 2/3P.

   The results showed that:

1) Filling a quarter of the length has none effect on the discharge coefficient, whatever the height filling (Fig. 11).

Figure 10.   Labyrinth weir with filling.

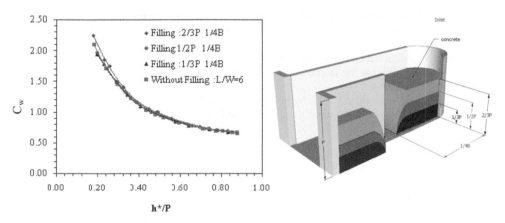

Figure 11.   Variation of the discharge coefficient Cw with ¼ B filling.

2) With half length filling, the discharge begins to be affected if the height filling is greater than P/3 (Fig. 12).

4   CONCLUSION

Following the analysis of the hydraulic behavior of labyrinth weirs, the number of dimensionless parameters governing the flow over the labyrinth weir upstream with rounded rectangular shape has been fixed at three.

These order parameters and secondary parameters (filling) have been experiments on six types of labyrinth weir, allowing to determine the impact of different dimensionless parameters on the hydraulic performance and to contribute to the optimization of this type of spillway. The tests concerned the ratio between the input and output a/b, the influence of the ratio L/W, the impact on the flow of lateral contraction of the channel Wc/W and the impact of alveoli filling.

The results showed that:

– The ratio 1.5 is the best among the tested values. The evidence of these tests is that the hydraulic performance increases with the ratio a/b for low and medium heads ($h^*/P < 0,5$).

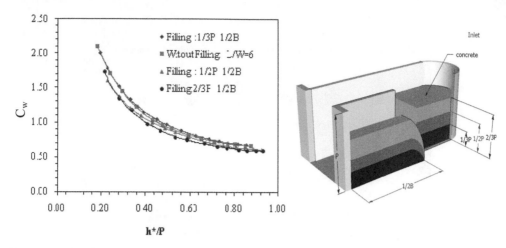

Figure 12.   Variation of the discharge coefficient C_W with ½ B filling.

– The ratio L/W has to be considered as the most important parameter in relation with the performance of this spillway type. This performance decreases rapidly with increasing upstream head.
– Experiments conducted on two types of dispositions (with lateral contraction and without lateral contraction) showed that the lateral contraction affects significantly the hydraulic performance of the labyrinth weir.
– One solution to reduce the cost of labyrinth weirs is to reduce the height of the walls while maintaining the same height of the weir; it may be possible with filling of upstream and downstream alveoli.

REFERENCES

Lux Iii, F., Hinchliffe, D.L. 1985 Design and Construction of Labyrinth Spillway, 15th Congress on Large Dams, ICOLD, Vol. IV, Q59, R 15, pp. 249–274, Lausanne.
Tullis, J.P., Amanian, N. Waldron, D. 1995. Design of Labyrinth Spillway, Journal of Hydraulic Engineering, Vol. 121, No.3, pp. 247–255.
Yildiz, D. Üzücek, E. 1996. Modeling the performance of labyrinth spillway, the International Journal on Hydropower & Dams Issue Three.

Labyrinth and Piano Key Weirs – PKW 2011 – Erpicum et al. (eds)
© 2011 Taylor & Francis Group, London, ISBN 978-0-415-68282-4

Influence of Piano Key Weir geometry on discharge

R.M. Anderson & B.P. Tullis
Utah Water Research Laboratory, Utah State University, Logan, Utah, USA

ABSTRACT: Due to the relatively recent introduction of the piano key weir (PKW) in both research and practice, no generally accepted standardized design procedure is currently available. This is due, in part, to the large number of PKW geometric parameters, and a limited understanding of their influences on the head-discharge relationship. This fact partially explains why recent PKW studies have included such a wide variety of geometric variations. However, despite the nonexistence of a standard design procedure, Hydrocoop (France), a non-profit dam spillways association, has reported the development of PKW configuration that they consider to be near optimal.

In an effort to develop a better understanding of how various geometric characteristics of the PKW affect discharge, a laboratory-scale sectional PKW model was fabricated using the geometry guidelines recommended by Hydrocoop. For comparative purposes, a rectangular labyrinth weir RLW with the same total crest length, weir height, inlet and outlet key widths, total weir width, wall thickness, and crest shape as the PKW was fabricated with removable ramped floors (same slope as the PKW ramped floors, with ramps installed the RLW essentially models the PKW with no overhangs). The PKW and RLW (with various ramp configurations) were tested over a wide range of discharges. Testing the RLW with and without the ramped floors was done in an effort to isolate of the influences of PKW ramped floors and overhangs on the head-discharge relationship. The paper presents the results of this study.

1 INTRODUCTION

1.1 *Non-linear weirs*

Many existing spillways are currently undersized and in need of replacement due to increasing magnitudes of probable maximum storm events, rising demands for reservoir water storage, and the continuing need to improve dam safety. Spillways commonly use weirs as the flow control structure. In the classical weir head-discharge relationship, Equation (1), the weir's discharge capacity (Q) is proportional to the weir length (L).

$$Q = \frac{2}{3} C_{dL} L \sqrt{2gH_t^{\frac{3}{2}}}$$

(1)

In Equation (1), Q is the discharge, C_{dL} is the discharge coefficient, g is the gravitational constant, L is the crest length, and H is the total head [piezometric head (h) plus velocity head ($V^2/2g$)] measured relative to the weir crest.

Non-linear weirs (e.g., labyrinth-type weirs) increase the discharge efficiency (larger Q for a given H), relative to linear weirs, by increasing L within the given footprint (longitudinal and transverse dimensions of the weir base) restrictions. A traditional labyrinth weir consists of a sequence of linear weirs, which has been folded in a zigzag fashion. Despite the fact that labyrinth weir discharge coefficient (C_{dL}) values, which are geometry and discharge dependent, are lower that linear weir discharge coefficients, the increase in L can increase Q, relative to a linear weir, by 3 to 4 times (Tullis et al. 1995) at a constant H value and channel width. For some spillway applications, the longitudinal and transverse dimensions of the control structure footprint may be limited (e.g., narrow dam crest). For such cases, sufficient room may not be available for a traditional trapezoidal labyrinth weir design and alternative non-linear weir designs, such as the piano key weir (PKW) should be considered.

1.2 *Piano key weir*

The PKW was developed to facilitate and improve the performance of labyrinth-type weirs applications with limited footprints. As shown in Figure 1, The PKW has a simple rectangular crest layout (in plan view) with inclined inlet and outlet keys. The inclined or ramped floors are cantilevered beyond the spillway footprint providing the PKW with a longer crest length, relative to a traditional labyrinth weir (vertical wall) with the same footprint.

Some of the important PKW geometric parameters used in this study, as shown in Figure 1, include the weir height (P), the weir wall height at the middle of the weir structure (P_m), crest centerline length (L), total weir width (W), slope of the inlet and outlet key floors (S_o and S_i respectively), inlet key width (W_i), outlet key width (W_o), upstream overhang length (B_o), downstream overhang length (B_i), wall thickness (T_s), and number of units (N_u; where one unit is equivalent to ($W_i + W_o + 2T_s$)/2).

A consequence of the PKW ramped floors is the upstream and/or downstream overhanging key apexes. The overall effect of PKW overhangs on discharge efficiency is not found in current literature. Generally, PKW geometry follows the geometry shown in Figure 1 with equal upstream and downstream overhang lengths (B_o and B_i respectively), and has been referred to as PKW geometry *Type-A* (Lempérière & Ouamane 2003). Variations of this geometry have been studied that include only larger upstream overhangs and no downstream overhangs, which have been referred to PKW geometry *Type-B*. As found in a study conducted by Lempérière & Ouamane (2003), PKW geometry *Type-B* results in an increase in discharge efficiency, relative to PKW geometry *Type-A* with all geometric parameters held constant except B_o, B_i, S_o and S_i. Changing the shape of the weir beneath the upstream overhangs to a more hydraulic shape by installing noses has also been shown to increase discharge efficiency (Lempérière & Ouamane 2003). Additional understanding of PKW overhangs and ramped floors would, in part, explain differences in discharge efficiency relative to traditional non-linear weirs (vertical walls), as well as help to explain, in part, some of the general hydraulic characteristics resulting from the geometry of the PKW structure.

Due to the relatively recent introduction of the PKW in both research and practice, no generally accepted standardized design procedure is currently available. This is due, in part, to the large number of geometric parameters and a limited understanding of their influences on the head-discharge relationship. Despite the absence of a standard design procedure, Hydrocoop (France),

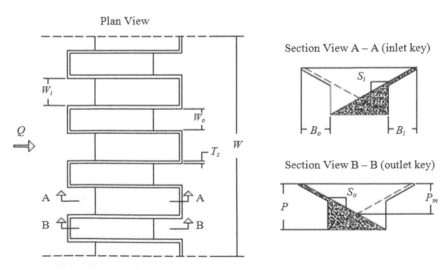

Figure 1.   PKW Parameters.

a non-profit dam spillways association, has more recently developed a PKW geometry they consider to be near optimal (Lempérière 2009).

1.3 Research objectives

In an effort to develop a better understanding of how some of geometric characteristics of the PKW affect discharge efficiency, a laboratory-scale sectional PKW model was fabricated using the geometry guidelines recommended by Lempérière (2009). For comparative purposes, a rectangular labyrinth weir (RLW) with the same L, P, W_i, W_o, W, N_u, T_s, and crest shape as the PKW was fabricated with removable ramped floors [same slope (S_o and S_i) as the PKW ramped floors]. The RLW with ramped floors is essentially equivalent to the PKW without the overhangs. The PKW and RLW with and without the removable ramps were tested over a wide range of discharges. Testing the RLW with and without the sloping false floors was done in an effort to isolate the influences of PKW ramped floors and overhangs on discharge efficiency.

2 EXPERIMENTAL SETUP

2.1 Testing facilities

Tests were conducted in a 93.4-cm wide by 61.0-cm deep by 7.4-m long rectangular flume. Water enters the flume through the head box containing a flow-distributing manifold followed by a baffle wall (to improve approach flow uniformity) and a floating surface wave suppressor. Acrylic flume sidewalls facilitated visual observations. A point gauge (readability 0.015-cm) mounted in a stilling well, hydraulically connected to flume a distance equal to 2-times the weir height (i.e., 39.4-cm) upstream of the test weir, was utilized to measure head (h). The flume has a maximum flow capacity of approximately 0.24 cubic meters per second (cms). Calibrated orifice meters ($\pm 0.2\%$ average uncertainty) and control valves, located in the parallel 10.2-cm and 30.5-cm supply lines enabled accurate flow rate measurements over a broad range of discharges. A schematic of the test flume is shown in Figure 2.

2.2 Model weir design, construction, and setup

All laboratory-scale weirs were designed to be fabricated using 1.27-cm thick acrylic sheeting. The material thickness was geometrically consistent with the concrete wall thickness (scale = 15.75) for one prototype structure [e.g., Goulours Dam (Laugier, 2007)]. All models were designed with $N_u = 8$, and feature a flat-top crest type. Weir pieces were cut out and assembled using acrylic glue. Following assembly, the weir crests were machined level using a CNC mill. The weirs were installed on an adjustable base and leveled using surveying equipment. A ramp was installed upstream of the weir to gradually transition the approach flow from the floor of the channel to the base of the test weir. A venting device was built to enable the testing of vented and non-vented nappe conditions. An overview the test weirs and the flume setups are shown in Figure 3.

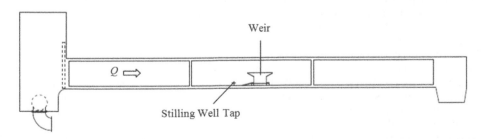

Figure 2.   Testing flume side view.

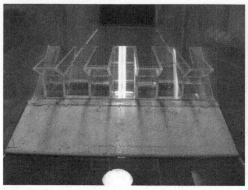

Figure 3. Setup of PKW (left) and RLW (right).

3 TESTING PROCEDURE

After the weirs were constructed, physical measurements of the crest lengths and weir heights were made; good agreement was found between the design and as-built weir dimensions. Following installation, a leak test was conducted to insure all joints were watertight.

Test flow rates ranged from 0.0071- to 0.024-cms for vented and non-vented nappe conditions utilizing a removable venting device located downstream of the weir assemblies. The upstream piezometric head (h) data, measured relative to the weir crest, were determined using the point gauge/stilling well assembly after the water level had been allowed to stabilized upstream of the weirs for a minimum of 5 minutes. To verify that stable flow conditions had been achieved, a minimum of 3 point-gauge readings were taken consecutively for each flow rate. If the h data were not consistent, the flow condition was allowed more stabilization time. Using the Q and h data, H and C_{dL} values were calculated using Equation (1).

4 EXPERIMENTAL RESULTS AND DISCUSSION

4.1 *Nappe venting and data uncertainty*

Preliminary weir test showed that the head-discharge data proved to be independent of the nappe conditions (vented and non-vented). Consequently, the majority of the tests were limited to the vented nappe condition. The measurement uncertainty in the data collection was calculated using the method outlined by Kline and McClintoch (1953); the maximum and average measurement uncertainty for all data sets was 3.20% and 2.09% respectively.

4.2 *Overhang effects*

PKW overhang effects were somewhat isolated in comparing the PKW and the RLW with the ramped floors installed. As shown in Figure 4, the PKW is more efficient (higher C_{dL} values) than the RLW weir with ramped floors over most of the range of H/P tested.

The effect of the PKW upstream overhangs on weir discharge efficiency is likely related to the nature of the inlet flow contraction and subsequent energy loss associated with flow entering the inlet key. The PKW overhang geometry increases the inlet flow area and wetter perimeter, relative to the RLW key inlet, resulting in a reduction of inlet velocities, flow contraction, and energy loss. Figure 5 presents a section side view of the PKW and RLW with ramped false floors at $H/P = 0.3$; the drop in the water surface profile is more pronounced on RLW with ramped floors, indicating a more significant flow contraction energy loss condition, relative to the PKW, which produced a relatively horizontal water surface profile.

Due to the downstream overhangs, the downstream end of the PKW outlet key has a larger area and wetted perimeter, relative to the RLW with ramped floors outlet key, resulting in more efficient

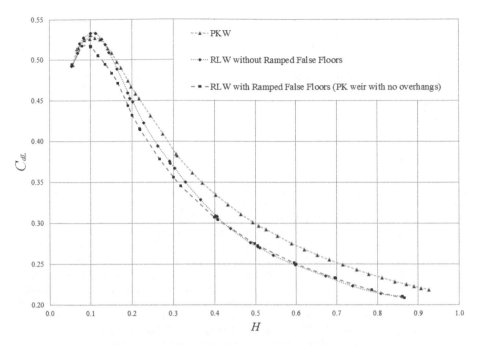

Figure 4.   Data for the RLW with and without ramped false floors and the PKW.

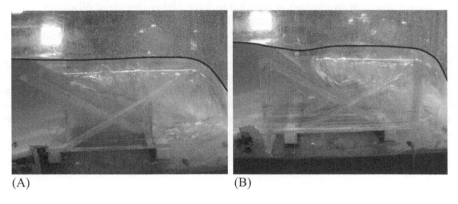

(A)   (B)

Figure 5.   PKW (A) and RLW with ramped false floors (B) side view at $H/P = 0.3$.

outlet key discharge exit conditions. This explains why it was observed that the outlet keys of the PKW did not fill with water as fast as the RLW outlet keys with ramped floors at similar flow conditions. As flow increases, an increase in PKW discharge efficiency was also likely influenced by a reduction in local submergence effects in the outlet keys, relative to RLW with ramped floors. Both upstream and downstream overhangs result in an increase in discharge efficiency.

4.3 Sloped floors

Testing the RLW with and without removable ramped floors does not entirely isolate the sloped floor effects of the PKW because of overhang effects, but the results do provide some general understanding of PKW ramped floor effects. As seen in Figure 4, the RLW without ramped floors is less efficient than PKW over most of the H/P range tested. In addition the RLW with and without ramps performed very similarly at higher H/P values ($H/P > 0.4$). This suggests that the ramp configuration for the PKW is not a significant factor when considering weir discharge efficiency, relative to the geometry of the PKW overhangs.

It was observed while testing the RLW with and without ramped floors, that the outlet cycles of the RLW without ramped floors filled with water at lower values of H/P (0.3–0.4), whereas the outlet cells of RLW with ramped floors did not fill until higher values of H/P (0.6–0.7). Ramped floors in the outlet cycles seem to reduce local submergence by helping to evacuate water out of the outlet cells (inducing supercritical flow out of the outlet cells). As shown in Figure 4, ramped floors in the inlet and outlet cycles result in a decrease in weir performance for $H/P < 0.4$, and no change in weir performance at $H/P > 0.4$, relative to having no ramps; it is expected that PKW ramped floors have a similar effect.

5 CONCLUSIONS

A laboratory-scale sectional piano key weir PKW model was fabricated using the geometry guidelines recommended by Lempérière (2009). For comparative purposes, a rectangular labyrinth weir (RLW) with the same total crest length, weir height, inlet key widths, outlet key widths, total weir width, wall thickness, and crest shape as the PKW was fabricated with removable ramped floors (same slope as the PKW ramped floors, with ramped floors installed the RLW essentially models the PKW with no overhangs). The PKW and RLW (with and without ramped floors installed) were tested over a wide range of discharges; aerated nappe conditions were considered. Testing the RLW with and without the sloping false floors was done in an effort to isolate of the influences of PKW ramped floors and overhangs on discharge efficiency. Following are the findings of this study.

PKW overhangs result in a significant increase in discharge efficiency, relative to the RLW with ramped floors (equivalent to a PKW with no overhangs). The PKW upstream overhang geometry increases the inlet flow area and wetter perimeter of the inlet key resulting in a reduction of inlet velocities, flow contraction, and energy loss. This may explain, in part, why the PKW geometry *Type-B* (larger upstream overhangs) is reported to have higher discharge efficiency than PKW geometry *Type-A* (smaller upstream overhangs). The PKW downstream overhang geometry results in a larger area and wetted perimeter in the outlet key exit, relative to the RLW with ramped floors, resulting in more efficient outlet key discharge exit conditions.

The RLW with and without ramped floors performed very similarly at higher H/P values ($H/P > 0.4$). This suggests that the ramp configuration for the PKW is not likely a significant factor when considering weir discharge efficiency, relative to the geometry of the PKW overhangs, at higher H/P values (H/P > 0.4). RLW ramped floors in the outlet cells aid in reducing local submergence by helping to evacuate water out of the outlet cells (inducing supercritical flow out of the outlet cells). It is expected that PKW ramped floors have a similar effect.

Additional research continues at the Utah Water Research Laboratory to investigate PKW geometry and various PKW configurations on discharge. PKW sloped floor studies also continue.

REFERENCES

Kline, S.J., and McClintock F.A. (1953). "Describing Uncertainties in single-sample Experiments." MechanI cal Engineering, 75(1), 3–8.
Lempérière, F. (2009). "New Labyrinth weirs triple the spillways discharge." <http://www.hydrocoop.org> (Feb. 8, 2010).
Lempérière, F., and Ouamane, A. (2003). "The Piano Keys weir: a new cost-effective solution for spillways." The international journal on Hydropower and Dams. 10(5).
Tullis, J. P., Amanian, N., and Waldron. D. (1995). "Design of Labyrinth Spillways." Journal of Hydraulic Engineering, 121(3), 247–255.
Laugier, F. (2007). "Design and construction of the first Piano Key Weir spillway at Goulours dam." The International Journal on Hydropower & Dams,14(5), 94–100.

Labyrinth and Piano Key Weirs – PKW 2011 – Erpicum et al. (eds)
© 2011 Taylor & Francis Group, London, ISBN 978-0-415-68282-4

Study of a piano-key morning glory to increase the spillway capacity of the Bage dam

G.M. Cicero, M. Barcouda & M. Luck
National Hydraulic and Environmental Laboratory, Chatou, France

E. Vettori
EDF CIH, Le Bourget, France

ABSTRACT: The thin arch dam of Bage was initially equipped with a morning glory spillway designed to discharge 34 m³/s at the Maximum Water Level. The updated evaluation of the design flood required to increase the discharge capacity to 73 m³/s. Many tests were thus performed on a physical model at 1/20 scale, with different spillway geometries at various crest levels but keeping the initial diameter of the vertical shaft. Among the alternatives, an innovative one, based on the fusion of a traditional morning glory with a piano-key weir has been developed and tested. This solution increased the overflow length of 85% without changing the global size of the spillway. This innovative geometry showed very good hydraulic performances. The discharge measured at MWL was 70% higher than with the current morning glory spillway. This paper presents the test results of this Piano Key Morning Glory, comparing the hydraulic performances to traditional morning glory profiles of similar diameter.

1 INTRODUCTION

The thin arch dam of Bage (Aveyron) was initially equipped with a morning glory spillway located on the right bank of the reservoir close to the dam. Its total discharge capacity at the Maximum Water Level (715.7 NGF) was designed to be 34 m³/s. The updated evaluation of the design flood (return period of 1000 years) required to increase the discharge capacity to 73 m³/s.

The objective of this framework is to optimize a new alternative to increase the spillway capacity. The design of the spillway can be changed and the current normal level (715 NGF) is allowed to be decreased while the characteristics of the discharge circuit have to be kept.

Many tests were thus performed at EDF R&D LNHE on a physical model at 1/20 scale, with different spillway geometries, various heads and water supply conditions.

The first tests were carried out to check the release capacity of the current spillway. Then various geometries increasing significantly the developed crest length were tested. Among the alternatives, an innovative one, based on the fusion of a traditional morning glory with a piano-key weir has been developed and tested. Although this solution showed very good hydraulic performances a classical shape of morning glory was preferred with respect to floating debris aspects.

This paper focus to the results of the tests performed on this Piano Key Morning Glory, comparing the hydraulic performances (flow observations and discharge characteristics) to the traditional morning glory of the current spillway which has a similar diameter.

2 EXPERIMENTAL SETUP

The experimental setup at scale 1/20 represents the reservoir, the thin arch dam (at crest 716.48 m NGF) and the morning glory spillway with a discharge circuit leading to a stilling basin. The morning glory (Fig. 1) is located on the right bank at 5.95 m upstream from the dam. After the first tests, an area was excavated around the morning glory to improve the supply of the spillway and

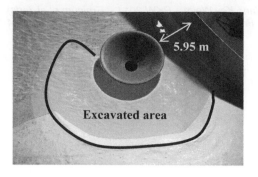
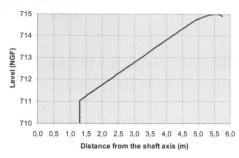

Figure 1.   View of the current morning glory (left) and profile (right).

avoid vortex formation. The crest diameter of the morning glory is 11.1 m. The outer rim diameter of the cup shaped inlet is 11.5 m (at 0.08 m under the crest).

The discharge circuit is composed of a vertical shaft and a horizontal gallery in concrete of the same diameter (2.6 m). The vertical shaft is 20 m long with a top level at 711.02 NGF and the horizontal gallery is 65.74 m long. The shaft and the gallery are in plexiglas to observe the flow regimes and air entrainment inside. Furthermore this material allows to respect the similitude on the Strickler coefficient ($K_{model} = 20^{1/6} K_{nature}$). The stilling basin is 20 m long with a bottom level at 688.35 NGF.

We measured the flow discharge with an electromagnetic flow-meter (1% accuracy) and the level of the reservoir (+/−1 mm accuracy).

3   EXPERIMENTAL RESULTS

3.1   Results of the current morning glory

Tests of the current spillway were carried out for various flow discharges from 9 to 80 m³/s. At 9 m³/s, we observe a bubbling flow in the bottom of the morning glory and air bubbles in the sides of the shaft. The bend and the horizontal gallery are not under pressure. With the discharge increase (Fig. 2 left), (i) the bubbling flow limit goes upwards (ii) air entrainment increases in the shaft and the gallery (iii) the gallery, the bend and the shaft goes successively under pressure.

For the initial design flood (34 m³/s) the reservoir reaches the maximum water level (Fig. 3). For the new design flood (73 m³/s), the reservoir level becomes unstable with quick oscillations up to the top of the model.

At 80 m³/s (Fig. 2 right), the transition to submerged flow occurs with the formation of a counterclockwise vortex in the morning glory. The reservoir level becomes very unstable and increases very quickly (Fig 3).

For free surface flow, the theoretical values were calculated according to Vischer & Hager (1998):

$$Q = C_{dR} 2\pi R \sqrt{2g} H^{1.5} \tag{1}$$

For 0.2 < H/R < 0.5:

$$C_{dR} = 0.515(1 - 0.2\frac{H}{R}) \tag{2}$$

For H/R < 0.2:

$$C_{dR} = 0.495 \tag{3}$$

The measured discharges are about 20% lower than the theoretical values (Fig. 3) but the water supply was not homogeneous due to the proximity of the boundaries and it was improved by an excavation around the morning glory (Fig. 1 left).

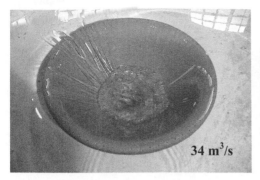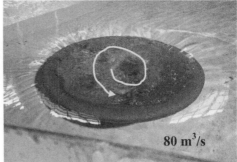

34 m³/s   80 m³/s

Figure 2.   Free surface flow (left) and submerged flow (right) over the current spillway.

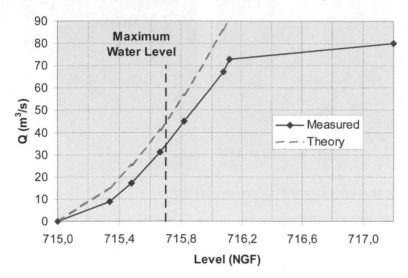

Figure 3.   Discharge capacity of the current morning glory.

3.2   *Results of the PK-Weir morning glory*

For the new design flood (73 m³/s), the reservoir level was 716.12 NGF (Fig. 3). Thus to keep the current spillway, it should be necessary to lower the crest level of about 0.50 m.

M. Barcouda proposed to test an innovative solution called "Papaya spillway" mixing the PK-Weir principles on a morning glory spillway (Barcouda et al. 2006). The circular spillway (Fig. 4) is divided into 12 inlet and outlet keys alternatively collecting the flow to the vertical shaft. In the inlet keys, the flow goes upwards on inclined planes. In the outlet keys, the flow goes downwards on shapes similar to the Creager profiles. The outer diameter is 10 m and the inner diameter is 2.6 m as the vertical shaft. This circular labyrinth increases the developed crest length (64.2 m) of 85% compared to the current morning glory (crest diameter = 11.1 m) and of 100% compared to a morning glory of the same diameter.

The tests carried out for the same discharges as the current spillway showed a better release capacity of the Papaya spillway. Up to 80 m³/s, the spillway is not submerged (Fig. 4). The flow enters radially in the inlet and outlet keys without any vortex formation. The air entrainment in the shaft does not increase with the discharge like with the current spillway (Fig. 5). Thus the Papaya spillway operates at higher discharges without being submerged.

At 90 m³/s, the spillway is submerged and there is a vortex like with the current spillway but the level in the reservoir is much lower than with the current spillway.

The rating curves (Fig. 6 left) show that the discharge capacity increases of about 20 m³/s with the Papaya spillway, for each reservoir level. For the new design flood (73 m³/s), the level of the

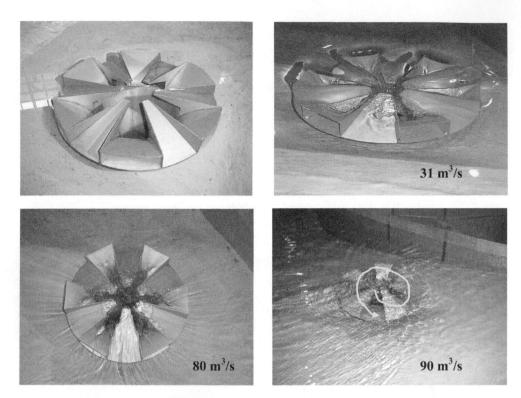

Figure 4.   Views of the "Papaya spillway" for various discharges.

Figure 5.   Views of air entrainment in the shaft with the Papaya spillway.

reservoir is about 20 cm higher than the maximum water level (instead of 50 cm with the current spillway).

If we consider the discharge coefficient C_{dR} versus the relative head H/R (Fig. 6 right), we can see that the Papaya spillway increases the release capacity especially at low heads where it can be 4 times higher than a traditional morning glory of the same diameter. The improvement of the release capacity decreases with the head but remains greater than 30%.

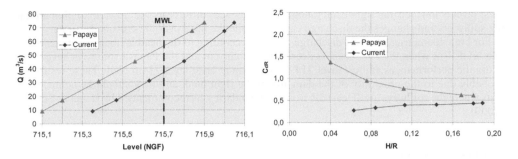

Figure 6.   Release capacity of the current and Papaya spillways at crest level 715 NGF.

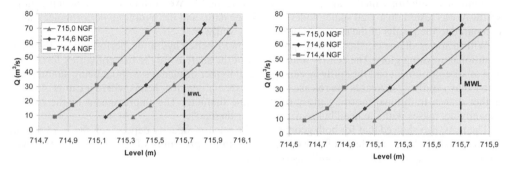

Figure 7.   Effect of the crest level on the rating curves (discharge versus level) of the current (left) and Papaya (right) spillways.

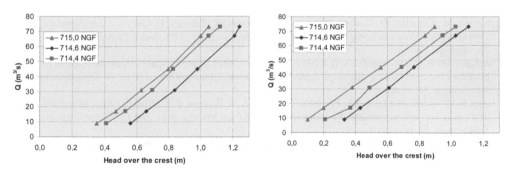

Figure 8.   Effect of the crest level on the rating curves (discharge versus head) of the current (left) and Papaya (right) spillways.

3.3   *Crest level effects*

New tests were performed at lower crest levels with both spillways to discharge the new design flood at the maximum water level. The crest level was changed easily from 20 to 60 cm by means of flanges set up or removed from the top of the shaft.

Figure 7 presents the rating curves measured at 3 crest levels. To discharge the new design flood (73 m³/s) at the maximum water level (715.7 NGF), the crest level should be lowered of 40 cm with the Papaya spillway (Fig. 7 right), and of 50 cm with the current spillway (Fig. 7 left).

The rating curves of the discharge versus the head over the crest (Fig. 8), show that the crest level influences the release capacity of both spillways. This is not the case of the linear spillways and of the rectangular PK-Weirs (Cicero et al. 2010) where the discharge depends mainly on the head over the crest. The hydraulic behaviour of the morning glory spillways is very influenced by the proximity of the boundaries which can induce vortex flow (Christodoulou et al. 2010). This is the case of the Bage spillway located rather close to the dam and to the right bank. Thus, when the crest level is lowered, the right bank topography influences differently the flow circulation around the spillways and changes the discharge capacity while the head over the crest is the same.

For both geometries, the best discharge capacity according to the head is obtained at crest level 715 NGF. But when the crest level is decreased, the current and the Papaya spillways have better hydraulic performances at crest level 714.4 NGF than at crest level 714.6 NGF. These results are probably due to the proximity of the dam and of the right bank and they should be compared to tests where the spillways are at least 2R from the boundaries, as recommended by Hager & Schleiss (2009).

4 CONCLUSIONS

An innovative solution called "Papaya spillway" mixing the PK-Weir principles on a morning glory spillway was tested on a 1/20 scale model of the Bage dam, and the hydraulic performances were compared to the results of the current morning glory spillway.

With a lower diameter, the Papaya spillway showed better hydraulic performances than the traditional morning glory. The central water supply of the shaft avoid the risks of vortex formation and of air entrainment and the spillway can operate at higher discharges without being submerged. The Papaya spillway increases the release capacity especially at low heads where it can be 4 times higher than a traditional morning glory of the same diameter The improvement of the release capacity decreases with the head but remains greater than 30%.

The tests also showed that for both geometries, the release capacity depends on the head and on the crest level. These crest level effects could be due to the proximity of the dam and of the right bank which are quite close to the spillway of Bage.

REFERENCES

Barcouda, M., Cazaillet, O., Cochet, P., Jones, B.A., Lacroix, S., Laugier, F., Odeyer, C. & Vigny, J.P. 2006. Cost effective increase in storage and safety of most dams fusegates of P.K. weirs. *22ème congrès des grands barrages, Barcelona.*

Christodoulou, A., Mavrommatis, A. & Papathanassiadis, T. 2010. Experimental study on the effect of piers and boundary proximity on the discharge capacity of a morning glory spillway. *1rst IAHR European congress, Edinbourgh, 04–06 mai.*

Cicero, G.M., Guene, C., Luck, M., Pinchard, T., Lochu, A. & Brousse, P.H. 2010. Experimental optimization of a piano key weir to increase the spillway capacity of the Malarce dam. *1rst IAHR European congress, Edinbourgh, 04–06 mai.*

Hager, W.H. & Schleiss, A.J. 2009. *Constructions hydrauliques – Ecoulements stationnaires –* Presses Polytechniques et universitaires romanes: Lausanne.

Vischer, D.L. & Hager, W.H. 1998. *Dam hydraulics.* Wiley: Chichester.

Physical modeling –
Downstream fittings

Labyrinth and Piano Key Weirs – PKW 2011 – Erpicum et al. (eds)
© 2011 Taylor & Francis Group, London, ISBN 978-0-415-68282-4

Contribution to the study of the Piano Key Weirs submerged by the downstream level

F. Belaabed & A. Ouamane

Laboratory of Hydraulic planning and Environment, University Mohamed Khider, Biskra, Algeria

ABSTRACT: Dam reservoirs are confronted with two fundamental problems. The first is the master of floods and the second is related to the loss of storage capacity due to the silting of the reservoir. One possible solution to these problems is to raise the sill level of the existing weir by its renovation in PK-Weir.

The PK-Weir is built across the rivers and on dams where the limitation of discharge or water level must be ensured in normal or submerged flow conditions. Various studies and researches carried out on the PK-Weir have focused only in the normal flow conditions. The present paper presents a study on the effect of the downstream level on the performance of the PK-Weir.

This study will be based primarily on model experiments and following on the approaches of other authors for rectilinear or labyrinth weirs submerged from downstream.

1   INTRODUCTION

Generally, the weirs are conceived to operate under free flow conditions, this means that the downstream water level lies below the crest level of the weir. However, when the downstream water level exceeds the level of the crest, the weir is not any more in a free flow condition but rather submerged.

In the case of a weir installed in an irrigation channel or through a natural stream, working in submerged conditions, a higher upstream head is required to assure a discharge equal to the one evacuated under free flow condition. In case the section upstream of the weir is significantly larger than the width of the weir; that engenders the formation of a storage reservoir upstream, the downstream submersion will then cause a decrease of the evacuated discharge.

Research works in the field of labyrinth weirs submerged by the downstream level are essentially related to the work of Tullis et al. (2006) and to that of Lopes et al. (2009). These works showed that the submersion does not impact before the downstream water level exceeds the weir crest. If the downstream water level continues to increase, it tends finally to equal the upstream water level and the structure will not work any more as a control structure.

According to the work of Tullis et al. (2006), the performance of a submerged labyrinth weir can be exactly described with a small error, with regard to experimental data. The ratio which expresses the relative downstream head (Hd/Ho), seems to be relatively independent from the side wall angle of the labyrinth weir. When the angle of the wall increases, the performance of the submerged labyrinth weir approaches that of a linear weir. The work of Lopes et al. (2009) showed that the shape of the labyrinth weir crest and the angle of the side walls have only a small effect on the relative head of the submerged flow.

The functioning of the PK-Weir under submerged conditions was the object of only few preliminary studies, so it became necessary to verify experimentally the effect of submersion on the performance of this type of weir. This study was realized at the university of Biskra on a PK-Weir model of type A (Fig. 1), whose geometrical characteristics are given in Table 1.

The experimental study focused on two typical cases of submersions, the first considering a sill downstream of the weir and the second with the presence of a flow under a gate (Fig. 2).

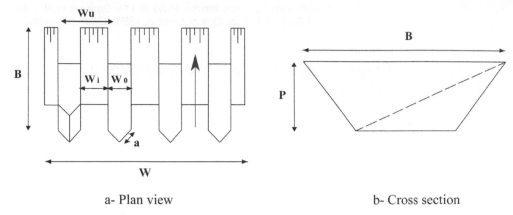

a- Plan view   b- Cross section

Figure 1.   Geometrical parameters of the experimented model.

Table 1.   Geometrical characteristics of the experimented PK-Weir model.

PK-Weir model A	n°	n	L cm	W cm	P cm	B cm	W_0 cm	W_i cm	L/W	W_u/P	W_i/W_0
	A1m	4	498	99.2	20	48.48	11.3	13.5	5.02	1.24	1.2

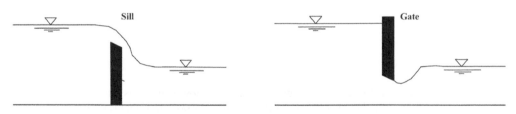

Figure 2.   Typical cases of downstream flow.

Figure 3.   Experimental channel with the PK-Weir, downstream view.

2   EXPERIMENTAL PROGRAM

The work was realized in an experimental facility composed of a set of open channels allowing to simulate flow conditions upstream and downstream of hydraulic structures (Fig. 3).

The upstream and downstream flow conditions are described in Figure 4 for free and submerged cases.

90

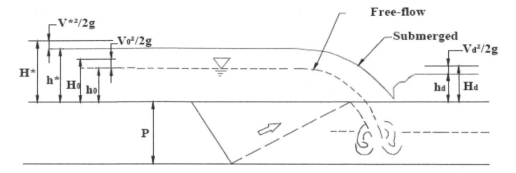

Figure 4. Definition of hydraulic parameters for the weir under free and submerged flow conditions.
Submerged flow
– H*: upstream total head;   – h*: upstream piezometric head;
– H_d: downstream total head;   – h_d: downstream piezometric head.
Free flow
– H_o: upstream total head;   – h_o: upstream piezometric head.

3 EXPERIMENTAL RESULTS

Tests made on the model of PK-Weir were realized in two phases, the first with a free flow and the second with a submerged flow. These tests were realized to verify the effect of the submersion on the performance of the PK-Weir. So, several cases were considered, according to the configuration of the outlet slab, the magnitude of the evacuated discharge and the flow controlled by an obstacle downstream (gate and sill).

3.1 Influence of the shape of the outlet slab (filling of outlet)

Three rates of filling of outlets were tested (the slab of outlet as steps). The obtained results showed that the effect of submersion is independent from the filling of outlets. This shows that the arrangement of the slab in outlets does not affect the evolution of the upstream head with regard to the downstream head.

3.2 Influence of the discharge

To estimate the effect of the submersion according to the discharge, tests were realized for three values of discharge (Q = 38 l/s, 60 l/s and 80 l/s). The obtained results point out that the variation of the upstream level with regard to the downstream level is independent from the discharge (Fig. 6).

3.3 Influence of the type of control structure downstream of the weir (gate and sill)

The increase of the downstream water level can appear due to the influence of an obstacle downstream of the weir. Two cases can be considered in relation with the type of control structure, the first concentrates flow on the bottom (flow under gate) and the second imposes a free surface flow (over a sill).
    The obtained results show, that flow on sill or under gate downstream from the PK-Weir has basically the same effect on the upstream part of the weir. According to Figure 7, the relative upstream total head for submerged conditions is not influenced by the type of control structure downstream of the weir.

3.4 Effect of the submersion on the discharge coefficient

The effect of submersion on the capacity of the PK-Weir can be determined by the variation of the discharge coefficient according to the rate of submersion. Tests made for various discharge

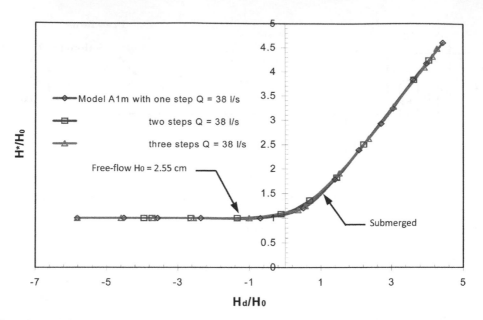

Figure 5. Relative upstream total head for submerged conditions: influence of the outlet slab configuration.

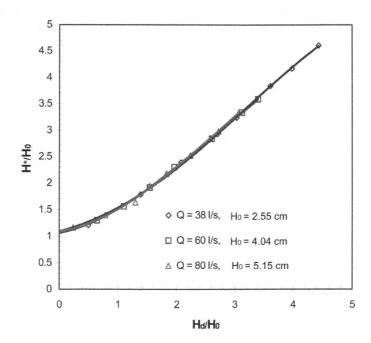

Figure 6. Relative upstream total head for submerged conditions according to the discharge.

values (Q = 38 l/s, 50 l/s, 60 l/s, 70 l/s, 80 l/s, 91 l/s, 100 l/s, 111 l/s and 122 l/s) under submerged flow and for Q = 30 l/s to 170 l/s for free flow. The results (Fig. 8) show that the PK-Weir under free flow conditions distinguishes by values of the discharge coefficient forming a superior envelope curve to the curves of submerged flow obtained for various discharges. The performance of submerged flow is strongly reduced for small heads and decreases more slowly for high relative heads.

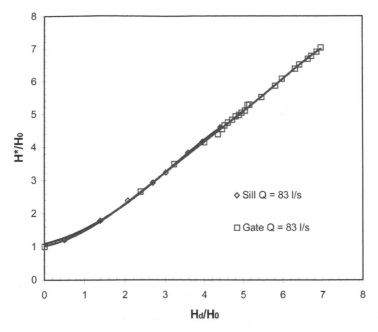

Figure 7. Relative upstream total head under submerged conditions, according to the control structure downstream of the weir.

The curves of submerged flow start from the envelope curve corresponding to the free flow, confirming that their initial conditions are in normal flow conditions.

When increasing the depth of the downstream water level, the weir becomes submerged. This submersion reveals itself by a fast decrease of the discharge coefficient for the small values of the relative upstream total head H^*/P $(0.12 < H^*/P < 0.6)$, this decrease losses of importance from a value H^*/P about 0,6 and the curves of discharge coefficient become more flattened.

One may notice that the curves of submerged flow are shifted in an ascending way (from small to high discharge). The gap between the various curves is almost constant essentially for the values of $H^*/P < 0,6$. This gap which corresponds to an increase of discharge in the order of 10 l/s engenders an increase of the discharge coefficient of about 0.1 for the values of $H^*/P < 0.6$.

One may notice that for the small discharges the influence of the downstream level is more pronounced. This can be explained by the small upstream heads when the weir works with small discharges.

Figure 8 shows that the influence of submergence is more significant when the upstream head is small. Figure 9 shows clearly this effect. For example, with a discharge of 38 l/s and a downstream total head $H_d = 5$ cm the variation of the upstream total head is of 3.5 cm, whereas for a higher discharge of 122 l/s and with the same value of H_d, the variation of the upstream total head is only 1.5 cm, about 50 % of the value corresponding to the former discharge. This observation is not true for the high heads where the variation of H^* is proportional to the variation of H_d.

Finally, it can be said that for the small values of H^*/P, the effect of submersion is very important for small discharges, however, this effect decreases gradually with the increase of discharge what implies an increase of the upstream total head H^*.

Results obtained experimentally are represented on Figure 10. This expresses the discharge coefficient related to the ratio of the downstream head reported to the upstream head. (H_d/H^*). Figure 10 shows that for values of $H_d/H^* < 0.35$, the decrease of the discharge coefficient is small. On the other hand for values of the relative downstream head $H_d/H^* > 0.35$, the discharge coefficient decreases gradually and the distance between curves is reduced until the

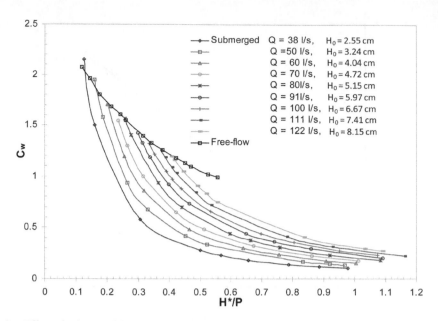

Figure 8.   Effect of submerged flow conditions on the discharge coefficient.

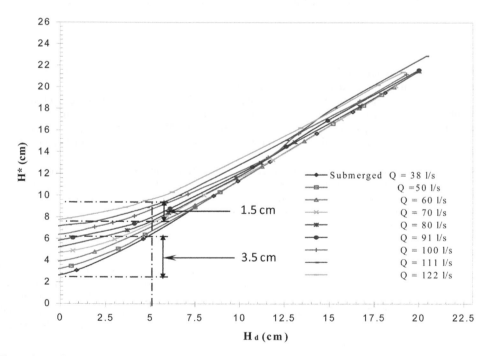

Figure 9.   Effect of the downstream total head and discharge on the submerged total upstream head.

various curves converge and tend to become a unique curve for a value of Hd/H* close to unity (Hd/H* = 1).

These results show that for the small values of Hd/H*, the PK-Weir is influenced by the downstream flow only weakly, on the other hand for the large values of Hd/H* the effect of the submersion on the upstream discharge coefficient is visible.

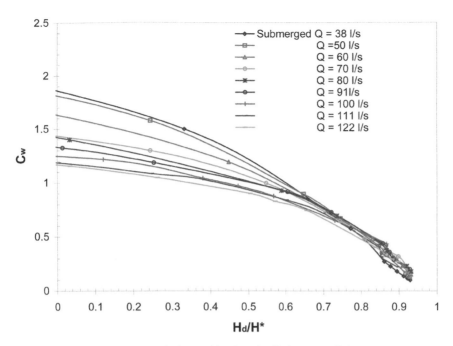

Figure 10.   Effect of the downstream relative total head on the discharge coefficient.

4   CONCLUSION

The PK-Weir can be designed to operate with a downstream water level higher than the crest of the weir. The results obtained for several cases of functioning show that the reduction of the performance is noticeable only when the downstream water level is higher than the level of the weir crest. When the downstream water level exceeds the weir crest, the downstream conditions can influence the flow on the PK-Weir and consequently its performance is affected. This influence was found to be independent of the shape of the PK-Weir. The upstream water depth varies in a proportional way with the downstream water depth. It was noticed that the variation of the upstream level with regard to the downstream level makes in a proportional constant way for various discharges. Tests realized for two types of control structures downstream (gate and sill) revealed that for a same given discharge, the submergence effect on the relative upstream total lead is not influenced by the type of structure.

REFERENCES

Tullis, P. (2006). Predicting submergence effects for labyrinth weirs. *International Symposium on Dams in the Societies of the XXI Century*. Barcelona.

Ouamane, A. Lempérière, F. (2010). "Study of various alternatives of shape of piano key weirs", *HYDRO 2010 – Meeting Demands in a Changing World,* Congress Centre, Lisbon, Portugal.

Lopes, R. Matos, J. Melo, J.F. (2009). Discharge capacity for free flow and submerged labyrinth weirs. Proc. 33rd IAHR congress, Vancouver, Canada.

Labyrinth and Piano Key Weirs – PKW 2011 – Erpicum et al. (eds)
© 2011 Taylor & Francis Group, London, ISBN 978-0-415-68282-4

Flow properties and residual energy downstream of labyrinth weirs

R. Lopes & J. Matos
Instituto Superior Técnico, Technical University of Lisbon, Lisbon, Portugal

J.F. Melo
National Laboratory of Civil Engineering, Lisbon, Portugal

ABSTRACT: Labyrinth weirs are an interesting option for the rehabilitation of existing spillways, requiring increased capacity, as well as for providing large spillway capacity in sites with limited width. In order to evaluate the flow properties on the chute downstream of the weir, including air entrainment, characteristic depths, and the energy dissipation, an experimental study was carried out on a flume assembled at the National Laboratory of Civil Engineering, Lisbon. The results show that the chute flow is basically three-dimensional in the vicinity of the weir, with air entrainment and with occurrence of shockwaves for supercritical flows. Quasi two-dimensional flow conditions occur at some cross-section further downstream. The respective normalized distance from the weir, along with the main characteristic depths, are evaluated as a function of the relative total upstream head over the weir crest and the magnification ratio. The residual energy is also reviewed and compared with previously published results.

1 INTRODUCTION

A labyrinth spillway is an overflow weir folded in plan view to provide a longer total effective length for a given overall spillway width. Therefore, labyrinth spillways may provide a higher discharge capacity than that of a straight overflow weir, for a given total upstream head and total width.

Several site-specific model studies and experimental research have been carried out in the last decades regarding the hydraulic performance of labyrinth weirs. A detailed review of past studies and hydraulic design guidelines are included in Falvey (2003). Additional research has been conducted since then, based on experimental studies (e.g., Tullis et al. 2007, Crookston and Tullis 2008), or including a numerical modelling approach (e.g., Paxson and Savage 2006). However, few studies have focused on the flow characteristics on the chute immediately downstream of labyrinth weirs. In Lopes et al. (2008), new experimental data was acquired on a flume assembled at the National Laboratory of Civil Engineering (LNEC), Lisbon, where trapezoidal labyrinth weirs were installed, for a single sidewall angle (30°). The results showed that the flow is basically three-dimensional up to a certain distance downstream of the weir, with formation of shockwaves and spray, except for high relative total upstream head, where the flow may be subcritical. It was found that two-dimensional flow conditions occur at some distance downstream of the weir, where the main flow properties, such as the mean air concentration, the equivalent clear water depth and the characteristic bulked depth become similar, regardless of the transverse coordinate. Such distance, normalized by the total energy upstream of the labyrinth weir, in relation to the chute bottom, was found to decrease slightly with the relative upstream head. The residual energy at the base of the labyrinth weir suits well the results presented by Magalhães and Lorena (1994), for similar sidewall angle.

The objective of the present paper is to analyse the location and the flow properties at the boundary cross-section which separates distinct flow regions on the chute (3-D versus 2-D), as well as the residual energy.

2 EXPERIMENTAL SETUP

The experimental study was conducted on a 41.0 m long, 2.0 m wide and 0.8 m high chute assembled at the National Laboratory of Civil Engineering (LNEC), where trapezoidal labyrinth weirs with the following geometric characteristics were installed: quarter-round crest profile; four or two cycles; height (p) equal to 0.25 m; sidewall angle (α) equal to 30° and 12° (i.e., total effective length over total width, or magnification ratio L/W equal to 1.8, and 4.0, respectively); cycle width over weir height, i.e. aspect ratio (w/p) equal to 2. The labyrinth was set over an horizontal slab.

The water discharge was measured with an electromagnetic flow meter located in the supply lines, with discharge measurement accuracy of ±0.5%. Discharges up to 0.31 m³/s (H/p ≤ 0.9) were investigated. Clear-water depths were measured with point gauges. The accuracy of the point gauges is ±0.05 mm. In order to obtain values of relative total upstream head (H/p) greater than 0.5, the width of the channel cross-section was reduced to 1.0 m.

The local air concentration was measured with a conductivity probe developed by the U.S. Bureau of Reclamation (Matos & Frizell 2000). The instrumentation was mounted on a trolley system that enabled longitudinal and transverse translation. The error on the vertical position of the instrumentation was less than 0.05 mm. The accuracy on the longitudinal and transverse position was estimated to be less than 2.0 mm. The conductivity probe data was checked with data obtained with a double tip fiber-optical probe developed by RBI Instrumentation, Grenoble, France, for identical flow conditions.

Air concentration measurements were acquired at various cross-sections along the chute (x ≤ 2.80 m, where x is the longitudinal distance from the labyrinth downstream apex) and at three different transverse positions, z, where z is the distance from the channel left sidewall: a) $z_1 = 0.5$ m (0.25 m for the 1.0 m wide channel); b) z_2 (channel centerline) = 1.0 m (0.5 m for the 1.0 m wide channel); c) $z_3 = 1.50$ m (0.75 m for the 1.0 m wide channel).

The flow downstream of the labyrinth weir with $\alpha = 30°$ was supercritical for H/p = 0.3 and 0.6 and subcritical for H/p = 0.8, whereas for $\alpha = 12°$, the flow was subcritical for H/p = 0.6 and 0.8. The air cavity beneath the nappe was not artificially ventilated. The Reynolds number varied between $0.7*10^5$ and $3.8*10^5$. According to previous findings on scale effects on air-water flows (e.g., Boes and Hager 2003, Chanson 2009), it is believed that the results may be exempted from major scale effects regarding the main flow properties. However, identical conclusion is not expected concerning other flow properties such as the number and sizes of entrained air bubbles or the air–water mass transfer, which were not addressed in the present study.

3 AIR-WATER FLOW PROPERTIES

3.1 Definitions

The local air concentration (C) is defined as the time-averaged value of the volume of air per unit volume of air and water.

The free-surface is defined herein as the air concentration line where C = 0.9. For self-aerated flow in a rectangular open channel, the characteristic depth (d) is defined as:

$$d = \int_0^{Y_{90}} (1 - C)dY \tag{1}$$

where (Y) is measured perpendicular to the channel and (Y_{90}) is the depth where the local air concentration is 90%.

The mean (depth averaged) air concentration (C_{mean}) is defined as:

$$C_{mean} = \frac{\int_0^{Y_{90}} C\, dY}{Y_{90}} \tag{2}$$

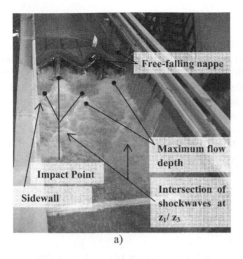

a)   b)

Figure 1.   Labyrinth weirs assembled at the LNEC flume: a) $\alpha = 30°$ and H/p $= 0.3$; b) $\alpha = 12°$ and H/p $= 0.6$.

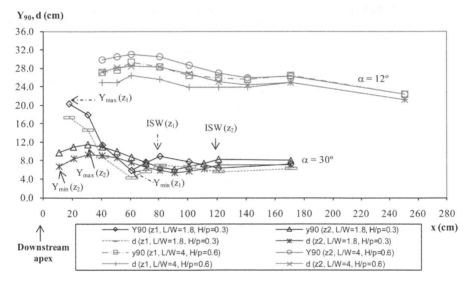

Figure 2.   Characteristic depths Y_{90} and d, down the chute. Note: ISW – Intersection of shockwaves; Y_{max} – Local maximum flow depth; Y_{min} – Local minimum flow depth.

Based on equations (1) and (2), the relationship between the characteristic depth (d) and the mean air concentration (C_{mean}) is given by

$$d = \left(1 - C_{mean}\right) Y_{90} \tag{3}$$

3.2   Basic flow patterns

The flow upstream of the weir is essentially two-dimensional, whereas in the chute further downstream, the flow is markedly three-dimensional, with air entrainment (Fig. 1). For supercritical flow conditions, the nappe interference and impact induce significant water splashing and jet deflection, followed by the propagation of shockwaves intersecting further downstream at z_1, z_2 and z_3 (Fig. 1a). For subcritical flows, no shockwaves are formed, and the free-surface exhibits a slightly wavy profile along the chute (Fig. 1b).

Typical flow patterns downstream of labyrinth weirs are shown in Figure 2, namely for $\alpha = 30°$ and H/p $= 0.3$ (supercritical flow) and for $\alpha = 12°$ and H/p $= 0.6$ (subcritical flow). The locations where local minimum and maximum flow depths occur, along with the intersection of shockwaves, are indicated in Figure 2, for $\alpha = 30°$ and H/p = 0.3.

For the labyrinth weir with $\alpha = 30°$ and H/p = 0.3 (Fig. 1 and 2), a local maximum flow depth occurs downstream of the nappe impact at z_1 and z_3 (Fig. 2, Y_{max} at z_1) followed by a region with strong streamline curvature and local minimum flow depths (Fig. 2, Y_{min} at z_1). Downstream of this region, another local maximum flow depth is observed when shockwaves intersect (Fig. 2, ISW at z_1). Further downstream, free-surface waves are observed. For the entire range of H/p tested, the flow pattern is similar at z_1 and z_3, regardless of the distance to the downstream apex of the labyrinth weir. At the channel centerline, z_2, a local minimum flow depth occurs (Fig. 2, Y_{min} at z_2) followed by a local maximum flow depth (Fig. 2, Y_{max} at z_2) that coincides with the origin of the shockwave at z_2 (Fig. 1). Further downstream shockwaves intersect again and another local maximum flow depth occurs (Fig. 2, ISW at z_2). Downstream of this region, free-surface waves of small amplitude take place.

3.3 Two-dimensional flow: location and flow properties

At some distance downstream of the labyrinth weir the flow becomes basically two-dimensional and the air concentration and velocity profiles, as well as the characteristic depths, become similar, regardless of the transverse coordinate. For defining such boundary condition, relative differences confined to ±5% were considered for the main variables at z_1 ($/z_3$) and z_2, namely the mean air concentration and the characteristic depths.

Figure 3 presents the dimensionless location where two-dimensional flow conditions occurred (x_{TDF}/H_0), as a function of the relative total upstream head (H/p), for labyrinth weirs with L/W = 1.8 ($\alpha = 30°$) and L/W = 4.0 ($\alpha = 12°$). Therein, H_0 is the total energy upstream of the labyrinth weir, in relation to the chute bottom. The results show that x_{TDF}/H_0 tends to decrease with H/p, approaching a constant value for large H/p. The influence of the magnification ratio (or the sidewall angle) seems to be relatively small. Even though the data set is limited, particularly for the larger magnification ratio, the following regression equation was obtained:

$$\frac{x_{TDF}}{H_0} = 9.850 + \frac{0.326}{(H/p)^2} - \frac{1.136}{L/W} \tag{4}$$

Figure 4 includes the dimensionless characteristic depths (d/H_0 and Y_{90}/H_0) at the location where two-dimensional flow conditions occur, as a function of H/p, along with the following

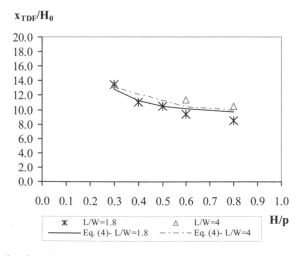

Figure 3.   Dimensionless location where two-dimensional flow (TDF) is reached.

regression equations:

$$\frac{d}{H_0} = -0.167 + 0.529\left(\frac{H}{p}\right) + 0.108\left(\frac{L}{W}\right) \qquad (5)$$

$$\frac{Y_{90}}{H_0} = -0.03 + 0.390\left(\frac{H}{p}\right) + 0.100\left(\frac{L}{W}\right) \qquad (6)$$

Both d/H_0 and Y_{90}/H_0 seem to increase linearly with H/p. For identical H/p, the characteristic depths also increase with the magnification ratio (L/W).

3.4 *Characteristic depths along the chute*

Figure 5 shows the characteristics depths obtained 3.0 cm distant from the left sidewall of the chute, as well as at the centerline (z_2), downstream of the labyrinth weir with L/W = 1.8 and H/p = 0.3.

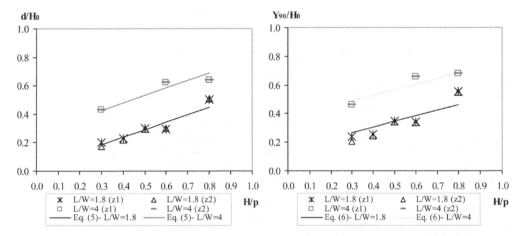

Figure 4.   Dimensionless characteristic depths at the location where two-dimensional flow (TDF) is reached.

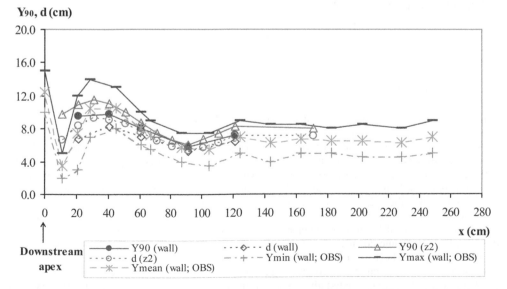

Figure 5.   Characteristic depths on the chute centerline and near the sidewall: L/W = 1.8 ($\alpha = 30°$), and H/p = 0.3. Note: Y_{max}, Y_{min}, Y_{mean} – maximum, minimum and mean flow depth at the sidewall.

101

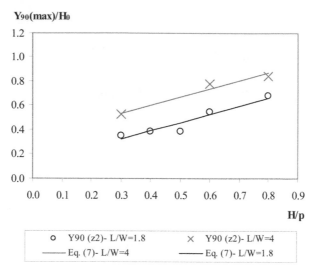

Figure 6. Dimensionless bulked depth on the chute centerline: $L/W = 1.8$ ($\alpha = 30°$); $L/W = 4.0$ ($\alpha = 12°$).

Figure 5 also presents flow depths observed at the sidewall of the chute, measured using sidewall rulers. In general, the mean flow depth at the sidewall (Y_{mean}) is similar to the bulked flow depth, Y_{90} (wall), obtained near the sidewall. Relative differences were confined to $\pm7\%$, except in the vicinity of the downstream apex of the labyrinth weir. The equivalent clear water depths fall between the minimum and the maximum flow depths measured at the sidewall. It is also shown that the characteristic depths obtained near the sidewall are similar to those obtained on the channel centerline, except in the vicinity of the downstream apex. These results suggest that the bulked depths on the chute centerline may also be taken as rough estimates of the sidewall bulked depths, downstream of the nappe impact.

In Figure 6, the maximum values of the dimensionless characteristic bulk depth on the chute centerline are plotted as a function of the relative total upstream head (H/p), for labyrinth weirs with magnification ratios of 1.8 and 4.0 (i.e., sidewall angles of 30° and 12°, respectively). The following equation indicates that $Y_{90(max)}/H_0$ increases almost linearly with H/p, and it also increases slightly with the magnification ratio (L/W):

$$\frac{Y_{90(max)}}{H_0} = -0.052 + 0.670\left(\frac{H}{p}\right) + 0.096\left(\frac{L}{W}\right) \tag{7}$$

4 RESIDUAL ENERGY

Magalhães & Lorena (1994) carried out an experimental study on one cycle trapezoidal labyrinth weirs with WES-type crest profiles. Based on experimental data, an empirical chart was presented to estimate the relative residual energy (H_1/H_0), where H_1/H_0 is given in function of the relative total upstream head (H/p) and the magnification ratio (L/W). The discharge was measured with rectangular and triangular shaped Basin weirs, whereas two point gauges were used to measure water levels in the flume: one was located 1.2 m upstream of the upstream apex of the labyrinth weir and the other about 2.0 m downstream from the first point gauge.

In the present study, water levels in the channel upstream of the labyrinth weir were measured at a cross-section located 1.0 m from the upstream apex, by using a point gauge. In order to estimate flow depths in the chute downstream of the labyrinth weir, distinct experimental procedures were adopted: Procedure 1) (EP1) Flow depths were measured downstream of the labyrinth weir at a cross-section where air bubbles were no longer visually observed within the flow; Procedure 2) (EP2) Flow depths were estimated after establishing an hydraulic jump at some specific location downstream of the labyrinth weir; two situations were tested, depending on the location of the upstream end of the hydraulic jump: a) at the same location as that adopted for EP1 (EP2a), for

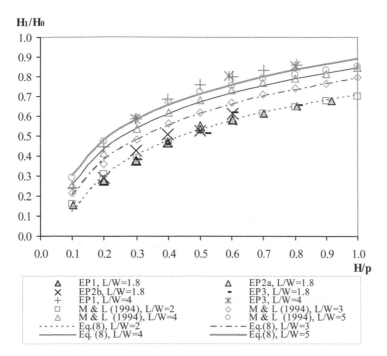

H₁/H₀ axis (vertical), H/p axis (horizontal)

Legend:
- △ EP1, L/W=1.8
- × EP2b, L/W=1.8
- + EP1, L/W=4
- □ M & L (1994), L/W=2
- △ M & L (1994), L/W=4
- ······ Eq.(8), L/W=2
- —— Eq. (8), L/W=4
- △ EP2a, L/W=1.8
- ▪ EP3, L/W=1.8
- ✳ EP3, L/W=4
- ◇ M & L (1994), L/W=3
- ○ M & L (1994), L/W=5
- —·—· Eq.(8), L/W=3
- —— Eq.(8), L/W=5

Figure 7.   Relative residual energy at the base of labyrinth weirs.

supercritical flow conditions, or b) at the location where minimum flow depths were observed at z_1 and z_3, corresponding to a cross-section within the three-dimensional flow region, absent from major bulked depths induced by cross-wave intersection (EP2b); Procedure 3) (EP3) Equivalent clear water depths obtained from the air concentration profiles (Eq. 3) at the section where two-dimensional flow occurred.

For the experimental procedure 1), clear-water depths were measured with point gauges mounted at a distance varying from 0.5 to 2.6 m downstream of the labyrinth weir, depending on the total upstream head ($0.1 \leq H/p \leq 0.9$). The mean value of the flow depths gathered at z_1, z_2 and z_3 was used to estimate the residual energy. Regarding the experimental procedure 2), the location of the hydraulic jump was controlled by a vertical-lift gate installed at the downstream end of the channel. For both procedures EP2a and EP2b, the downstream jump depths at z_1, z_2 and z_3 were measured using a point gauge. The upstream conjugate depths were then estimated using those average depth values and applying the hydraulic jump momentum equation. For some flow conditions ($L/W = 1.8$ and $H/p \geq 0.6$ for EP2a; $L/W = 1.8$ and $H/p > 0.7$ for EP2b) the hydraulic jump could not be established.

The values of the relative residual energy estimated by EP1 and EP2a are quite similar, as expected. The relative differences were lower than 2%. In addition, the relative differences between the relative residual energy estimated by methods EP2b and EP3, in relation to the results obtained via method EP1, were found to be smaller than 10%.

The experimental results obtained in the present study and those by Magalhães & Lorena (1994) are plotted in Figure 7, along with the formula developed by Lopes et al. (2006), based on the review of the data gathered by Magalhães & Lorena (1994) (Eq. 8):

$$\frac{H_1}{H_0} = 0.571 + 0.254 \ln\left(\frac{H}{p}\right) + 0.199 \ln\left(\frac{L}{B}\right) \tag{8}$$

The results presented in Figure 7 show that:

- The relative residual energy increases with the relative total upstream head (H/p), and with the magnification ratio (L/W), similarly as observed by Magalhães and Lorena (1994).

103

- The relative residual energy obtained in the present study, for $L/W = 1.8$ ($\alpha = 30°$), is within 5% of that presented by Magalhães and Lorena (1994) for $H/p \geq 0.4$, regardless of the experimental method. Slightly larger differences, up to $\pm 10\%$, occur for $H/p < 0.4$, as noted in Lopes et al. (2008).
- The relative residual energy obtained in the present study for $L/W = 4.0$ ($\alpha = 12°$), is slightly larger than that presented by Magalhães and Lorena (1994) for $H/p \geq 0.2$. However, the relative differences are confined to $\pm 10\%$, irrespective of the experimental method.

The results plotted in Figure 7 suggest that the residual energy is not expected to be significantly influenced by the number of cycles or by the crest shape.

5 CONCLUSIONS

The present investigation provides additional information on the flow properties downstream of labyrinth weirs, on a horizontal channel. The study has highlighted the following conclusions: 1) the flow is basically three-dimensional in the vicinity of the labyrinth weir, with air entrainment and shockwaves for supercritical flow conditions; 2) two-dimensional flow conditions (TDF) take place shortly downstream of the weir, where the main flow properties are basically independent of the transverse coordinate; 3) the normalized distance where TDF occurs tends to decrease with increasing H/p, approaching a constant value for large H/p; 4) at the TDF cross-section, the characteristic depths increase with H/p and L/W; 5) the maximum bulked depths increase with H/p and with L/W; 6) the relative residual energy obtained in the present study is within 10% of the that obtained from Magalhães and Lorena (1994), for the tested range of H/p and L/W.

ACKNOWLEDGMENTS

The authors acknowledge the support given by the National Laboratory of Civil Engineering (LNEC), and by the National Science Foundation (FCT), via Project PTDC/ECM/108128/2008 and the Ph.D. scholarship SFRH/BD/16333/2004, granted to the first author.

REFERENCES

Boes, R.H. & Hager, W.H. 2003. Two-phase flow characteristics of stepped spillways. *Journal of Hydraulic Engineering* 129(9): 661–670.

Chanson, H. 2009. Turbulent air-water flows in hydraulic structures: dynamic similarity and scale effects. *Environmental Fluid Mechanics* 9(2): 125–142.

Crookston, B. & Tullis, B. 2008. Labyrinth weirs. In S. Pagliara (ed.), *Proc. 2nd Int. Junior Researcher and Engineer Workshop on Hydraulic Structures, Pisa, Italy*: 59–63. Pisa: Edizioni Plus, University of Pisa.

Falvey, H.T. 2003. *Hydraulic Design of Labyrinth Weirs*. Reston: ASCE Press.

Lopes, R., Matos, J. & Melo, J.F. 2006. Discharge capacity and residual energy of labyrinth weirs. In J. Matos and H. Chanson (eds), *Proc. Int. Junior Researcher and Engineer Workshop on Hydraulic Structures, Montemor-o-Novo, Portugal*: 47–55. Brisbane: Report CH61/06, Div. of Civil Engineering, The University of Queensland.

Lopes, R., Matos, J. & Melo, J.F. 2008. Characteristic depths and energy dissipation downstream of a labyrinth weir. In S. Pagliara (ed.), *Proc. 2nd Int. Junior Researcher and Engineer Workshop on Hydraulic Structures, Pisa, Italy*: 51–58. Pisa: Edizioni Plus, University of Pisa.

Magalhães, A.P. & Lorena, M. 1994. Perdas de energia do Escoamento sobre Soleiras em Labirinto. *Proc. 6º SILUSB/1º SILUSBA, Lisbon, Portugal*: 203–211 (*in Portuguese*).

Matos, J. & Frizell, K.H. 2000. Air concentration and velocity measurements on self-aerated flow down stepped chutes.*Proc. Joint Conference on Water Resources Engineering and Water Resources Planning & Management, ASCE, CD-ROM, Minneapolis. USA*.

Paxson, G. & Savage, B. 2006. Labyrinth spillways: comparison of two popular U.S.A. design methods and consideration of non-standard approach conditions and geometries. In J. Matos and H. Chanson (eds), *Proc. Int. Junior Researcher and Engineer Workshop on Hydraulic Structures, Montemor-o-Novo, Portugal*: 37–46. Brisbane: Report CH61/06, Div. of Civil Eng., The University of Queensland.

Tullis, B.P, Young, J.C. & Chandler, M.A. 2007. Head-discharge relationship for submerged labyrinth weirs. *Journal of Hydraulic Engineering* 133(3): 248–254.

Labyrinth and Piano Key Weirs – PKW 2011 – Erpicum et al. (eds)
© 2011 Taylor & Francis Group, London, ISBN 978-0-415-68282-4

Energy dissipation on a stepped spillway downstream of a Piano Key Weir – Experimental study

S. Erpicum, O. Machiels*, P. Archambeau, B. Dewals & M. Pirotton
Laboratory of Hydrology, Applied Hydrodynamics and Hydraulic Constructions (HACH), ArGEnCo Department, Liège University, Liège, Belgium
*Fund for Education to Industrial and Agricultural Research, F.R.I.A.

C. Daux
Coyne et Bellier – Tractebel Eng., Gennevilliers, France

ABSTRACT: Piano Key Weir (PKW) is a new type of weir showing appealing hydraulic capacities. Its specific geometric features create different interacting flows and jets downstream of the structure. This suggests that the use of a PKW to control the flow upstream of a stepped spillway may help in enhancing energy dissipation on the downstream channel. The first results of an experimental study carried out at the Laboratory of Engineering Hydraulics – HACH in the scope of a Coyne & Bellier – Tractebel Engineering project are presented in the paper. The goal of the study was to compare the energy dissipation on a stepped spillway downstream of a PKW with the one which takes place on the same spillway equipped with a standard ogee-crested weir. Despite similar global energy dissipation rates whatever the weir type due to the relative important height of the spillway, significant differences in the aeration of the flow along the spillway have been shown. This paves the way to further investigations.

1 INTRODUCTION

Piano Key Weir (PKW) specific geometric features create different interacting flows and jets downstream of the structure (Fig. 1), and thus energy dissipation and air entrainment. These observations

Figure 1. Interacting flows over a scale model of a PKW.

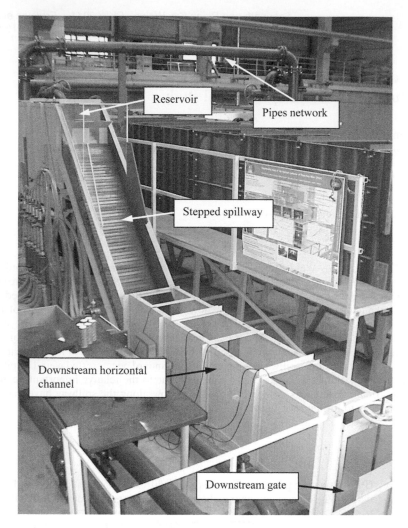

Figure 2.   Global view of the experimental facility.

suggest that the use of a PKW to control the flow upstream of a stepped spillway may help in enhancing energy dissipation on the downstream channel.

In order to verify this assumption, an experimental study has been initiated by Coyne et Bellier – Tractebel Engineering and the Laboratory of Engineering Hydraulics – HACH of the University of Liège. The goal of the tests depicted in this paper was to compare, in an idealized environment, the energy dissipation on a stepped spillway downstream of a PKW with the one which takes place on the same spillway equipped with a standard ogee-crested weir.

The experimental facility is depicted in section 2, the test procedure is presented in section 3 and the first results are analyzed in section 4.

2   EXPERIMENTAL FACILITY

An existing facility at the HACH laboratory (Dewals et al., 2004) has been used for this study. It is made of a 2 m high and 0.494 m wide stepped spillway linked to an upstream reservoir and a 3 m long downstream horizontal channel (Fig. 2). The spillway slope is 52° (78%) with 3 cm high regular steps. Bank wall are made of steel, PVC or Plexiglas plates to limit friction effects.

Figure 3.   PKW 1 model (left) and PKW 2 one (right).

Table 1.   General dimensions and aspect ratios of PKW 1 and 2.

	W_i [cm]	W_o [cm]	T_s [mm]	P [cm]	B [cm]	B_i [cm]	B_o [cm]	W_i/W_o [-]	P/W_i [-]	T_s/W_i [-]	L_u/W_u [-]
PKW1	16.9	12.3	15	26.2	62.3	6.8	8.7	1.37	1.55	0.09	4.78
PKW2	9.8	7.7	10	16.3	38.8	11	13.9	1.27	1.66	0.10	4.88

An adjustable vertical gate is at the downstream extremity of the horizontal channel. The reservoir is supplied in water by a regulated pump connected to a pressurized pipes network.

Discharge is measured on the upstream pipe using an electromagnetic flow meter (accuracy of 1%). Ultrasonic probes (resolution of 0.025 mm, accuracy of 2%) enable to measure the water depth in the reservoir and along the downstream horizontal channel. Water depth measurements have been done on a period of 120 s with a frequency of 10 Hz.

Three different models of weir have been considered upstream of the stepped spillway. In a first time, tests have been done with a standard ogee-crested weir (Dewals et al., 2004). Global spillway height is 2.039 m in that case. In a second time, two varied layouts of a PKW have been placed on the top of the stepped spillway (Fig. 3). The first one (PKW 1) represents 1.5 inlets and 1.5 outlets while the second one (PKW 2) is made of 2.5 inlets and 2.5 outlets. In both cases, the aspect ratios of the PKW are almost identical (Table 1). Only the scale of the models varies. PKW 2 dimensions are generally 1.6 times smaller than PKW 1 ones, except the global width which is kept constant to the experimental facility one, and the walls thickness which as to match the commercial dimensions of PVC plates. PKW are of a modified type A (upstream and downstream overhangs but with varied lengths) and steps have been built in the outlets to improve energy dissipation, as already tested by Leite Ribeiro et al. (2007). With PKW 1, the global spillway height is 2.258 m and it is 2.156 m with PKW 2.

3   METHODOLOGY

The tests consisted in calculating, for several constant discharges, the energy dissipation on the weir and along the spillway. This has been done by comparison of the head in the reservoir H_{up} and the head at the spillway toe H_{dw} (Fig. 4). For the first one, it can be easily calculated from the reservoir level measurements h_{up}. For the second one, because of air entrainment and flow velocity, it is difficult to measure accurately the flow depth directly at the spillway toe. An indirect method has thus been applied (Shvainshtein, 1999).

By means of regulating the gate downstream of the horizontal channel, a hydraulic jump has been created at the middle of the channel. The first conjugate depth (supercritical – upstream) has been calculated from the measured depth downstream of the jump, which is the second conjugate depth (subcritical) and is less varied and aerated.

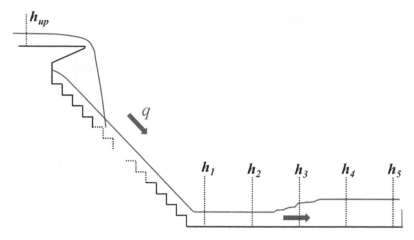

Figure 4.   Sketch of the experimental facility and tests principle.

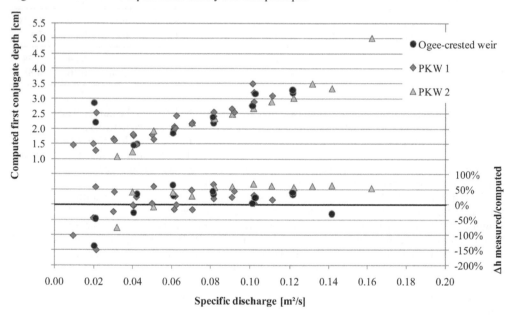

Figure 5.   Computed first conjugate depths and relative difference in between measured depths at probes 1 and 2 and computed one as function of the specific discharge in the spillway.

Water depth has been measured to probes 4 and 5 (Fig. 4) to define the second conjugate depth of the hydraulic jump and then calculate the first conjugate depth, which has been considered to compute the residual flow energy at the dam toe. In addition, direct measurements of the water depth at probes 1 and 2 (Fig. 4) enable to compute a rough direct evaluation of the residual energy downstream of the spillway. This shows a general overestimation of around 50% of the measured downstream water depth compared to the calculated one (Fig. 5). This is in agreement with the observation of a strong aeration of the flow at the spillway toe.

4   RESULTS

The results of the tests carried out with the three weir configurations for discharges ranging from 10 to 80 l/s are summarized on figure 6 in the way used by Boes (Boes & Hager, 2003), where H_{dam} is the dam height and h_c the critical depth computed from the spillway and weir width.

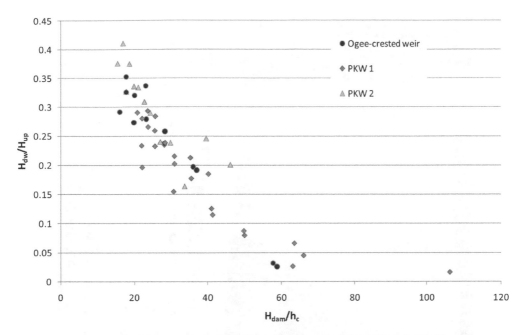

Figure 6. Relative downstream head ratio H_{dw}/H_{max} as function of relative spillway height H_{dam}/h_c.

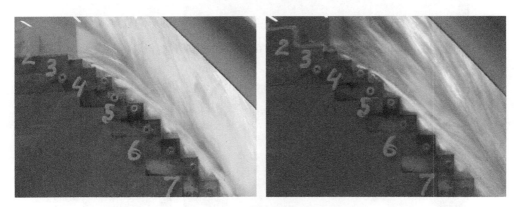

Figure 7. Aeration of the flow on the first steps – PKW 1, $Q = 50$ l/s (q = 0.1 m²/s – left) and PKW 2, $Q = 70$ l/s (q = 0.14 m²/s – right).

Regarding energy dissipation, no clear difference appears in between the tested configurations. This can be explained by the spillway length which is certainly sufficient to reach near uniform flow conditions at the spillway toe in the range of discharges considered. Indeed, according to Boes & Hager (2003), for the spillway slope considered in this paper, the minimum relative dam height $H_{dam,u}/h_c$ to attain uniform flow is 20.5.

However, for the same specific discharges in the spillway, important differences in the flow have been observed depending on the type of weir. With both PKW geometries, the flow is fully aerated at the beginning of the spillway (Fig. 8 and Fig. 9), while it is not the case with the ogee crested weir (Fig. 10). These observations are consistent with the ones of Ho Ta Khanh et al. (2011).

The aeration of the flow on the first steps of the spillway downstream of a PKW is underlined on the pictures of figure 7 for rather high discharges. No air can be seen on the steps in the outlet, but air entrainment begins just downstream of the PKW toe, upstream of the impact point of the jet coming out from the inlet apex.

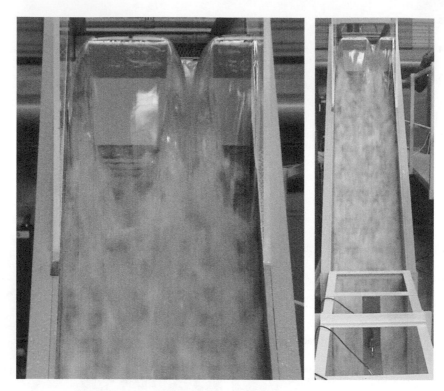

Figure 8.   PKW 1 – Aeration of the flow – $Q = 20\,\text{l/s}$ ($q = 0.04\,\text{m}^2/\text{s}$).

Figure 9.   PKW 2 – Aeration of the flow – $Q = 40\,\text{l/s}$ ($q = 0.08\,\text{m}^2/\text{s}$ – left), $Q = 60\,\text{l/s}$ ($q = 0.12\,\text{m}^2/\text{s}$ – right).

Figure 10.   Ogee crested weir – Variation of the inception point location along the spillway – $Q = 20\,l/s$ ($q = 0.04\,m^2/s$ – left), $Q = 40\,l/s$ ($q = 0.08\,m^2/s$ – center), $Q = 60\,l/s$ ($q = 0.12\,m^2/s$ – right).

5   CONCLUSION

Experimental tests carried out on a 52° slope stepped spillway for relative dam height $H_{dam,u}/h_c$ close or higher than 20 showed that the energy dissipation along the structure is equivalent if the structure is equipped with an ogee-crested weir or two different layouts of a PKW.

For the same specific discharges, significant differences have been observed in the flow upstream of the spillway depending on the type of weir. With both PKW geometries, the flow is fully aerated at the beginning of the spillway, while it is not the case with the ogee-crested weir.

These observations suggest quicker effective energy dissipation downstream of a PKW than downstream of an ogee-crested weir. This assumption will be validated by further tests considering a reduced height of the stepped spillway.

Finally, improvement of the calculation accuracy might be provided by considering a modified conjugate depths relation taking into account the high volume of air in the flow upstream of the hydraulic jump.

REFERENCES

Boes R.M. & Hager W.H. 2003. Hydraulic design of stepped spillways. *Journal of hydraulic engineering ASCE*. 129(9): 671–679.
Dewals, B.J., André S., Schleiss A. & Pirotton M. 2004. Validation of a quasi-2D model for aerated flows over mild and steep stepped spillways. In Liong, Phoon & Babovic (eds). *Proc. of the 6th Int. Conf. on Hydroinformatics, Singapore*. Vol I, 63–70. World Scientific Publishing Company.
Ho Ta Khanh M., Truong Chi Hien & Nguyen Thanh Hai. 2011. Main results of the PK Weir model tests in Vietnam (2004 to 2010). *International Workshop on Labyrinth and Piano Key weirs*, Liège, Belgium.
Leite Ribeiro M., Boillat J.-L., Schleiss A., Laugier F. & Albalat C. 2007. Rehabilitation of St-Marc Dam – Experimental Optimization of a Piano Key Weir. *Proceedings of the 32nd Congress of IAHR. Venice, Italy*.
Shvainshtein A.M. 1999. Stepped spillways and energy dissipation. *Hydrotechnical construction*, 33(5): 275–282.

Labyrinth and Piano Key Weirs – PKW 2011 – Erpicum et al. (eds)
© 2011 Taylor & Francis Group, London, ISBN 978-0-415-68282-4

Coupled spillway devices and energy dissipation system at St-Marc Dam (France)

M. Leite Ribeiro, J.-L. Boillat & A.J. Schleiss
Laboratory of Hydraulic Constructions (LCH), Ecole Polytechnique Fédérale de Lausanne (EPFL), Switzerland

F. Laugier
Electricité de France (EDF), Centre d'Ingénierie Hydraulique, Savoie Technolac, Le Bourget-du-Lac, France

ABSTRACT: The physical modeling tests for the rehabilitation of St-Marc Dam with a PKW are presented. A particular focus is put on the energy dissipation downstream from the PKW. The adopted solution is a leaned "ski-jump gutter" placed at the contact line between the downstream face of the dam and the natural foundation rock. It consists of a cylindrical profile, developed around an inclined axis and closed by a horizontal reach at the end of the structure. The aim of this solution is to guide the flow issued from the PKW to the stilling basin of the left existing spillway. The experimental tests consider various operation conditions which required pressure measurements at different impact zones.

Structural design of the PKW is impacted by the fact that the concrete of the dam is subjected to a noteworthy blowing reaction. For that reason, the new structure could not be anchored in the existing dam. Thus, the spillway behaves as a gravity structure.

1   INTRODUCTION

The Saint-Marc dam, property of EDF (France) is a 40 m high concrete gravity dam with an overall crest length of 170 m. It was built between 1926 and 1930 and is located 20 km upstream of Limoges, France (Fig. 1 left). The created reservoir covers an area of 150 ha, for a retention volume of 20 mio.m³. The power plant is located at the toe of the dam providing 13.5 MW through three Francis turbines. The original spillways include 3 sluices with Creager type crests: one 7.5 m wide equipped with a radial gate on the right bank and two identical 10 m wide also fit out with

Figure 1.   Downstream view of St-Marc dam with the original spillway devices (left), and with the additional PKW (right).

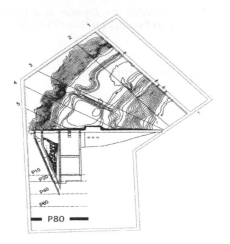

Figure 2.   Limits of the hydraulic model (left) and upstream view of the spillway devices, including the PKW (right).

radial gates. Normal and maximum water levels are fixed at elevation 282.00 and 283.50 m NGF respectively, while spillway crests level lies at 278.50 m NGF. Before the construction of the PKW, the maximum spillway capacity was 623 m³/s.

The update of hydrological studies according to the Gradex method (CFGB, 1994) has shown that the existing discharge capacity is smaller than the 1'000 years return period design flood (750 m³/s). In order to satisfy the required dam safety, EDF-CIH undertook a feasibility study to provide an additional discharge work to the existing spillways. A technical-economical comparison of several designs was carried out, among them the Piano Key Weir (PKW) turned out to be the best solution (Fig. 1 right). This alternative presents a high performance regarding risk analysis (reliability of free surface flow spillway) and maintenance costs (no mechanical or electrical device).

The PKW introduces however particular hydraulic behaviors requiring special attention. Among them, aeration of the jet napes issued from the weir crest as well as reception and guidance on the downstream part of the dam have to be mentioned. The overflow also requires an adapted energy dissipation device, in good agreement with existing works at the toe of the dam.

2   DESIGN CRITERIA AND EXPERIMENTAL RESULTS

2.1   *Experimental setup and device*

The experimental model of the St-Marc's Dam constructed at LCH includes part of the reservoir, the dam body, the existing spillways and the designed PKW (crest at elevation 282.15 m NGF). On the downstream side, the model reproduces the stilling basin of the left spillway, the channel of the right bank spillway and the current topography of the valley. The model extends approximately 130 m upstream and downstream of the dam (Fig. 2 left).

The existing gated spillways and the PKW, as well as the stilling basin and channel are made of PVC. The topography of the reservoir and the reach of the Taurion River downstream of the dam are reproduced with a cement cover. The experimental setup is constructed with a geometrical scale factor 1:30 (Fig. 2 right). The model operates with respect to Froude similarity.

2.2   *Upstream and downsteam flow conditions*

The upstream boundary conditions of the spillway devices revealed to be positively influenced by the PKW operation during floods. This impact could be put in evidence by the analysis of the superficial flow field velocities based on LSPIV (Large Scale Particle Image Velocimetry)

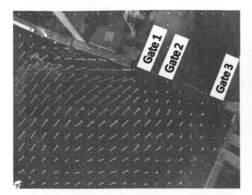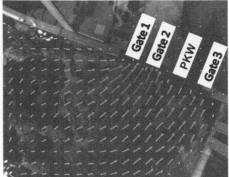

Figure 3.   Average flow pattern in the reservoir without (left) and with PKW (right) at maximum operation level.

measurements (Kantoush et al., 2008). Tests were conducted for different spillway operations, with and without PKW. When comparing the average flow fields it could be noticed that before the construction of the PKW, the main flow is deflected to the left side of the basin (Fig. 3 left). Moreover, the approach velocity is reduced near the gates due to a circulation cell forming in the right corner. As a result of the PKW construction, the flow patterns completely change. The flow orientates more straight and perpendicular to the gates, the right corner circulation is suppressed and the approach velocity is more uniform (Fig. 3 right).

The downstream hydraulic boundary conditions regulated on the model have a significant influence on the energy dissipation at the toe of the dam. For that reason, the hydraulic profiles of the Taurion River were computed by numerical *Hec-Ras* 1D-Simulation (USACE, 2002) over the reach comprised between the next reservoir of Chauvan and cross section P80 corresponding to the downstream limit of the model (Fig. 2 left). Different river roughness coefficients were considered in the frame of a sensitivity analysis, varying the Manning's coefficients in a ±20% range. Computations confirm that the water level at dam of Chauvan does not modify the flow conditions at the outlet of St-Marc. However, the bridge St-Martin Teressus located in the downstream vicinity influences the rating curve at section P80. The roughness sensitivity analysis reveals a maximal difference of 30 cm in the range of highest discharges. For the hydraulic tests, the smoothest configuration was adopted, placing the results on the safe side concerning the energy dissipation behavior at the toe of the dam.

2.3   *Spillway capacity*

2.3.1   *PKW capacity*

The PKW design was constrained by the available space on the dam crest as only a 20 m wide pass was affordable. Two different shapes of PKW, a trapezoidal and a rectangular, were tested in the model (Leite Ribeiro et al., 2009; Laugier et al., 2009). The final rectangular design includes two wide inlet keys and three smaller upstream ones (Fig. 2 right). The bottom part of the PKW is 5 m wide. The PKW sidewalls are 12.05 m long and 4 m high. The structure is cantilevered upstream and downstream equally. It is thus well self-balanced. The minimum thickness of vertical walls is not less than 25 cm, principally for durability reasons and construction terms. The PKW has a capacity of 134 m³/s at 283.50 m NGF, which corresponds to an upstream hydraulic head o 1.35 m.

The adjustment of the rating curves for the tested PKW was based on the classical equation (1) for the linear crest spillways, as presented in Leite Ribeiro et al (2007a, b). In this fitting, the PKW is considered as a linear crest spillway with the effective length (L_{eff}) decreasing with the increase of head (H).

$$Q = C_{dL} L_{eff} \sqrt{2g} H^{\frac{3}{2}}$$ (1)

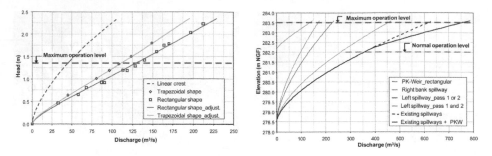

Figure 4. Rating curves of the tested PKW (left) and of the overall spillway devices (right).

The discharge coefficient C_{dL} was assumed for a sharp-crested weir over the total developed length of the PKW. This coefficient is based on the equation of SIA (Carlier, 1972) as follows:

$$C_{dL} = 0.410\left(1 + \frac{1}{1000H + 1.6}\right)\left[1 + 0.5\left(\frac{H}{H + P}\right)^2\right]$$ (2)

It is a function of the hydraulic head (H) and the height of the vertical walls of a linear weir (P), For the PKW, P varies along the crest and was defined as a first approximation, by its mean value considered equal to $P/2$, for $H = 0$, P being the height of PKW at inlet key entrance. For computation of L_{eff}, the proposed model defines the effective length in function of the total developed length of the PKW (L), the width of the pass (W), the total head upstream of the PKW (H) (Pralong et al., 2011) and two coefficients, k and n depending on the geometric parameters of the structure.

$$L_{eff} = W + \frac{1}{\left(kH + \frac{1}{\sqrt[n]{L - W}}\right)^n}$$ (3)

In the present case, the rating curve for the PKW was fitted with $k = 0.055$ and $n = 5$ for the rectangular crest shape, respectively $k = 0.055$ and $n = 7$ for the trapezoidal shape. Figure 4 (left) presents the rating curve for both tested PKWs by comparison with the theoretical rating curve of a linear Creager shape with crest length equal to the width of the pass $W = 14.50$ m. This result demonstrates the efficiency of the PKW under low heads. The fitted curve agrees with measurements and denotes a coherent convergence towards the classical broad crested spillway of length W for increasing H.

2.3.2 Total overflow capacity

The curves presented on Figure 4 (right) for the 3 gated weirs were obtained experimentally and correspond very well to the computed ones according to classical theory of standard weirs (Hager & Schleiss, 2009). The total overflow capacity of the St-Marc dam with the PKW is 757 m³/s at the maximum operation level. This value is higher than the required capacity.

2.4 Energy dissipation

The trajectories of the jets issued from both inlet and outlet keys of the PKW impact the downstream slab of the dam before pursuing over the natural topography. In order to dissipate the energy of the flow, six different solutions were designed and tested on the model, including protection of the natural gneiss, baffle blocks on the slab, deflector walls and shifted ski jumps. Figure 5 illustrates the situation without any device and the adopted solution (inclined gutter).

For geological reasons related to the safety of existing structures and to excavated volumes, the best option consisted to divert the flow towards the existing stilling basin. The adopted solution is a leaning 'ski-jump gutter' placed at the contact line between the downstream face of the dam and the natural rock, as shown in Figure 5.

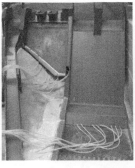

Figure 5. Spillway devices with the natural rock apron downstream from the PKW (left); with the inclined gutter and pressure measuring connections (middle); flow pattern at maximum operation level (right).

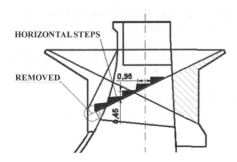

Figure 6. PKW optimization measures: Steps insertion and length shortening of the outlet key. Schematic profile (left) and downstream view of the model (right).

The gutter consists of a cylindrical profile with a constant diameter placed after along an inclined axis with a horizontal reach at the downstream end of the structure. The aim of this solution is to guide the flow from the PKW to the stilling basin of the left spillway.

To improve the energy dissipation and to mitigate the impact length of the flow on the downstream face of the dam, additional measures had to be taken at the outlets keys of the PKW. Some steps were constructed on the inclined slabs in order to contribute to energy dissipation (André et al., 2008a,b) and consequently to shortening the jet's trajectories (Fig. 6).

These additional measures have been quite efficient in improving the performance of the ski-jump gutter. The design of the steps was done taking care not to reduce the hydraulic capacity of the PKW over the operating range.

2.5 Pressure measurements

To analyze the behavior of the energy dissipating structure downstream of the PKW, pressure measurements have been made at several points in the inlet of the PKW, on the downstream face of the dam, in the gutter, on the right guide wall of the left spillway and in the stilling basin. An electronic transducer with a precision higher than 0.1 mm and a sampling frequency of 100 Hz was used.

The signal statistical analysis, considered the mean, minimum and maximum values, the standard deviation, the energy spectrum density and the distribution function curve. Following comments can be made, on these measurements:

– Inlet key of the PKW: Vibrations occur on the structure with 0.8 Hz frequencies at prototype scale, if the jet is not aerated. These vibrations disappear with complete aeration of the under nape of the jet.
– Downstream face of the dam: Mean pressure values from 0 to 2 m water column (w.c.) and maximum standard deviation value of 0.5 m w.c. at the point with the maximum mean value.

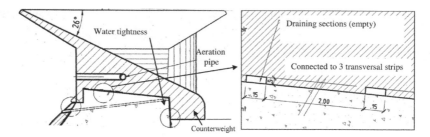

Figure 7. PKW schematic section with details of the aeration and drainage systems.

- Inclined gutter: Mean pressure values from 3.2 to 8.5 m w.c. and a maximum standard deviation of 1.1 m w.c.
- Right guide wall of the left spillway: Mean pressure values lay close to zero. This means that no collision occur between the jet coming from the inclined gutter and the wall. The standard deviation varies from 0.30 to 0.5 m w.c.
- Stilling basin of the left spillway: Pressure measurements have been compared between the existing situation and the project design. Results show that the PKW will not lead to significant variations of the mean values and standard deviations for most points. The highest difference is observed in the shaded zone of the pier dividing the two existing passes. At this location, pressure values are actually close to 0 m w.c. and will increase to 2.5 m w.c. with the PKW.

3 STRUCTURAL ASPECTS AND CONSTRUCTION

3.1 *Concrete expansion pathology*

The PKW structural design is strongly impacted by the fact that the dam's concrete is subjected to a noteworthy expansion (20 up to 40 μm/m/year), mainly due to alkali-aggregate reaction. Moreover, this pathology affects the resistance to hydraulic erosion of the concrete located on the downstream face of the dam, which is exposed to the PKW outflow. Since the characteristic duration of the floods is sufficiently long to enable severe damages (48 hours), a protection of the downstream face of the dam with a concrete slab is required.

Anchoring classically the new structures in the expansive existing concrete would on the one hand generate very high stresses in structures, and on the other hand require using tremendous quantities of connectors. Besides, the question of long term integrity of an anchor's sealing made in an expansive concrete revealed to be quite complicated. Consequently, it was decided not to anchor the new structure on the dam, but only to ensure its stability by means of compressive and friction forces. Stresses generated by dam expansion are then limited to friction capacity between both materials.

3.2 *PKW stability*

In order to provide a gravity stability, a counterweight beam was added upstream of the inlet keys. In addition, water tightness (by redundant geomembrane and bituminous joint) and drainage systems are designed to prevent any uplift forces development as presented in Figure 7.

The PKW stability analysis was conducted according to the usual bi-dimensional gravity dams' calculation method and safety standards in force for EDF's dams (Table 1). Safety factors were also checked in a deteriorated tightness configuration, with a linear uplift diagram hypothesis. In this table, values in parentheses correspond to residual safety factors calculated without water tightness in case of both geomembrane and bituminous joint defect.

The seismic effects were taken into account as pseudo-static forces, but with a ground acceleration amplification which was evaluated through a temporal analysis applied to the unmodified dam's profile. As the fundamental mode of the dam matches the seismic spectrum's peak, the 0.1 g ground acceleration reaches 0.35 g at PKW foundation level.

Table 1. PKW stability factors.

Conditions	Upstream effective stress	Downstream effective stress	Uncompressed length	HSF*	OSF*
Normal Level + Ice	49 (0) kPa	166 (166) kPa	0 (0.01) m	2.64 (2.04)	4.17 (1.91)
EDF criterium	$0 < \sigma < 5\,MPa$	$0 < \sigma < 5\,MPa$	*0 m*	*>1.33*	*>1.40*
PMF	0 (0) kPa	231 (335) kPa	0.04 (3.06) m	2.08 (1.17)	3.18 (1.18)
EDF criterium	$0 < \sigma < 7.5\,MPa$	$0 < \sigma < 7.5\,MPa$		*>1.10*	*>1.30*
MCE	0 (0) kPa	215 (248) kPa	0.20 (1.81) m	1.09 (0.83)	2.71 (1.47)
EDF criterium	$0 < \sigma < 15\,MPa$	$0 < \sigma < 15\,MPa$		*>1.05*	*>1.20*

* HSF = Horizontal Safety Factor; OSF = Overturning Safety Factor – (with water tightness defect)

Resulting safety factors are all correct. Nevertheless, a lack of horizontal resistance appears during MCE (Maximum Credible Earthquake) in case of severe tightness defect. For this reason, a stop beam was added downstream of the outlets' ends. The latter and the upstream counterweight beam are designed to resist unaided the maximum horizontal resulting force. In order to allow concrete expansion, a 3 cm clearance will be kept between upstream counterweight beam and dam. The impact of the PKW on global dam's stability is less than 2%. It's worth noticing too that for durability improvement, walls thickness have been increased to 25 cm/min.

3.3 Downstream face protection slab

Water flow on the downstream face of the dam may last for more than 48 hours and reaches velocities above 20 m/s. The nature of this flow is rather different from the ones which occur in the existing spillway (no water impact on the concrete versus an average impact angle of 25° for the PKW and flow direction parallel to the spillway).

This outflow could thus be linked with cavitations, cracks, development of turbulences and uplift pressures, potential sources of erosion of the concrete on the downstream face. In addition, this concrete is more than 75 years old, with small cement content, affected by expansion pathology and behaves a poor surface quality.

It therefore appears difficult to assess what could be the consequences of a long lasting water flow on the existing downstream face concrete. It was consequently decided to construct it with a 50 cm thick concrete slab. This protection layer shall be drained.

The stability of the downstream face protection slab is ensured by its dead load and by clamping to the gutter. A draining geogrid will prevent any uplift forces development, in addition with draining walls. Nevertheless, the slab will be subjected to negative pressures due to the flow. Final design is carried out according to the pressure measurements presented above. A preliminary technico-economical study leaded to a slab thickness of 0.5 m, which was considered a comfortable dimension with regards to the behavior under vibrations.

3.4 "Ski jump" gutter

As for the PKW and the downstream face protection slab, the gutter shall not be anchored to the dam. In fact it will be simply resting on steps cut at the foot of downstream face of the dam. The interface between gutter and dam will be drained to avoid uplift pressures.

Three types of loads are therefore applied to the gutter: self weight, hydraulic loads coming from the water flow deviation and loads transmitted by the downstream face protection slab which is clamped to the gutter.

The component of the gutter self weight which is parallel to the gutter slope forms a 31° angle with the horizontal. This load is taken by the friction reaction at the interface considering a 30° friction angle between new concrete and existing concrete. The standard safety margin is given by the friction and abutment of the horizontal part of the gutter. Hydraulic loads add "weight" to the gutter because of their direction and the way loads are transmitted to the gutter.

The horizontal component of loads transmitted by the downstream face protection slab will be supported by the abutment to the rock and the moment will be taken by the gutter self weight. There is no existing rock to support the horizontal loads applied to the bottom horizontal part of the gutter. This area will be supported by the upper sloped part of the gutter and will behave as a cantilever beam with regard to horizontal loads. Design calculations were carried out with extreme pressure values measured on the model.

3.5 Aeration

In order to aerate the flow on the PKW, two air intakes (400 mm diameter) were implemented (Fig. 7). Both structures start below outlet keys and come out to the dam crest, both sides of the PKW. They are protected by stainless steel covers and connected to each other in order to allow redundancy of air supply in case of blockage of one pipe.

4 CONCLUSIONS

The Saint-Marc hydraulic scheme presented a lack of spillway capacity after updating the 1'000 years return period flood. In order to provide the capacity in deficit, a PKW was designed. The flow guidance and energy dissipation behind the new structure required original and innovative solutions due to delicate geological context and existing works. The experimental study carried out at LCH on a physical model with scale factor 1/30 allowed the validation of the proposed PKW and the design of an energy dissipation structure for the new spillway.

Among the tested shapes of the PKW, the rectangular one has been adopted to the rehabilitation project. Its maximum capacity of 134 m³/s allows increasing the total spillway capacity up to 757 m³/s, slightly higher than the design flood. Experimental tests also demonstrated the efficiency of the PKW under lower heads. LSPIV measurements of the approaching field velocities in the reservoir confirmed that no negative interactions occur for different gate operations, with and without PKW.

For the energy dissipation of the jet issued from the PKW, the adopted solution is a leaned "ski-jump gutter" placed at the contact line between the downstream face of the dam and the natural rock with the objective to guide the flow from the PKW to the existing stilling basin of the left spillway.

Dynamic pressure measurements performed on the physical model gave additional information needed to the final design of the rehabilitation project, particularly concerning static and dynamic solicitation of the structures and also the jet aeration.

The structural design of all hydraulic structures (PKW, downstream protection slab and "ski-jump" gutter) is strongly impacted by the concrete expanding pathology. It was decided not to anchor the new structures to the dam, but only ensure their stability by means of compressive and friction forces.

REFERENCES

André, S., Boillat, J.-L. & Schleiss, A. 2008a. Ecoulements aérés sur évacuateurs en marches d'escalier équipées de macro-rugosités – Partie I: caractéristiques hydrauliques. *La Houille Blanche – Revue internationale de l'eau* (N° 1): 91–100.
André, S., Boillat, J.-L. & Schleiss, A. 2008b. Ecoulements aérés sur évacuateurs en marches d'escalier équipées de macro-rugosités – Partie II: dissipation d'énergie. *La Houille Blanche – Revue internationale de l'eau* (N° 1): 101–108.
Carlier, M. 1972. Hydraulique Générale et Appliquée. *Ed. Eyrolles*, France.
CFGB, Comité Français des Grands Barrages. 1994. Les crues de projet des barrages: Méthode du gradex, 18ème *congrès du Comité International des grands barrages*, Durban, Afrique du Sud.
Hager, W.H. & Schleiss, A.J. 2009. *Traité de Génie Civil Volume 15 – Constructions hydrauliques, Ecoulements stationnaires*. Lausanne: PPUR – Presses Polytechniques Romandes.
Kantoush, S. A., et al. 2008. Flow field investigation in a rectangular shallow reservoir using UVP, LSPIV and numerical modelling, *Flow Measurement and Instrumentation*, 19, 139–144.

Laugier, F., Lochu, A., Gille, C., Leite Ribeiro, M. & Boillat, J-L. 2009. Design and construction of a labyrinth PKW spillway at Saint-Marc dam, France. *Hydropower & Dams*, Issue Five, pp. 100–107.

Leite Ribeiro, M., Bieri, M., Boillat, J.-L., Schleiss, A., Delorme, F. & Laugier, F. 2009. Hydraulic capacity improvement of existing spillways – design of a piano key weirs. *23rd Congress of the Int. Commission on Large Dams CIGB-ICOLD;* Proc. (Volume III, Q.90-R.43), 25–29 May 2009, Brasilia.

Leite Ribeiro M., Boillat J.-L., Schleiss A., Laugier F. & Albalat C. 2007(a). Rehabilitation of St-Marc Dam – Experimental Optimization of a Piano Key Weir. *Proceedings of the 32nd Congress of IAHR. Venice, Italy*.

Leite Ribeiro M., Boillat J.-L., Kantoush S., Albalat C., Laugier F. & Lochu A. 2007(b). Rehabilitation of St-Marc dam – Model studies for the spillways. *Proceedings of Hydro 2007 "New approaches for a new era"*. Granada, Spain.

Pralong, J., Blancher, B., Laugier, F., Machiels, O., Erpicum, S., Pirotton, M., Leite Ribeiro, M., Boillat, J-L., & Schleiss, A.J. 2011. Proposal of a naming convention for the Piano Key Weir geometrical parameters. *International Workshop on Labyrinth and Piano Key weirs,* Liège, Belgium.

USACE, US ARMY Corps of Engineers (2002). HEC-RAS River Analysis System, User's and Reference Manuals; Version 3.1, *Hydrologic Engineering Center*, Davis, CA, November 2002.

Labyrinth and Piano Key Weirs – PKW 2011 – Erpicum et al. (eds)
© *2011 Taylor & Francis Group, London, ISBN 978-0-415-68282-4*

Energy dissipation downstream of Piano Key Weirs – Case study of Gloriettes Dam (France)

M. Bieri, M. Federspiel & J.-L. Boillat
Ecole Polytechnique Fédérale de Lausanne (EPFL), Laboratory of Hydraulic Constructions (LCH), Lausanne, Switzerland

B. Houdant, L. Faramond & F. Delorme
Electricité de France (EDF), Centre d'Ingénierie Hydraulique (CIH), Le Bourget-du-Lac, France

ABSTRACT: The Gloriettes concrete arch dam in the French Pyrenees showed a deficit of 80 m^3/s for the new design flood of 150 m^3/s. To address the problem, a Piano Key Weir (PKW) as an additional spillway was designed. The existence of a geotechnically unstable zone downstream of the PKW site, as well as requirements for environmental integration, did not allow a direct trajectory of the tailrace channel, but imposed an abrupt change in direction halfway. A stepped channel including an intermediate stilling basin was conceived for this purpose. The energy dissipation over the total length of the restitution channel for this condition was found to be about 90 per cent. Since the construction works could only be carried out in summer, the works were split into two stages: summer 2009 (excavation of the stepped reaches and the stilling basin) and summer 2010 (construction of the PKW and completion).

1 INTRODUCTION

The Gloriettes dam, a concrete arch structure on the Gave d'Estaubé River, operated by Electricité de France (EDF) is located in the French Pyrenees (Fig. 1). It was constructed between 1949 and 1951. The initial spillway consists of four free-overflow sluices on the dam crest at 1667 m NGF. Its capacity is about 70 m^3/s at the maximum operating level of 1667.8 m NGF. For the new design flood of a 1000 year return period, a peak discharge of 150 m^3/s was defined. In order to compensate for the deficit of 80 m^3/s, a complementary spillway on the right bank had to be constructed. Two different shapes of Piano Key Weir (PKW) with weir crests at the same level as the existing spillway have been designed and evaluated by physical model tests. The results have been published in Bieri et al. (2010).

The existence of a geotechnically unstable zone downstream of the new PKW site, as well as requirements for environmental integration, did not allow for a simple and direct trajectory of the tailrace channel, but instead there must be an abrupt change in direction. The main design purpose was to produce continuous and maximal energy dissipation, by avoiding overflows.

The first design of the tailrace channel involved two straight reaches, interconnected by a 120° curve. The preliminary evaluation was based on an analysis of the water level, flow velocity and energy dissipation. The energy grade line was computed by applying the Bernoulli equation on steep slopes and the energy dissipation was based on the classical Colebrook-White equation. The original design allowed for only a low level of energy dissipation, leading to high flow velocities of about 25 m/s. The cavitation index σ at the bottom was computed, showing that the threshold value for cavitation risk of 0.2 was exceeded in several locations.

On the other hand, it was shown by using the empirical Knapp formula (Hager & Schleiss 2009), that the initial incurved part of the channel with its small radius could generate water levels of up to 6 m as a result of the centrifugal acceleration and stationary waves. This hydraulic behaviour led to a completely new design for the restitution channel with two stepped reaches and an intermediate stilling basin, allowing the 120° change of direction halfway.

Figure 1. Downstream view of Gloriettes dam with existing spillway and construction site of the PKW.

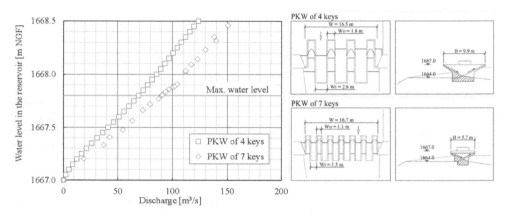

Figure 2. Capacity and geometrical characteristics of the two tested PKW.

2 DESIGN CRITERIA

2.1 *Boundary conditions*

The upstream boundary conditions of the restitution channel are imposed by the PKW. Several tests were carried out to evaluate the hydraulic capacity of two types of PKW in relation to the existing spillway on the dam crest (Fig. 2). The results relating to the hydraulic aspects of the weir system are discussed in Leite et al. (2009). These tests led to a maximum discharge of $80\,\text{m}^3/\text{s}$ in the restitution channel for the design flood of $150\,\text{m}^3/\text{s}$.

2.2 *Stepped channel reaches*

To achieve maximum energy dissipation on the two channel reaches, steps were provided along the profile. The preliminary design was based on a research project on stepped spillways (André et al. 2008a, b). Using Froude similarity, scale effects related to air transport and velocity profiles are negligible for scale factors λ between 5 and 15 (Boes 2000). In models exceeding this factor, air bubbles are proportionally too large and air transport capacity tends to be lower than for an equivalent prototype. Furthermore, scale effects are small for Reynolds numbers over 10^5 and Weber numbers higher than 100 (Boes & Hager 2003), which was the case for the conducted tests. For quasi-uniform flow, low discharges generate nappe flow and high discharges skimming flow conditions. Nappe flow dissipates energy by flow impact on the steps. Skimming flow is less efficient, due to the development of entrapped flow cells between the steps. When both flow regimes are partly present, the regime is called transitional flow. The onset can be defined by Equation 1 for transitional flow and by Equation 2 for skimming flow, where h_c (Equation 3) is the critical water depth, h_s the step height, l_s the step length, q_w the unit flow discharge and g the gravity acceleration (Fig. 3a):

$$(h_c / h_s) = 0.743 \cdot (h_s / l_s)^{-0.244} \qquad (1)$$

$$(h_c / h_s) = 0.939 \cdot (h_s / l_s)^{-0.367} \qquad (2)$$

$$h_c = (q_w^2 / g)^{1/3} \qquad (3)$$

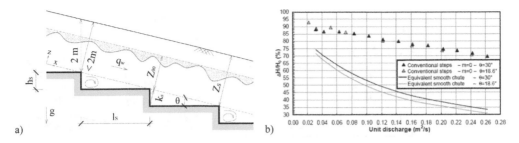

a)   b)

Figure 3. Longitudinal profile of the channel with skimming flow conditions (on the left) and relative energy loss $\Delta H/H_0$ (on the right) for conventional stepped (without macro-roughness elements $m = 0$) and smooth chutes for two chute slopes θ (André 2004).

The dissipated energy ΔH for different height to length of step ratios is generally between 85 and 95 per cent of the available head H_0. The head loss ΔH (Fig. 3b) is estimated based on an experimentally developed relationship between unit discharge and relative energy loss $\Delta H/H_0$. These results were obtained for 0.06 m high steps in the test flume. For 1 m steps, the scale factor λ is 16.67, and for 2 m, it becomes 33.33. The corresponding discharge scale factors ($\lambda^{5/2}$) are 1.13 and 6.42, respectively. For a discharge of 80 m³/s and a channel width of 7.5 m, the skimming flow regime establishes and the energy dissipation is about 90 per cent for both step heights.

In both cases, the main design criteria was avoiding overflow and minimizing the excavation volume, which led to variable step lengths. By knowing the average chute slopes θ of the two channel reaches, 19° and 18° respectively, the normal height of the steps k_s (Equation 4) and Froude number for steep slope F_θ (Equation 5) could be defined, and consequently the depth of uniform flow mixture $Z_{90,u}$ (Equation 6). This flow depth corresponds to the position where the air concentration is 90 per cent:

$$k_s = h_s \cdot \cos\theta \qquad (4)$$

$$F_\theta = q_w / \sqrt{g \cdot k_s^3 \cdot \cos\theta} \qquad (5)$$

$$Z_{90,u} / k_s = 0.58 \cdot F_\theta^{0.6\cos\theta} \qquad (6)$$

For $h_s = 1$ m and 2 m, $Z_{90,u}$ is about 1.2 m, and 1.3 m respectively. By applying a safety factor of 1.5, the water depth for the maximum discharge of 80 m³/s is 2 m. This depth is required to define the offset Z_s of the channel ground from the surrounding topography. It leads to the lower limit for the channel bottom, which should not be overtopped by the steps. Because of the very long steps, requiring a significant excavation volume, the 2 m high steps were refused.

2.3 Stilling basin

The intermediate stilling basin had to be able to dissipate the residual energy of the flow at the downstream end of the upper reach of the restitution channel. It had to be integrated in the existing topography so as to provide critical flow conditions at the beginning of the second channel reach. To ensure satisfactory performance of the sudden expansion of the stilling basin, a central sill has proved to be effective (Bremen 1990). The optimal design of the sill leads to symmetrical and stable jumps with significant contribution of the lateral eddies to the energy dissipation process, and almost uniform tailwater velocity distribution (Fig. 4a).

For a given discharge, the inflow depth h_1, the flow velocity v_1 and the Froude number F_1 at the outflow of the first reach of the channel can be estimated by applying Equation 6. By using the Bélanger formula, the conjugated downstream flow depth h_2 is calculated. For the design flood of 80 m³/s, $h_1 = 1.1$ m, $v_1 = 9.7$ m/s, $Fr_1 = 3.0$ and as a result $h_2 = 4.1$ m.

According to Bremen (1990), the length of the basin x_j is defined by Equation 7. The calculated value of $x_j = 16$ m was increased here by 5 m, to guide the water in the curved part of the basin towards the outlet:

$$x_j = h_1 \cdot (6.29 \cdot Fr_1 - 3.59) \cong 4.5 \cdot h_2 \qquad (7)$$

125

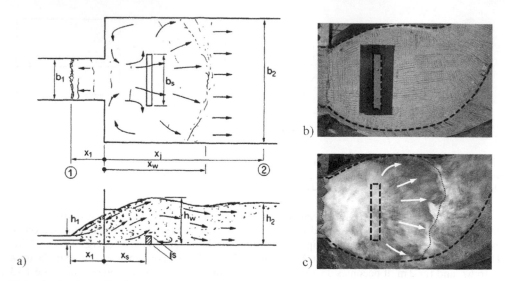

Figure 4. Notation for sill-controlled jumps in sudden expansion (Bremen 1990) (a) and final design of the modelled basin, empty (b) and with overall flow pattern (c).

The optimal sill geometry mainly depends on Fr_1, the expansion ratio between the basin width b_2 and the channel width b_1 as well as h_1 and b_1. In the present case $b_2 = 15$ m was chosen twice as $b_1 = 7.5$ m. The best flow conditions are obtained by a non-dimensional sill position X_p which is higher than 0.8 (Bremen 1990). Considering the progressive enlargement of the inflow part X_p was fixed at 1.75. The optimal position of the sill related to the channel outlet $x_s = 6.5$ m can be defined by Equation 8:

$$X_p = \frac{x_s}{(b_2 - b_1)/2} \tag{8}$$

The sill height $s = 1.6$ m is obtained by Equation 9:

$$s = \frac{x_s / x_j + 0.0116 \cdot Fr_1 - 0.225}{0.155 - 0.008 \cdot Fr_1} \tag{9}$$

The sill width $b_s = 9.4$ m can be calculated with Equation 10:

$$b_s = b_1 \cdot (1 + 0.25 \cdot (b_2 / b_1 - 1)) \tag{10}$$

For reasons of environmental integration, the smoothly rounded side walls of the basin (Fig. 4b) are the only difference between the model configuration and the theoretical case. Flow patterns remained therefore quite similar (Fig. 4c).

3 PHYSICAL MODEL TESTS

3.1 Experimental setup and device

For the physical modelling a geometrical scale factor λ of 30 and Froude similarity were applied. The scale factor is out of range for modelling two-phase flow. The possibly lower air transport capacity would reduce the air transport in the model and increase the flow resistance. Therefore energy dissipation would be overestimated. However, quasi-uniform flow is only achieved on the last 15 m of the reaches (Boes & Hager 2003) and the effect of air entrainment can therefore be considered as not relevant.

The experimental setup consisted of two connected parts. The first, reproducing the reservoir, the arch dam and the spillways, was placed in a square steel tank. The second was dedicated to the

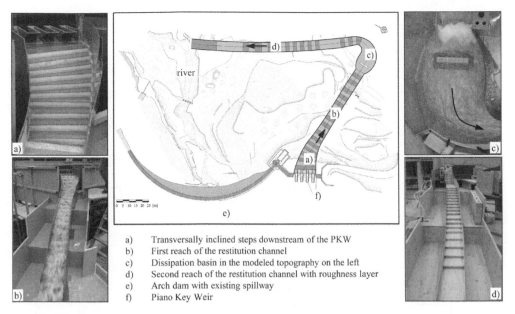

a) Transversally inclined steps downstream of the PKW
b) First reach of the restitution channel
c) Dissipation basin in the modeled topography on the left
d) Second reach of the restitution channel with roughness layer
e) Arch dam with existing spillway
f) Piano Key Weir

Figure 5.   Plan view of the final design of Piano Key Weir and the restitution channel.

restitution channel, which was reproduced within surrounding walls, allowing extensions in all three dimensions. The channel reaches and the intermediate stilling basin were made of PVC. The water level in the tank was controlled by two ultrasonic sensors and the discharge by an electromagnetic flowmeter. Flow velocities were measured by a micropropeller with 1 mm/s accuracy. The static and dynamic pressures on the sill in the dissipation basin were measured by piezometric tubes and a piezoresistive sensor at 100 Hz sampling frequency.

3.2   Test programme and optimisation procedure

For the performance and optimization tests an iterative and systematic procedure was applied. All the experiments were carried out originally with the design flood of $80 \, \text{m}^3/\text{s}$. The optimized configuration was verified afterwards, with lower discharges of 20, 40 and $60 \, \text{m}^3/\text{s}$.

In a first step, only the first reach of the restitution channel was modelled, tested and improved. To check the transversal flow distribution, the water levels on both lateral walls were measured manually. The flow behaviour in the incurved part immediately downstream of the PKW required particular attention, two complementary measures for flow equilibration and overflow obviation were developed and tested. On the one hand, the steps were laterally inclined by reducing the interior height and increasing the external one. In addition, the longitudinal step configuration was adapted. On the other hand, a guide wall at the outer bank was provided, to reduce the risk of overflow and erosion of the nearby foundation of the arch dam. Not only was the height of this structure, but also its shape part of the optimization procedure (Fig. 5a).

The main function of the stilling basin downstream of the first reach of the channel is to allow for a 120° change of direction under subcritical flow conditions. The easily adaptable experimental installation facilitated the optimization of several elements of the basin. The bottom level, the surrounding form, the in- and outflow connections of the basin, the height and form of the sill, as well as the adjoining natural topography were systematically adapted and qualitatively evaluated (Fig. 5c). The static and dynamic pressures on the sill under flood conditions were measured and the energetic spectral density analysis aimed to define the structurally problematic frequencies.

To simulate the roughness of uncoated excavated rock, with estimated irregularities of 5 to 10 cm on the prototype, a rough cement layer was added at the corresponding surfaces of the channel. Strips of approximately 2 mm high and 10 mm wide were applied (Fig. 5d).

For the final configuration the water levels were measured again, with and without roughness layer. The outflow velocities at the end of the two channel reaches were measured by micropropeller

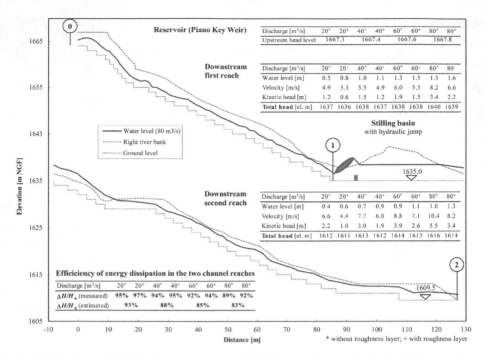

Figure 6. Velocity and water level measurements with efficiency of energy dissipation in the two channel reaches, without and with roughness layer for four different discharges.

in a 10 mm regular grid. The integrated average velocity of the subcritical flow made it possible to estimate the kinetic as well as the total head, and finally the corresponding energy dissipation.

3.3 Results

The optimization of the first reach of the channel led, concerning the upper part, to a particular design of transversally inclined concrete steps and a side wall to avoid overflow. The second reach of the restitution channel is smoothly curved and has longer steps than the first one. At the outlet of the basin, lateral guide walls allow for uniform flow distribution. At the outlet of the second reach, the flow is guided securely towards the axis of the natural river.

The velocities measured at the outlet of the two channel reaches make it possible to estimate the energy dissipation, expressed as the ratio between head loss and total head (Fig. 6). As expected the efficiency decreases with increasing discharge. The preliminary computation shows good agreement with the experimental results for the estimation of head losses for low discharges, related to nappe flow conditions. For skimming flow conditions, the discrepancy increases with the discharge. The irregular step disposition as well as the channel curvature may be causes of these differences. For the rough layer, the measured velocities at the channel outlets are lower, and as a result the head losses and efficiency are higher. To take into account uncertainties about air entrainment, a safety factor of 1.3 was applied. The optimized design of the stilling basin allows for the required 120° change in direction. It consists of an excavated round pool, enforced by lateral walls on the left and downstream parts. To stabilize the water depth in the basin between 4 and 5 m for the design flood, the outlet must be 6.2 m wide and made of concrete. The maximum jump height is about 9 m. The sill is located 5 m downstream of the upper channel outlet, and the preliminary defined height $s = 1.6$ m was confirmed by the tests. The round shape of the basin avoids lateral recirculation zones, but reduces the flow section. For that reason, the sill width b_s has been shortened to 7.5 m, corresponding to the channel width. The measured pressures on the front part of the sill are about 5.3 m water column (Tab. 1) and slightly asymmetrical, which can be explained by the asymmetrical shape of the basin. Standard deviations between 1.3 and 1.9 m indicate highly varying flow with a risk of negative values. The measures also show that the back of the sill is about 50 per cent less loaded.

128

Table 1. Static and dynamic pressures on front and back parts of the sill for the design flood of 80 m³/s.

Water column [m]		Static pressure	Dynamic pressure				Test configuration
		μ	μ	max	min	σ	
Front part	Left	5.1	5.2	13.7	−2.3	1.9	
	Center	5.1	5.4	13.2	0.5	1.3	
	Right	5.4	5.4	12.4	−0.6	1.6	
Back part	Left	2.1	2.2	8.1	−2.1	0.8	
	Center	2.3	2.4	5.2	−0.7	0.8	
	Right	2.1	2.3	6.4	−1.2	0.8	

Figure 7. Gloriettes dam area: overall view of the final project (a), upstream view of PKW (b), inclined steps downstream of the PKW (c) and stilling basin with lateral walls and sill (d).

4 STRUCTURAL ASPECTS AND REALISATION

The Gloriettes dam is located at high altitude and in a wild part of the French Pyrenees. The dam site is inaccessible throughout the whole winter because the road is covered by snow and crosses several avalanche areas. Even in summer, the dam site is a long way from all utilities. It takes 1.5 h to drive to the next concrete factory, for example. Since the construction could only be carried out in summer (from the beginning of May until the end of October), it was decided to split the project into two stages:

- **Stage 1**, during summer 2009, consisted of rock blasting and excavation for the second reach of the channel (Fig. 7a), the stilling basin (Fig. 7d) and the lower part of the first reach, as well as constructing the side walls of the stilling basin and the channel reaches.
- **Stage 2**, during summer 2010, consisted of rock blasting and digging the upper part of the first channel reach with the side walls (Fig. 7c), sawing the concrete wall and constructing the PKW (Fig. 7b), which was realized in less than four months.

Since the arch dam is located close to the excavation area, rock blasting was strictly monitored by accelerometers located on the dam and around its foundation. Two blasts produced vibrations with a maximal speed higher than 10 mm/s, but lower than the limit value of 15 mm/s.

Table 2. Main quantities of material for the construction works.

Excavations	Explosives	Riprap	Concrete	Shotcrete	Reinforcement	Formwork
$9000\,m^3$	$2100\,kg$	$800\,m^3$	$1150\,m^3$	$60\,m^3$	$33,000\,kg$	$1300\,m^2$

From a geological point of view, the right bank of the river, where the new spillway and the channel reaches were built, is formed of gneiss with strong foliation oriented perpendicular to the valley axis and with a high dip. The rock is fractured, ground by the former glacier, but not very degraded. Some geological difficulties were encountered particularly during the construction of the stilling basin: clay faults had to be treated with riprap filled with concrete. The main quantities of material for the construction works are given in Table 2. The total project cost was about €1.5 million. The main part of the cost was related to the energy dissipation devices.

5 CONCLUSIONS

Piano Key Weirs are compact and adaptive elements for increasing the flood discharge capacity of existing dams. The evacuation of water downstream of these structures requires original and innovative solutions. The Gloriettes dam spillway upgrade project has involved particularly complex boundary conditions, including a change of direction by 120° of the restitution channel halfway. The channel configuration, with two stepped reaches and an intermediate stilling basin, made it possible to guide the water to the downstream river. Uniform flow distribution on the spillway chute, as well as energy dissipation up to 90 per cent could finally be achieved, minimizing the excavation volume and also meeting integration requirements into the mountainous environment.

The chosen approach, which involved a preliminary design based on a theoretical relations and simple numerical computations followed by systematic tests on a physical model, provided satisfying results in terms of both the technical and economic aspects. Meteorological, geological and environmental issues made the Gloriettes dam rehabilitation project especially ambitious. Nevertheless, the construction works could be completed in the planned time schedule.

REFERENCES

André, S. 2004. High velocity aerated flow on stepped chutes with macro-roughness elements. *Thèse No 2993*, Ecole Polytechnique Fédérale de Lausanne.

André, S., Boillat, J.-L. & Schleiss, A. 2008a. Ecoulements aérés sur évacuateurs en marches d'escalier équipées de macro-rugosités – Partie I: caractéristiques hydrauliques. *La Houille Blanche – Revue internationale de l'eau* (N° 1): 91–100.

André, S., Boillat, J.-L. & Schleiss, A. 2008b. Ecoulements aérés sur évacuateurs en marches d'escalier équipées de macro-rugosités – Partie II: dissipation d'énergie. *La Houille Blanche – Revue internationale de l'eau* (N° 1): 101–108.

Bieri, M., Federspiel, M., Boillat, J.-L., Houdant, B. & Delorme, F. 2010. Spillway discharge capacity upgrade at Gloriettes dam. *International Journal on Hydropower & Dams* 17(5): 88–93.

Boes, R. 2000. Scale effects in modelling two-phase stepped spillway flow. In E. Minor & W. Hager (eds), *Hydraulics of Stepped Spillways*: 53–60. Rotterdam: Balkema.

Boes, R. & Hager, W. 2003. Two-phase flow characteristics of stepped spillways. *J. of Hydraulic Engineering* 129(9): 661–670.

Bremen, R. 1990. Expanding Stilling Basin, *Thèse No 850*, Ecole Polytechnique Fédérale de Lausanne.

Hager, W.H. & Schleiss, A.J. 2009. *Traité de Génie Civil Volume 15 – Constructions hydrauliques, Ecoulements stationnaires*. Lausanne: PPUR – Presses Polytechniques Romandes.

Leite Ribeiro, M., Bieri, M., Boillat, J.-L., Schleiss, A., Delorme, F. & Laugier, F. 2009. Hydraulic capacity improvement of existing spillways – design of a piano key weir. *23rd Congress of the Int. Commission on Large Dams; Proc. (Volume III, Q.90-R.43)*, 25–29 May 2009, Brasilia.

Numerical modeling

Labyrinth and Piano Key Weirs – PKW 2011 – Erpicum et al. (eds)
© *2011 Taylor & Francis Group, London, ISBN 978-0-415-68282-4*

A sensitivity analysis of Piano Key Weirs geometrical parameters based on 3D numerical modeling

J. Pralong, F. Montarros, B. Blancher & F. Laugier
EDF – Hydro Engineering Center, France

ABSTRACT: Piano Key Weirs (PKW) are new cost-effective free-flow spillways, designed to substantially increase the discharge capacity of dams. Four have already been achieved on EDF dams to meet the new hydrology requirements. However, their structure involves a lot of geometrical parameters and the hydraulic behaviour is complex. A better understanding of flow patterns in PKW and an optimisation of their shape are becoming real issues. A 3D numerical model has been developed at EDF on the CFD-Software Flow-3D®. Calibrated on the physical model results of Lempérière et al. (2003) and validated on other EDF physical model studies data, the numerical model has been used to carry out a sensitivity analysis for geometrical parameters of PKW. Several geometrical parameters (alveoli width ratios, deflectors, etc.) have been tested on the PKW model A of Lempérière et al. (2003) for a large range of flowrates. This paper presents the results of the sensitivity analysis and makes recommendations for these parameters according to the results of more than a hundred numerical simulations. These optima are likely to be valid only for the given configuration and dimensions of the PKW model A, but they give a first idea of the influence of the tested parameters.

1 INTRODUCTION

Recent computational power increase has enabled numerical modeling to become competitive in the frame of hydraulic engineering studies. EDF – Hydro Engineering Center has purchased *Flow-3D®*, a commercially available CFD-software, four years ago and numerous discharge capacity studies have been carried out since using numerical modeling.

In the mean time, EDF – Hydro Engineering Center has implemented Piano Key Weir spillways on some of its existing dams and others are currently designed in the frame of rehabilitation works. A need for an optimisation of these spillays has risen. However, the complexities of both the PKW structure and the flow patterns make it a hard task.

3D numerical modeling turned out to be an interesting opportunity as it could enable to model various geometries within a reasonable delay. Therefore, a *Flow-3D®* model was developed two years ago, calibrated on experimental data from Lempérière et al. (2003) before being validated on data collected from physical model studies. The model has demonstrated a good correlation with the experimental results.

A sensitivity analysis has been carried out based on this model to assess optima for several geometrical parameters of PKW. The initial configuration is the PKW model A (Lempérière et al., 2003) defined by equal upstream and downstream overhangs and equal inlets and outlets widths (Figure 1).

Sensitivity analyses have been carried out for the inlet/outlet width ratio, for the upstream offset of the structure, for the sidewall thickness, for the entrance section of the inlet area and for parapet walls. Starting from the configuration model A (Lempérière et al., 2003), each analysis relies on four to six different geometries which have been tested for a range of upstream head going from 0.5 m to 4 m.

Figure 1. PKW model A physical model (Lempérière et al., 2003) (EDF – Hydraulic laboratory).

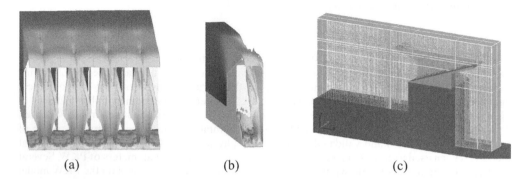

(a)   (b)   (c)

Figure 2. Reconstituted (a) and modeled (b) flows (PKW model A (Lempérière et al., 2003), 1.5 m head) and 3D view of the mesh and (c).

Table 1. PKW model A geometrical properties (Lempérière et al., 2003).

Model A piano keys weir

Geometrical paremeter	Value m	Geometrical paremeter	Value m
W_i	2.4	B	12
W_o	2.4	B_i	3
T_s	0.02 (Steel sidewall)	B_o	3
P	4	B_b	6
P_p	0	B_n	0

2 NUMERICAL MODEL CALIBRATION AND VALIDATION

The 3D numerical model of PKW relies on the following main assumptions: only half of a PKW unit (half an inlet, half an outlet and a sidewall) is modeled using symmetry lateral boundary conditions without turbulence model. The modeling domain is reduced as much as possible, and modeling without turbulence model activated enables to decrease the calculation times. Sensitivity analysis has proven that turbulence modeling does not impact discharge capacity results. The model solves the Navier-Stockes equations on a 1 million cells mesh (Figure 2) which requires around 8 hours in calculation time (8 cores, 2.5 Ghz) to model 30 seconds of flow (convergence toward steady state accomplished).

The calibration data provided by Lempérière et al. (2003) have enabled the comparison between numerical and physical results on a large range of flowrates (Table 1). The efficiency is expressed as a specific flowrate, i.e. the total flowrate divided by the width of the weir modeled.

All the deviations and gains calculated in this paper are assessed as follows:

$$deviation = \frac{tested - reference}{reference}, \quad \text{where reference is the physical model result} \qquad (1)$$

Table 2. Comparison between physical and numerical model specific discharges.

Upstream Head H m	Ratio H/P	Model A piano keys weir		Deviation %
		Physical model q_{sW} m³/s/m	Numerical model q_{sW} m³/s/m	
0.5	0.125	3.5	3.6	+2.9
1.0	0.250	8.2	7.7	−6.1
1.5	0.375	11.5	12.5	−8.3
2.0	0.500	15.6	14.9	−4.5
4.0	1.000	28.7	28.3	−1.4

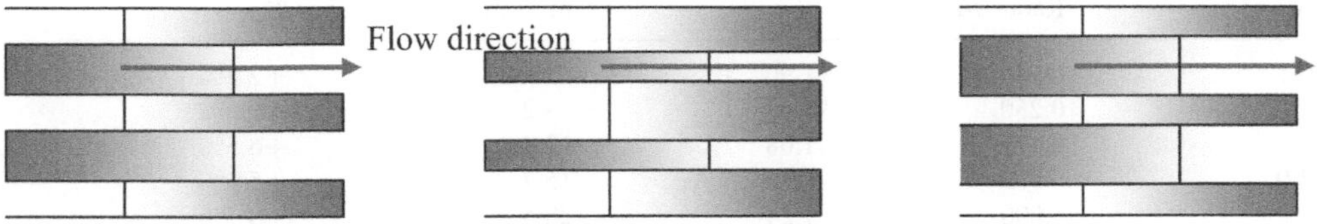

Figure 3. Variation of the inlet/outlet width ratio.

The *Flow-3D®* model, once calibrated, gives results with a mean relative deviation matching 4.5% in terms of specific flowrates. Comparisons with data available on EDF PKW physical models have also led to encouraging results, with a mean relative deviation in specific flowrates not exceeding 5%. The model has been validated, and is considered reliable for further sensitivity analysis on PKW geometrical parameters.

3 SENSITIVITY ANALYSIS

3.1 *Inlet/outlet widths ratio*

The flow patterns in the inlet and outlet are very different. The flow enters the structure within the inlet before rising up to the crest to be spilled either at the downstream extremity of the inlet or in the outlet over the sidewalls. If the dimensions of the inlet define how much water can feed the sidewalls, the dimensions of the outlet are likely to limit the discharge capacity through submergence effects. There is no obvious reasons why inlets and outlets should have the same width (W_i end W_o respectively). In order to compare PKW with the same total width W, it is the ratio W_i/W_o that has been studied, keeping the sum of the two constant (Figure 3).

The results have been analyzed and polynomial regressions have been performed to determine optima for each upstream head tested. The optimal ratio is always above 1 for the tested geometries. This shows that the outlets do not need to be as wide as the inlet to be able to spill the incoming water.

The optimum ratio is dependent on the upstream head and is lower for higher heads. Parallel results analyses have shown that the higher the upstream head, the more sensitive the discharge capacity is to the W_i/W_o ratio. Indeed, for low heads, only the developed crest length (constant here) determine the flowrate as no submergence effect happens. Discharge capacities versus W_i/W_o ratios curves are very flat at their apex, which means that the optima are quite stable.

This sensitivity analysis tends to recommend inlets wider than outlets as submergence effects in the outlets are less restrictive than flow feeding limitation in the inlets.

3.2 *Upstream overhang length*

PKW structure can be self-balancely set on the crest of a dam, which encourages research in global upstream/downstream mass distribution design. A sensitivity analysis has been carried out from

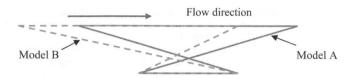

Figure 4.   Variation of the PKW cross section.

Table 3.   Optima, specific discharge associated and gain due to optimal configurations.

Upstream Head H m	Ratio H/P	Optimum W_i/W_o	Specific flowrate q_{sW} m^3/s/m	Gain in q_{sW} comparing to PKW model A (numerical results) %
0.5	0.125	1.90	3.9	+7.4%
1.0	0.250	1.68	8.2	+6.3%
1.5	0.375	1.68	12.2	+6.5%
2.0	0.500	1.63	15.7	+5.1%
4.0	1.000	1.24	28.5	+0.7%

Figure 5.   Variation in entrance section proportions (inlet sections in red, outlet sections in blue).

a discharge capacity optimisation point of view which results are detailed hereafter. Keeping a constant upstream/downstream length B of the structure, the upstream overhangs have been progressively extended up to PKW model B (Lempérière et al., 2003)] (without downstream overhangs) (Figure 4). No test has been carried to lengthen the downstream overhangs as such configurations would not be interesting from a self-balancing point of view.

The sensitivity analysis has shown that extending the upstream overhang can be interesting to improve the discharge capacity. However, the gains are not very significant, with a mean value around 2.5% and a maximum of 6.6% for proper PKW model B with 0.5 m head (Table 3). The configurations compared hereafter are defined by their B/B_o ratios.

Model B configurations seem more efficient according to the results. This parameter is very interesting, essentially for low upstream heads. Structural issues must be taken into consideration as construction implementation and structure stability are very dependant on the upstream overhang length.

3.3   Inlet entrance section area

The previous sensitivity analysis on inlet/outlet width ratio has shown how important it is to maximize the flow within the inlet. The control reach in the inlet is the entrance section area, which surface is defined by $W_i^* P$ (inlet width time the global PKW heigth). Then for a given surface area, different couples (W_i, P) are possible. The global height P can be increased as it is often possible to strike of the existing work. Reducing W_i is likely to increase the specific flow-rate geometrically, but friction phenomena could then counterbalance this benefit. Two configurations with a smaller W_i and a greater P have been tested (Figure 5). The first keeps the model A outlet width and the second reduces both inlet and outlet width equally.

Table 4. Optima, specific discharge associated and gain due to optimal configurations.

Upstream Head H m	Ratio H/P	Optimum configuration	Specific flowrate q_{sW} m^3/s/m	Gain comparing to PKW model A (numerical results) %
0.5	0.125	PKW model B ($B_o/B = 0.5$)	4.1	+13.8%
1.0	0.250	PKW model B ($B_o/B = 0.5$)	8.3	+7.8%
2.0	0.500	$B_o/B = 5/12$	15.7	+5.3%
4.0	1.000	$B_o/B = 5/12$	28.9	+2.1%

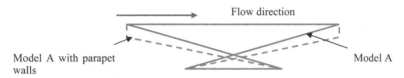

Figure 6. Implementation of parapet walls – cross section.

Table 5. Specific discharge associated to the configurations simulated.

Upstream Head H m	PKW model A Specific flowrate m^3/s/m	Inlet section modified only Specific flowrate m^3/s/m	Inlet and outlet section modified Specific flowrate m^3/s/m
0.5	4.1	4.1	4.9
1.0	8.3	8.6	10.2
2.0	15.7	16.4	19.9
4.0	28.9	32.8	35.7

The inlet width W_i has been divided by 3/2 and the total height P has been multiplied by the same value. In the last configuration, the outlet width W_o has also been divided by 3/2. Results show that decreasing W_i is interesting, even more W_o is also reduced (Table 4).

This sensitivity analysis has shown that increasing the total height P while keeping the alveoli (inlet and outlet) cross-sections constant enables to increase the PKW efficiency by a factor of more than 20%. When the available width on the dam crest is restricted, striking of few meters of work is an interesting option for designers. However, there must be a limit, as friction effects could become significant for too narrow keys.

3.4 Parapet walls

Parapet walls are optional features that can be placed on the crest of the PKW either to heighten the structure or to modify the cross sections in inlets and outlets keeping the global height P constant (Figure 6). Only the second case is considered hereafter. A qualitative analysis has been carried out to define where parapet walls could be interesting: all along the crest or only along some sections. A parapet wall was set all along the PKW crest, having the same height everywhere. Parapet walls heights P_p of 0.5, 1 and 2 m have been tested on PKW model A (Lempérière et al., 2003) ($P = 4$ m).

The parapet walls are similar all along the PKW crest. They have a rectangular cross-section. Simulations have enabled to calculate flowrates separately along the inlet and outlet extremities and along the sidewall crest. Results show that parapet walls are interesting upstream the outlet as they lead to an increase of the outlet volume, hence a reduction of submergence effects. However, a parapet wall downstream the inlet seems to disturb the flow and reduces the discharge capacity (Table 5).

Table 6. Gain in specific discharge comparing with PKW model A.

Upstream Head H m	Inlet			Outlet		
	$P_p = 0.5$ m %	$P_p = 1$ m %	$P_p = 2$ m %	$P_p = 0.5$ m %	$P_p = 1$ m %	$P_p = 2$ m %
0.5	+2.1	−2.2	−11.6	+22.0	+30.7	+46.6
1.0	−2.1	−5.0	−18.9	+8.7	+8.8	+22.1
2.0	−2.5	−6.3	−13.8	+7.1	+10.3	+5.9
4.0	+6.7	−4.3	−22.0	+3.7	+7.6	+7.1

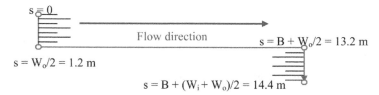

Figure 7. Curvilinear abscissa *s: plan view*.

The gain in the outlet is more important for low heads and raises up to 46.6% for 0.5 m water head ($P_p = 2$ m). A significant parapet wall (50% of the PKW height P) in the outlet is very advantageous and remains hydraulically interesting for higher heads. In the inlet, the higher the parapet walls, the worst the discharge capacities. Even the smaller parapet wall (0.5 m high) does not provide interesting results.

This analysis recommends the use of a significant parapet wall upstream the outlet but no parapet walls downstream the inlet. The implementation of parapet walls all along the PKW crest remains competitive with a 5% mean increase in discharge capacity: the gain from the outlet balances the loss of the inlet.

4 QUALITATIVE ANALYSIS – FLOW RATE DISTRIBUTION

The results presented previously are based on discharge capacity comparisons. Therefore, the optima are defined using quantitative considerations. The complexity of the flow makes uneasy the hydraulic interpretation of the results. There is an obvious need for a qualitative analysis of the flow within PKW. The numerical model which this study relies on enables to discretise the discharge along the PKW crest. A qualitative comparison has been performed between PKW model A and model B (Figure 4) regarding the discharge distribution along the PKW crest. To plot the distribution, the crest developed on half the PKW unit considered has been metered using the curvilinear abscissa *s* (Figure 7).

The comparison has been performed for an upstream head matching 1 meter, and thus for specific flow rates q_{sW} for PKW model A and model B of 7.7 m^3/s/m and 8.3 m^3/s/m respectively.

The local discharge flow rates are measured normally to planes that are 1 m wide along the crest. This explains the low specific flow rate at the sidewall upstream and downstream extremity where the flow is less normal to the crest (Figure 8).

The comparison between the two PKW shows that the feeding of both inlet and outlet extremity are very close. The PKW model B inlet is slightly less feeded due to a higher efficiency of the sidewall upstream. The PKW model A is more efficient along the first third of the sidewall while it is the model B that works better along the last two third. This gain for PKW model B counterbalances the loss happening upstream and makes it 10% more efficient than PKW model A (Figure 8).

This behavior can be explained by the reduction in inlet entrance section. It seems to enhance the vertical rising of the flow toward the sidewall crest. This interpretation may not be the only one

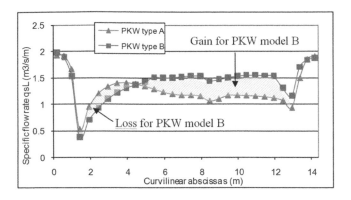

Figure 8.   Comparison between PKW model A and model B flow rate distribution along the crest.

but proves the interest of numerical modeling with such an analysis. Flow rate distribution analysis is likely to improve the hydraulic understanding of the flow within PKW.

5   CONCLUSIONS

The model implemented with *Flow-3D*® has proven its efficiency and relevance for sensitivity analysis of PKW geometrical parameters. It has enabled to perform sensitivity analysis to some parameters, and further tests are currently carried out at EDF – Hydro Engineering Center. The results discussed in this paper are valid for a PKW configuration close to model A (Lempérière et al., 2003). It seems that the upstream/downstream overhang length distribution is not a very sensitive parameter. However, this study has shown that using wider inlets than outlets is interesting. Increasing the PKW height is also improving the PKW efficiency, keeping the alveoli cross sections constant. Parapet walls can be cleverly set upstream the outlets to increase their volume and limit submergence effects.

Flow rate distribution analysis has proven to be complementary with sensibility analyses by adding a qualitative interpretation of the flow behaviour.

Parallel studies performed at *EPFL-LCH* (Le Doucen et al., 2009) and at the *Ulg-HACH* (Machiels et al., 2009) have confirmed some of the conclusions mentioned above.

REFERENCES

Machiels, O., Erpicum, S., Archambeau, P., Dewals, B., Pirotton, M., (2009). *Analyse expérimentale du fonctionnement hydrauliquedes déversoirs en touche de piano.* Colloque SHF-CFBR, Lyon.
Le Doucen, O., Leite Ribeiro, M., Boillat, J. L., Schleiss, A., Laugier, F., (2009). *Etude paramétrique de la capacité des PK-weirs.* Colloque SHF-CFBR, Lyon.
Lempérière, F., Ouamane, A. (2003). The Piano Keys weir: a new cost-effective solution for spillways. Hydropower & Dams, 144(6).
Pralong, J., Blancher, B., Laugier, F., Machiels, O., Erpicum, S., Pirotton, M., Leite Ribeiro, M., Boillat. J-L., and Schleiss, A.J. 2011. Proposal of a naming convention for the Piano Key Weir geometrical parameters. *International Workshop on Labyrinth and Piano Key weirs,* Liège, Belgium.

Labyrinth and Piano Key Weirs – PKW 2011 – Erpicum et al. (eds)
© 2011 Taylor & Francis Group, London, ISBN 978-0-415-68282-4

Hydraulic comparison between Piano Key Weirs and labyrinth spillways

B. Blancher, F. Montarros & F. Laugier
EDF Centre d'Ingénierie Hydraulique, Savoie Technolac, Le Bourget du Lac, France

ABSTRACT: Piano key weir is an evolution of traditional labyrinth, which are both considered as ones of the most efficient shapes. The corresponding significant increase in discharge has been evaluated using a commercially available CFD program, *FLOW-3D®* which has been validated completing physical model comparison studies with Paxson and Savage (2006) for classical labyrinth and Laugier et al. (2009) for piano key weir. This paper compares the hydraulic behavior of one cycle of each spillway using Reynolds-averaged Navier-Stokes numerical models. The results indicate that a basic piano key weir discharge performance is about 15 to 20% more efficient than the equivalent rectangular-shaped labyrinth one for total upstream head on weir ranging from 0.5 m to 4 m. Considering the CFD analysis, it seems that the efficiency of piano key weir, relatively to the labyrinth, comes firstly from the sidewall part of the spillway. The gap is slightly reduced by the piano key weir outlet part which is more sensitive to downstream submergence. Further tests describe the sensitivity of the labyrinth spillway and piano key weir hydraulic characteristics to approach flow conditions, through the approach depth effect, and moderate geometric changes, through the sidewall angle effect. The results confirm the fact that V-shaped (trapezoidal) labyrinth spillways are less efficient than rectangular ones for low upstream head. This trend inverts for higher heads.

1 INTRODUCTION

Labyrinth spillways are particularly well-suited for providing a large-capacity spillway in a site with restricted width. This is due to significant increase in crest length and unit discharge for a given width over conventional weirs for a given head (Darvas, 1971). Nevertheless, it also complicates the flow patterns and induces loss of efficiency caused by the flow interference from the jets of adjacent crests. A similar new cost effective solution (Lempérière & Ouamane, 2003), named piano key weir, tend to upgrade capacity and limit interferences. The major differences result from the overhanging structure, alternating spilling channels with receiving chutes, of the piano key weir spillway and the V-shapes opportunity of the labyrinth spillway.

Therefore, when designing a new spillway, the choice between a labyrinth one (vertical walls) and a piano key weir (overhanging structure) can be determined from the required flow rate and economic considerations. The result is clearly linked to a technical (structure and hydraulic) and economical compromise. At first sight, labyrinth spillway seems to be easier to build (vertical walls casing) and therefore less expensive. However, piano key weir seems to be more efficient (better specific discharge and structure stability) and needs a lower covered area, which can be attractive when the dam crest width or the water head are limited.

The aim of this study is to compare the hydraulic efficiency of a traditional rectangular-shaped labyrinth spillway with a piano key weir one for a same length of crest configuration.

2 LITERATURE REVIEW

2.1 *Labyrinth and piano key weir parameters*

The labyrinth weir is based on vertical walls and flat bottom. Its shape can be rectangular, trapezoidal or triangular. The discharge characteristics of labyrinth weirs are primarily a function of the weir height (P), the upstream head on weir (H), the weir width (W), the half apex width (a), the sidewall length (B), the sidewall angle (α). The developed length (L = 2B + 4a) of the labyrinth weir (one cycle) is generally defined as the total length along the crest.

The piano key weir can be seen as an evolution or an optimisation of the labyrinth spillway, which cannot be used on top of the usual concrete gravity dam sections because of its substantially flat required area. The shape of the piano key weir is based on a rectangular layout, a reduced width of the elements and an inclined bottom of the upstream and downstream part. The part where the flow enters is known as the inlet, and the other part the outlet. Therefore, the discharge characteristics are logically dependant on more parameters (Pralong et al., 2011), such as the inlet (W_i) and outlet (W_o) widths or the upstream (B_o) and downstream (B_i) overhangs.

2.2 *Labyrinth and piano key weir design methods*

Taylor (1968) started to study the hydraulic behaviour of labyrinth weirs, comparing it to that of a sharp-crested weir, before developing design criteria (Hay & Taylor, 1970). Darvas (1971) based his work on physical model studies to develop a family of curves evaluating spillway performance. Nowadays, the main methods (design and construction of labyrinth spillway) referred to Lux & Hinchliff (1985) and Tullis et al. (1995). The last one has been developed thanks to physical modelling results. Both methods suggest a quarter-round crest shape, a maximum H/P of about 0.7 and a W/P ratio greater than 2 (3 to 4 for Tullis et al.). Falvey (2003) summarized these methods (Fig. 1) and introduced the concept of interference at the apexes. In addition, he presented a spreadsheet that can be used to optimize the dimensions of the labyrinth while minimizing the structure cost. Ghare (2008) submit a methodology based on the previous works to determine the optimal design of trapezoidal labyrinth weirs. Nowadays, more than 40 labyrinth installations are noted in the literature, mainly situated in the U.S.A. or in Portugal.

Hydroccop has more than 10 years of experience designing piano key weirs. This began with the preliminary model in 1999 at the EDF National Hydraulic Laboratory and has continued with detailed tests in 2003 at Biskra University, taking care of the structural and construction facilities. Lempérière & Ouamane (2003) suggest 2 solutions. The first one, Model A with similar upstream and downstream overhangs, favours the use of pre-cast concrete elements and may be preferred to improve many existing spillways when the second one, Model B with only an upstream overhang, could be attractive for many future large dams. Some works are in progress to understand piano key weir hydraulic behaviour and to upgrade its efficiency. Le Doucen et al. (2009) complete a piano key weir capacity parametric study before Leite Ribeiro et al. (2009) suggest a piano key weir design to improve hydraulic capacity of existing spillway at the International Commission on Large Dams hold in Brasilia. Machiels et al. (2010) achieved an experimental study of the alveoli widths influence on the release capacity and elaborate a 1D numerical approach to model the flow over a piano key weir. Finally, Truong Chi Hien & Ho Ta Kanh (2009) studied hydraulic models of Model B piano key weir with stepped spillway. Since 2008, EDF experimented with several numerical models of piano key weirs to upgrade their design and built 4 of them (Goulours dam, Saint-Marc dam, L'Etroit dam and Gloriettes dam).

2.3 *Labyrinth and piano key weir hydraulic performance sensitivity*

The hydraulic performance is partially dependant on the geometrical parameters (crest shape, weir height, sidewall length and angle, half apex width) and is affected by other factors such as approach conditions and downstream conditions.

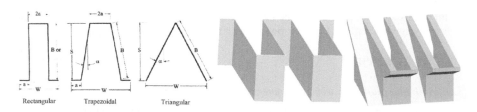

Figure 1.   Labyrinth weir shapes and 3D views of a labyrinth spillway and a piano keys weir.

According to the United States Army Corps of Engineers (1990), the labyrinth spillway hydraulic characteristics are extremely sensitive to approach flow conditions. This requires sitting the crest configuration as far upstream into the reservoir as possible in order to achieve approach flow nearly perpendicular to the labyrinth spillway crest as shown by Falvey (2003). The numerical results sensitivity to a modification of the depth of approach on piano key weir and labyrinth spillways is studied below.

2.4 Numerical modelling accuracy

Savage et al. (2004) point out that a CFD model can provide realistic results of discharge over labyrinth, especially for weirs that may be raised to an atypical height which current design methods are not appropriate for. Furthermore, this use allows the mapping of the flow field and a finer analysis comparing to physical model studies.

According to Paxson & Savage (2006), a RANS Computational Fluid Dynamics (CFD) model, using commercially available software (*FLOW-3D®*), for a given labyrinth geometry configuration is shown to provide results equivalent to those obtained using the previous design methods (5% deviation). Furthermore, non-standard labyrinth approach conditions or geometries have been tested on physical and numerical models to determine the design methods' applicability for lower than recommended values of the aspect ratio (W/P). These methods seem to be inappropriate for W/P < 2 even if hydraulic performance increase for these lower values.

Laugier et al. (2009) demonstrate the numerical model accuracy to determine the piano key weir discharge, using *FLOW-3D®* and comparing the 3D results with physical model data.

3 NUMERICAL MODELLING

To compare the piano key weir and the labyrinth spillways with the CFD model, 3D solid models having the geometric parameters shown in Table 1 were constructed and imported into the program. The numerical modeled piano key weir configuration is defined by Lempérière & Ouamane (2003) model A, i.e. a 4 m high spillway with similar inlet and outlet width and similar upstream and downstream overhangs. The hydraulic physical model tests were made with immaterial wall thickness. One should be aware that the considered wall thickness is not realistic for structures that have been built until now (too thin).

The numerical modeled rectangular-shaped labyrinth configuration is based on the same crest geometry (length and thickness) as the piano key weir. Only inclined bottoms are replaced by vertical walls. This inclination has a significant effect on the performance of all labyrinth weirs.

Regarding the calculation time limitation, numerical models of those spillways were restricted to a half part of an inlet and a half part of an outlet, assuming symmetry side boundary conditions. Some sensitivity tests to different turbulence models ended in the same results (specific discharge

Table 1. Considered piano key weir and labyrinth spillways configurations.

	Model A piano key weir	Rectangular-shaped labyrinth
Upstream-downstream length (B)	12 m	12 m
Upstream overhang crest length (B_o)	3 m	0 m
Downstream overhang crest length (B_i)	3 m	0 m
Base length (B_b)	6 m	12 m
Wall height (P)	4 m	4 m
Width of a unit (W_u)	4.8 m	4.8 m
Inlet key width (W_i)	2.4 m	2.4 m
Outlet key width (W_o)	2.4 m	2.4 m
Crests & wall thickness ($T_s = T_i = T_o$)	0.02 m ~ 0 m	0.02 m ~ 0 m
Total developed length of a unit (L_u)	28.8 m	28.8 m
Developed length ratio of a unit (n_u)	6	6
Aspect ratio (W_u/P)	1.2	1.2

Table 2. Piano key weir and labyrinth total specific discharge.

Upstream Head H	Ratio H/P	Model A piano key weir specific discharge referred to the total width (q_{sW})			Rectangular-shaped labyrinth specific discharge referred to the total width (q_{sW})			P.K. weir/ labyrinth
		Physical model	Numerical model	Deviation	Tullis design method (6°)	Numerical model (0°)	Deviation	
m		m³/s/m	m³/s/m	%	m³/s/m	m³/s/m	%	%
0.5	0.125	3.5	3.6	+2.9	2.8	3.1	+10.7	+17.7
1.0	0.250	8.2	7.7	−6.1	6.8	6.1	−10.3	+27.2
2.0	0.500	15.6	14.9	−4.5	13.9	12.6	−9.4	+18.0
4.0	1.000	28.7	28.3	−1.4	26.9	24.4	−9.3	+16.2

variance lower than 1%). The cell size of the mesh ranges from 0.06 m to 0.10 m and is slightly refined at the vicinity of the crest. The total number of cells is close to 1 million. It takes 20 to 40 hours to simulate 40 to 50 seconds of transient outflow. The upstream boundary length from the spillways was sufficient for a profile flow to be established.

4 RESULTS

4.1 Velocity flow field

Computed streamlines profiles and specific discharge distribution along the crest give the opportunity to observe phenomenon and to propose explanations for the numerical results. Thus, observing the velocity flow field in the inlet part of the labyrinth spillway demonstrate that the piano key weir inclined bottom is well adapted, preserving only the hydraulically efficient area.

4.2 Total specific discharge

The calculated total specific discharge of the piano key weir and the labyrinth FLOW-3D® numerical models are compared with the physical model results (Lempérière & Ouamane, 2003) for the first one and with the Tullis et al. (1995) design method for the second one, i.e. with a specific discharge $q_{sW} = (2/3).C_{dL}.(L/W).\sqrt{2g}.H^{3/2}$ with $C_{dL} = A + B(H/P) + C(H/P)^2 + D(H/P)^3 + E(H/P)^4$. The coefficients A, B, C, D and E are dependant from the sidewall angle. Unfortunately, they are not defined for a 0° sidewall angle which is our case. For the smaller defined sidewall angle (6°), A = 0.49, B = −0.24, C = −1.20, D = 2.17 and E = −1.03. This limitation explains the deviation (10%), shown in next table, between the design method and the numerical results, which besides have been validated by Paxson and Savage (2006). Nevertheless, it seems coherent that the 0° sidewall angle labyrinth is more efficient than the 6° sidewall angle labyrinth for lower upstream heads (because of its higher total developed crest length) and less efficient for higher upstream heads (because of its higher sensitivity to downstream submergence). Note that Falvey (2003) efficacity parameter analysis shows that the 6° sidewall angle was optimum for labyrinth weir.

The specific discharges of the piano key weir and the labyrinth follow a linear correlation for higher heads. The downstream submergence operating in the outlet parts for higher heads, which is observed for a 4 m upstream head, seems to have the same influence on both spillways.

The numerical models results suitable linear regressions are $q_{sW} = 28.0 \times (H/P) + 0.5$ for the piano key weir and $q_{sW} = 24.4 \times (H/P) + 0.1$ for the labyrinth. The power regressions are $q_{sW} = 29.0 \times (H/P)^{0.99}$ for the piano key weir and $q_{sW} = 24.7 \times (H/P)^{1.00}$ for the labyrinth spillway.

4.3 Specific discharge along the crest

Contrary to the physical model, the numerical one gives the opportunity to determine easily the specific discharge along the crest for the labyrinth and the piano key weir spillways and to identify

Table 3. Specific discharge of the inlet, outlet and sidewall of the piano key weir and labyrinth.

Upstream head (H)	Ratio (H/P)	Model A piano key weir specific discharge referred to the developed length			Rectangular-shaped labyrinth specific discharge referred to the developed length		
		Inlet (q_{si})	Outlet (q_{so})	Sidewall (q_{sh})	Inlet (q_{si})	Outlet (q_{so})	Sidewall (q_{sh})
m		$m^3/s/m$	$m^3/s/m$	$m^3/s/m$	$m^3/s/m$	$m^3/s/m$	$m^3/s/m$
0.5	0.125	0.7	0.7	0.6	0.7	0.7	0.5
1.0	0.250	1.8	2.0	1.2	1.7	2.0	0.8
2.0	0.500	4.4	5.3	2.0	4.0	5.8	1.5
4.0	1.000	9.1	14.4	3.3	8.7	14.6	2.5

Table 4. Percentage of total discharge in the inlet, outlet and sidewall of the piano key weir and labyrinth.

Upstream head H	Ratio H/P	Model A piano key weir percentage of total discharge			Rectangular-shaped labyrinth percentage of total discharge		
		Inlet	Outlet	Sidewall	Inlet	Outlet	Sidewall
m		%	%	%	%	%	%
0.5	0.125	9	10	81	11	12	77
1.0	0.250	12	13	76	14	16	70
2.0	0.500	15	18	67	16	23	61
4.0	1.000	16	25	59	18	30	52

efficient parts of the inlet, outlet and sidewall. Geometrically speaking, in our cases, the sidewall crest length is close to 83.3% $(B_h/(B_h + \frac{1}{2} (W_i + W_o)))$ of the total crest length whereas the inlet and outlet ones are about 8.3% $(\frac{1}{2} W_i/(B + \frac{1}{2} (W_i + W_o)))$.

The analysis of the discharge mapping reveals that:

- the inlet part of the piano key weir spillway is slightly more efficient ($<10\%$) than the inlet part of the labyrinth one, which may be explained by the recirculation phenomena occurring for the labyrinth,
- the outlet part of the piano key weir spillway is slightly less efficient ($<8\%$) than the outlet part of the labyrinth one, which may be explained by the downstream submergence facility due to the inclined bottom of the piano key weir,
- the sidewall part of the piano key weir spillway is significantly more efficient (14% to 21%) than the sidewall part of the labyrinth one,
- the distribution of total discharge between outlet, inlet and sidewall is close to the geometrical overflowing crest length distribution for lower heads,
- the percentage of total discharge passing through the sidewall decreases considerably when upstream head increases, from 80% for H = 0.5 m to 50% for H = 4.0 m.

4.4 Total discharge coefficient

The total discharge coefficient of the considered spillways have been calculated assuming that $C_{dW} = Q/W.\sqrt{2g}.H^{3/2}$. The piano key weir discharge coefficient is higher than a classical Creager spillway (design head of 2 m) from +450% for H = 0.5 m to +50% for H = 4 m.

4.5 Discharge coefficient along the crest

The discharge coefficient along the crest have been calculated assuming that $C_{dWi} = Q_i/W_i.\sqrt{2g}.H^{3/2}$, $C_{dWo} = Q_o/W_o.\sqrt{2g}.H^{3/2}$ and $C_{dWh} = Q_h/B_h.\sqrt{2g}.H^{3/2}$.

Table 5. Total discharge coefficient of a Creager, a piano key weir and a labyrinth.

Upstream head (H)	Ratio (H/P)	Discharge coefficient (C_{dW})		
		Creager ($H_d = 2$ m)	Piano key weir	Labyrinth
0.5 m	0.125	0.418	2.305	1.959
1.0 m	0.250	0.455	1.741	1.368
2.0 m	0.500	0.494	1.188	1.007
4.0 m	1.000	0.537	0.799	0.688

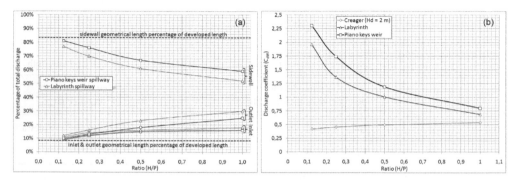

Figure 2. Inlet, outlet and sidewall discharge contribution (a) Total discharge coefficients spillways (b).

Table 6. Discharge coefficient of the inlet, outlet and sidewall parts of the piano key weir and labyrinth.

Upstream head H	Ratio H/P	Model A piano key weir discharge coefficient related to the developed length			Rectangular-shaped labyrinth discharge coefficient related to the developed length		
		Inlet (C_{di})	Outlet (C_{do})	Sidewall (C_{dh})	Inlet (C_{di})	Outlet (C_{do})	Sidewall (C_{dh})
0.5 m	0.125	0.432	0.440	0.375	0.432	0.466	0.303
1.0 m	0.250	0.405	0.451	0.263	0.383	0.445	0.192
2.0 m	0.500	0.356	0.425	0.160	0.326	0.467	0.123
4.0 m	1.000	0.260	0.409	0.094	0.249	0.416	0.072

5 SENSITIVITY

5.1 Approach conditions

Piano key weir spillways are generally built at the top of dam crest whereas labyrinth spillways are usually built directly near the ground. The depth of approach is different considering those two cases and may change the discharge results. That is why tests have been done with a depth of approach that equals the spillway height to determine the total specific discharge loss.

Contrary to the labyrinth spillway, the depth of approach's influence on the piano key weir spillway is significant. Nevertheless, the piano key weir is still more efficient.

5.2 Sidewall angle

In most cases, the labyrinth spillways' geometry is not rectangular-shaped but trapezoidal with a side wall angle ranging from 6° to 18°. This angle aims to preserve the flow from downstream

Table 7. Piano key weir and labyrinth total specific discharge with and without depth of approach.

	Model A piano key weir specific discharge referred to the total width (q_{sW})			Rectangular-shaped labyrinth specific discharge referred to the total width (q_{sW})		
Upstream head	Depth of approach	No depth of approach	Deviation	Depth of approach	No depth of approach	Deviation
m	$m^3/s/m$	$m^3/s/m$	%	$m^3/s/m$	$m^3/s/m$	%
1.0	7.7	7.2	−6.5	6.1	6.1	<1
2.0	14.9	13.6	−8.4	12.6	12.5	<1
4.0	28.3	25.8	−8.8	24.4	24.3	<1

Table 8. Considered rectangular and trapezoidal-shaped labyrinth spillways configurations.

	Rectangular-shaped labyrinth	Trapezoidal-shaped labyrinth
Upstream-downstream length (B)	12 m	12 m
Upstream overhang crest length (B_o)	0 m	0 m
Downstream overhang crest length (B_i)	0 m	0 m
Base length (B_b)	12 m	12 m
Sidewall overflowing crest length (B_h)	11.6 m	12.14 m
Wall height (P)	4 m	4 m
Width of a unit (W_u)	4.8 m	11.76 m
Inlet key width (W_i)	2 m	2 m (apex width)
Outlet key width (W_o)	2 m	2 m (apex width)
Crests & wall thickness ($T_s = T_i = T_o$)	0.4 m	0.4 m
Total developed length of a unit (L_u)	28 m	28.88 m
Developed length ratio of a unit (n_u)	5.83	2.46
Aspect ratio (W_u/P)	1.20	2.94
Sidewall angle (α)	0°	17.17°

submergence. The interactions between the outlet flow and the upstream part of the sidewall flow are reduced with this angle.

5.2.1 *Upstream-downstream length and apex width constant values*

First tests have been done with a P = 4 m high labyrinth spillway with a B = 12 m length of weir and a $W_i/2 = W_o/2 = 1$ m half apex width. The angle value is supposed to be about 17.2°, which means that sidewall length is about 12 m/cos(17.2°) = 12.55 m. This value has been studied for an EDF project.

Contrary to the first tests, wall thickness are equal to 0.4 m in order to avoid meshing resolution problems with the trapezoidal-shape. Tests for the rectangular-shaped labyrinth spillway have been done a second time with this thickness in order to compare relevant results. The specific discharges of these numerical models have been calculated (Table 9).

The main conclusions are:

- A constructible wall thickness (0.4 m) gives a specific discharge decrease from 6% to 12% compared to a thin and difficult to be built wall thickness (2 cm). These values are lower than those obtained in the same case for piano key weir (decrease from 10% to 20% with a 0.4 m wall thickness).
- The rectangular-shaped labyrinth is more efficient (in terms of specific discharge) for lower upstream heads than the trapezoidal-shaped one, thanks to its longest total crest length.
- The trapezoidal-shaped labyrinth is more efficient (in terms of specific discharge) for higher upstream heads than the rectangular-shaped one, thanks to its reduced sensitivity to downstream submergence.

147

Table 9. Rectangular and trapezoidal-shaped labyrinth spillways specific discharge.

Upstream head H	Ratio H/P	Rectangular-shaped labyrinth spillway specific discharge referred to the total width (q_{sW})			Trapezoidal-shaped labyrinth spillway specific discharge referred to the total width (q_{sW})	
		Wall thickness e = 2 cm	Wall thickness e = 0.4 m	Deviation	Wall thickness e = 0.4 m	Deviation
m		$m^3/s/m$	$m^3/s/m$	%	$m^3/s/m$	%
0.5	0.125	3.1	2.7	−12.7	1.6	−41
1.0	0.250	6.1	5.7	−5.8	4.7	−18
2.0	0.500	12.6	11.8	−6.8	10.9	−8
3.0	0.750	*	17.2	*	17.5	+2
4.0	1.000	24.4	22.7	−6.9	24.2	+7

Table 10. Considered rectangular and trapezoidal-shaped labyrinth spillways configurations.

	Trapezoidal-shaped labyrinth (n°1)	Rectangular-shaped labyrinth (n°2)	Trapezoidal-shaped labyrinth (n°3)
Upstream-downstream length (B)	12 m	12 m	12 m
Upstream overhang crest length (B_o)	0 m	0 m	0 m
Downstream overhang crest length (B_i)	0 m	0 m	0 m
Base length (B_b)	12 m	12 m	12 m
Sidewall overflowing crest length (B_h)	12.14 m	11.6 m	12.14 m
Wall height (P)	4 m	4 m	4 m
Width of a unit (W_u)	11.76 m	11.76 m	11.76 m
Inlet key width (W_i)	2 m (apex width)	5.48 m	8.92 m (apex width)
Outlet key width (W_o)	2 m (apex width)	5.48 m	8.92 m (apex width)
Crests & wall thickness ($T_s = T_i = T_o$)	0.4 m	0.4 m	0.4 m
Total developed length of a unit (L_u)	28.88 m	34.96 m	43.2 m
Developed length ratio of a unit (n_u)	2.46	2.97	3.67
Aspect ratio (W_u/P)	2.94	2.94	2.94
Sidewall angle (α)	17.17°	0 °	−17.17°

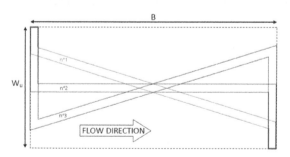

Figure 3. Labyrinth configurations.

Tests should determine the critical head which the rectangular-shaped and the trapezoidal-shaped labyrinths have the same specific discharge for, whereas their total crest length is different (lower for the trapezoidal one). This head is between 2 m and 3 m in our case.

5.2.2 Labyrinth length and width constant values
Second tests have been done with a 4 m high labyrinth spillway with a B = 12 m length of weir and a W = 5.881 m width. Wall thickness equals 0.4 m in order to avoid meshing resolution problems with the trapezoidal-shape.

Table 11. Rectangular and trapezoidal-shaped labyrinth spillways specific discharge.

| Upstream head (H) | Ratio (H/P) | Specific discharge referred to total width (q_{sW}) | | |
		Trapezoïdal-shaped labyrinth n°1	Rectangular-shaped labyrinth n°2	Trapezoïdal-shaped labyrinth n°3
m		$m^3/s/m$	$m^3/s/m$	$m^3/s/m$
0.5	0.125	1.6	1.9	1.6
1.0	0.250	4.7	4.6	3.6
2.0	0.500	10.9	10.7	7.2
3.0	0.750	17.5	16.1	12.1
4.0	1.000	24.2	22.0	16.9

The specific discharges of these numerical models have been calculated. The rectangular-shaped labyrinth (n°2) is more efficient than the other ones for lower upstream heads whereas the classical trapezoidal-shaped labyrinth (n°1) is more efficient for higher upstream heads. The last trapezoidal-shaped labyrinth (n°3) is less efficient than the first two ones instead of his total length along the crest is bigger. The inlet entry section and the outlet final section are not large enough to obtain correspondingly good approach conditions and good evacuating conditions.

6 CONCLUSIONS

In the case of the numerical models considered configurations, the piano key weir is close to 20% more efficient in terms of specific discharge than the labyrinth spillway. The specific discharge gap between the piano key weir and the labyrinth is mainly due to the sidewall flow efficiency. Contrary to the piano key weir, the depth of approach seems to have less influence on the labyrinth specific discharge. The specific discharge of the piano key weir decreases from about 6% to 12% when the walls thickness comes from 0.02 m to 0.4 m. In the case of labyrinths, the trapezoidal-shaped spillway is efficient for higher upstream heads when the rectangular-shaped one is for lower heads. A negative sidewall angle does not seem to improve specific discharge even for low heads.

Other piano key weir and labyrinth configurations will have to be tested to optimize the hydraulic capacity of these structures. Future works will be focused on a piano key weir sidewall angle or a combination of a piano key weir inlet part and a traditional labyrinth outlet part.

REFERENCES

Darvas, L.A. 1971. Discussion of Performance and Design of Labyrinth Weirs. *ASCE Journal of the Hydraulics Division* (HY8): 1246–1251.
Falvey, H.T. 2003. *Hydraulic Design of Labyrinth Weirs*. Reston, Virginia: ASCE Press.
Ghare, A.D., Mhaisalkar, V.A. & Porey, P.D. 2008. An Approach to Optimal Design of Trapezoidal Labyrinth Weirs. *World Applied Sciences Journal* 3(6): 934–938.
Hay, N. & Taylor, G. 1970. Performance and Design of Labyrinth Weirs. *ASCE Journal of the Hydraulics Division* 96(HY11): 2337–2357.
Hinchcliff, D. & Houston, K.L. 1984. Hydraulic Design and Application of Labyrinth Spillways. *USBR Concrete Dams and Hydraulics Branch Design Memorandum*.
Laugier, F., Blancher, B., Guyot, G. & Valette, E. 2009. Assessment of numerical flow model for standard and complex cases of free flow spillway discharge capacity. *Proc. of HYDRO 2009*. Lyon, France.
Le Doucen, O., Leite Ribeiro, M., Boillat, J.L., Schleiss, A.J. & Laugier, F. 2009. PK Weir capacity parametric study. *Proc. of the SHF Congress on hydraulic physical models*. Lyon, France.
Leite Ribeiro, M. Bieri, M. Boillat, J.L. Schleiss, A.J., Delorme, F. & Laugier, F. 2009. Hydraulic capacity improvement of existing spillways, piano key weir design. 23rd ICOLD. Brasilia.
Lempérière, F. & Ouamane, A. 2003. The piano keys weir: a new cost-effective solutions for spillways. *Hydropower & Dams* (5): 144–149.

Lux, F. 1984. Discharge Characteristics of Labyrinth Weirs. *Proc. of the Conference Water for Resource Development*: 385–389.

Lux III, F.L. & Hinchcliff, D. 1985. Design and Construction of Labyrinth Spillways. *Proc. of the 15th International Congress On Large Dams* 4 (Q.59 R.15): 249–274.

Machiels, O. 2010, Erpicum, S., Archambeau, P., Dewals, B. & Pirotton, M. 2010. Experimental study of the alveoli widths influence on the release capacity of Piano Key Weirs. *La Houille Blanche*: 22–28.

Paxson, G. & Savage, B. 2006. Labyrinth spillways: Comparison of two popular U.S.A. design methods and consideration of non-standard approach conditions and geometries. In J. Matos & H. Chanson, *International Junior Researcher and Engineer Workshop on Hydraulic Structures* CH61/06. Brisbane, Australia: Division of Civil Engineering, The University of Queensland.

Pralong, J., Vermeulen, J., Laugier, F., Erpicum, S. & Boillat, J.L. 2011. A naming convention for the Piano Key Weirs geometrical parameters. *Proc. of the International Workshop on Labyrinth and Piano Key Weirs*. Liège, Belgium, February 9–11, 2011.

Savage, B., Frizell, K. & Crowder, J. 2004. Brains versus Brawn: The Changing World of Hydraulic Model Studies. *ASDSO 2004 Annual Conference Proceedings* PAP-933. Hydraulic Investigations and Laboratory Services Group Reports.

Tullis, J. P., Armanian, N. & Waldron, D. 1995. Design of labyrinth spillways. *Journal of Hydraulic Engineering* 121(3): 247–255. ASCE Press.

Truong Chi Hien & Ho Ta Khanh, M. 2009. *Hydraulic model studies of the PK Weir (Type B) with stepped spillway.*

US Army Corps of Engineers. 1990. *Hydraulic design of spillways*, EM 1110-2-1603: 5-5.

Labyrinth and Piano Key Weirs – PKW 2011 – Erpicum et al. (eds)
© 2011 Taylor & Francis Group, London, ISBN 978-0-415-68282-4

1D numerical modeling of the flow over a Piano Key Weir

S. Erpicum, O. Machiels*, P. Archambeau, B. Dewals & M. Pirotton
Research unit of Hydrology, Applied Hydrodynamics and Hydraulic Constructions (HACH), ArGEnCo Department – MS²F – University of Liège (ULg), Liège, Belgium
* Belgian Fund for education to Industrial and Agricultural Research, FRIA

ABSTRACT: Because of the PK-Weir geometric specificities, its hydraulic capacity remains difficult to predict without using experimental techniques. A 1D numerical model of the flow over a PK-Weir has been developed at the Research unit HACH at the University of Liège. It is based on a 1D modeling of the inlet and the outlet separately, with a single upstream reservoir and interactions between both flows by exchange of mass and momentum along the lateral crest. The comparison of the numerical results with various experimental data showed the ability of the numerical model to predict with reasonable accuracy the release capacity of a PK-Weir, whatever its geometry. The tests of the solver enable to highlight the significant influence of the inlet and outlet width on the weir release capacity. They also suggest that improvements of the numerical model may lie in the evaluation of the discharge coefficient of the lateral crest.

1 INTRODUCTION

The Piano Key Weir (PK-Weir) is an original type of free weir developed by Lempérière and Ouamane (Blanc & Lempérière, 2001; Ouamane & Lempérière, 2003) to reduce the base length of labyrinth weirs by using overhangs and facilitate thus their location on dams crest. The first scale model studies showed that this new type of weir can be four times more efficient in terms of discharge release than a standard Creager at constant head and crest length on a dam (Ouamane & Lempérière, 2006). Several PK-Weir projects have already been studied and constructed in the world, especially by Electricité de France (EDF) (Laugier, 2007; Laugier et al., 2009), showing appealing performances. PK-Weir is today a true technical/economical solution to safely increase the discharge capacity of existing dams.

The PK-Weir shows geometric specificities such as up- and/or downstream overhangs with variable width, inlet and outlet bottom slopes, . . . involving a large set of parameters. The flow over a PK-Weir is similar to the one over a conventional labyrinth weir, with in addition a significant effect of the inlet slope on the lateral weirs supply and a possible strong interaction between the outlet and inlet flows. Experimental studies have been or are currently carried out in different laboratories to characterize the influence of a number of the PK-Weir geometric parameters on its discharge capacity [Le Doucen et al., 2009; Machiels et al., 2009; Truong Chi et al., 2006]. Regarding numerical modeling, a 3D approach is currently undertaken by EDF, showing promising results [Luck et al., 2010].

This paper shows that a simplified numerical model, based on a 1D hydraulic approach, can succeed in predicting the discharge capacity of a PK-Weir in free flow conditions. This model has been developed in the framework of a coupled numerical – experimental research currently undertaken at the Laboratory of Engineering Hydraulics of the University of Liège.

The numerical model is presented in section 2. It is based on a 1D modeling of the inlet and the outlet, taking into account the possible interaction between both flows by exchange of discharge along the lateral crest. The performance of the model, simple to apply and little time consuming, is discussed in section 4 on the basis of a comparison of the numerical results with varied experimental data detailed in section 3.

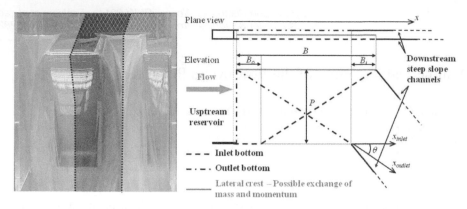

Figure 1.   Basic element of a PK-Weir (left) and numerical model layout with main geometric parameters (right).

2   FLOW MODEL

2.1   *Principles and objectives*

The main goal of the numerical model is to help in identifying the most relevant geometric parameters of the PK-Weir governing its release capacities and to assess their pertinent range of variation prior to undertaking experimental studies.

The numerical model considers the smallest hydraulic element of a PK-Weir, made of a lateral wall, half an inlet and half an outlet. The inlet and the outlet are modeled as parallel 1D channels, possibly interacting by exchange of mass and momentum along the lateral crest, and linked by an upstream reservoir (Fig. 1). To avoid the need for a downstream boundary condition and considering that a PK-Weir usually works as a free weir, i.e. without influence of the tailwater level, the inlet and the outlet are both extended by arbitrary steep slope channels, ensuring supercritical flow downstream of the simulation (Fig. 1).

According to experimental observations (Machiels et al., 2009), the main flow direction in the outlet follows the bottom slope. Consequently, the x-axis has been locally inclined in the numerical model. It is not the case in the inlet where the main flow direction is mostly horizontal, with hydrostatic pressure distribution along the vertical. The x-axis has thus been directed horizontally along the inlet (Fig. 1).

The geometric parameters needed to set up the numerical model are the weir height P, the lateral crest length B, the up- and downstream overhangs lengths B_o and B_i and the inlet and outlet widths W_i and W_o. In the model, the inlet and outlet channel width is respectively $W_i/2$ and $W_o/2$.

2.2   *Mathematical model*

The flow model, depicted in details by Erpicum (Erpicum et al., 2010c), is based on the one-dimensional cross-section-averaged equations of mass and momentum conservation, assuming a rectangular cross-section. In this standard 1D approach, it is basically assumed that velocities normal to the main flow direction are significantly smaller than those in this main flow direction. Consequently, the pressure field is almost hydrostatic everywhere and the free surface is horizontal along the transverse direction. Considering the x-axis inclination defined in subsection 2.1, this assumption is consistent with the problem considered in this paper.

Source terms are the topography, the friction terms and the lateral exchange terms. For the later, a α coefficient [0,1] enables to quantify the effect on momentum of the lateral discharge.

2.3   *Grid and numerical scheme*

The mesh size in both channels is constant to take advantage of the lower computation time and the gain in accuracy provided by regular grids.

To enable a direct computation of the lateral exchanges mesh by mesh without interpolation of the flow variables, i.e. to compute free surface level and velocity at the same absolute abscissa in both channels, the mesh size in the outlet is $\sin\theta\,\Delta x$ when it is Δx in the inlet.

The flow equations are discretized in space with a finite volume scheme. This ensures a correct mass and momentum conservation, which is needed for handling properly discontinuous solutions such as moving hydraulic jumps. As a consequence, no assumption is required regarding the smoothness of the solution. Reconstruction at cells interfaces is performed with a first order constant approach. The fluxes are computed by a Flux Vector Splitting (FVS) method, where the upwinding direction of each term of the fluxes is simply dictated by the sign of the flow velocity reconstructed at the cells interfaces (Erpicum et al., 2010a).

2.4 Source terms

The discretization of the topography gradients is always a challenging task when setting up a numerical flow solver based on depth- or cross-section-averaged equations. The bed slope appears as a source term in the momentum equations. As a driving force of the flow, it has however to be discretized consistently with the pressure and advective terms.

The first step to assess topography gradients discretization is to analyze the situation of still water on an irregular bottom. In this case, momentum equations simplify and there are only two remaining terms: the hydrostatic pressure variation and the topography gradient. A suitable treatment of the topography gradient source term is thus easy to set up (Erpicum et al., 2010b).

This approach fulfils the numerical compatibility conditions defined by Nujic (Nujic, 1995) regarding the stability of water at rest. Nevertheless, it constitutes only a first step towards an adequate form of the topography gradient discretization as it is not entirely suited regarding water in movement over an irregular bed. The effect of kinetic terms is not taken into account and, consequently, poor evaluation of the flow energy evolution can occur when modeling flow, even stationary, over an irregular topography (Erpicum, 2006). To overcome this problem on the upstream side of the outlet channel, where the topography gradient is locally the most important, the momentum equation has been locally replaced by the energy equation, using an approach depicted by Erpicum (Erpicum, 2006). This technique has not been applied on the whole inlet and outlet length as it is not suited to compute correctly shocks such as hydraulic jumps, which may occur in the outlet.

The friction terms are conventionally modeled with the Manning formula, where the Manning coefficient characterizes the surface roughness. To consider the reduced width of the inlet and the outlet, the hydraulic radius R is defined by equation 1, where z_s is the lateral weir elevation, z_b the bottom elevation, h the water depth, L the key width and Ω the cross section

$$R = \frac{\Omega}{L + \min(h, z_s - z_b)} \tag{1}$$

Finally, the lateral unit discharge in the momentum exchange terms is computed on each point of the lateral weir depending on the head difference ΔH between the inlet and the outlet, without considering the kinetic terms along the inlet and outlet axis:

$$\Delta H = \max\left(0, Z_{inlet} - z_s\right) - \max\left(0, Z_{outlet} - z_s\right) \tag{2}$$

$$q_l = \mu\sqrt{2g|\Delta H|^3}\,\mathrm{sgn}\left(\Delta H\right) \tag{3}$$

where Z is the free surface elevation and μ is the lateral weir discharge coefficient.

2.5 Upstream reservoir

The upstream reservoir is an important part of the numerical model as it distributes the discharge between the inlet and the outlet channels. It is also in the upstream reservoir that the value of the head on the PK-Weir is measured to characterize the release efficiency of the structure.

The upstream reservoir is modeled as two special twin 1D finite volumes, with distinct discharges $Q_{R,out}$ and $Q_{R,in}$ but a single cross-section value Ω_R, as depicted in figure 2. The reservoir width is

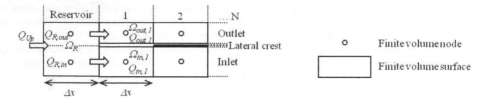

Figure 2. Modeling of the upstream reservoir and links with the inlet and the outlet.

$(W_i/2 + W_o/2)$ and it is only one space step in length. All the source terms are neglected to compute the time evolution of the three reservoir variables on the basis of mass and energy conservation. The mathematical model in the reservoir results merely in one mass balance equation and two energy equations, one for the velocity term related to each discharge (Erpicum et al., 2010c).

2.6 *Other features*

The value of the upstream discharge is the only value to be prescribed as a boundary condition. The steep slope of both channels in the downstream part of the model leads to supercritical flow and therefore no outflow boundary condition is needed.

Since the model is applied to compute steady-state solutions, the time integration is performed by means of a Runge-Kutta algorithm, providing adequate dissipation in time. For stability reasons, the time step is constrained by the Courant-Friedrichs-Levy condition.

A convergence criteria has been defined on the basis of the discharge evolution in the reservoir compared to the upstream discharge boundary condition. When the relative difference between both values is lower than a given tolerance during a fixed time period, the computation is assumed to be converged.

3 APPLICATION AND COMPARISON WITH EXPERIMENTAL DATA

To assess the model performance and accuracy, the numerical results have been compared with experimental data measured on a number of various scale models of PK-Weir tested in different hydraulic laboratories (Laboratoire National d'Hydraulique et Environnement – EDF, Chatou, France; Hydraulic Department of Biskra University, Biskra, Algeria and Laboratoire d'Hydraulique des Constructions – ULg, Liège, Belgium). The main geometric parameters of the scale models are summarized in table 1. Some experimental tests have been realized using a support under the PK-Weir to simulate the weir position on a dam crest. This support affects the water height in the inlet along the upstream overhang and thus the head/discharge curve. The support or dam height P_d has thus been considered for the numerical modeling. It has directly been accounted for in the inlet bathymetry.

The numerical model has been built with a space step in the inlet of 5 mm. The outlet axis inclination is given in table 1. The constant lateral discharge coefficient values (Eq. 3) used for the simulations are the ones of thin (Chatou, Biskra) or thick (Liege) crested weir. The Manning roughness coefficient value is chosen equal to 0.011 s/m$^{1/3}$ for both PVC and steel. α coefficient in momentum exchange terms has been assumed to be equal to 1 in the inlet and in the outlet (full exchange of momentum from the discharged water).

The numerical model has been used to compute the flow over the PK-Weir in the range of specific discharge, calculated with the width of the PK-Weir in the upstream reservoir, detailed in table 1. From these computations, the "numerical" head/discharge relations or the PK-Weir equivalent discharge coefficient (Ouamane & Lempérière, 2006) can be compared with the corresponding experimental data (Figs. 3 and 4).

4 DISCUSSION

The numerical results are generally in good agreement with the experimental ones, especially for moderate heads and head ratios. The shape of the numerical curves is close to the corresponding

Table 1. Geometric and numerical parameters of the scale models considered for comparison.

Laboratory	Model	W_i [m]	W_o [m]	B_o [m]	B_i [m]	B [m]	P [m]	T_s [mm]	P_d [m]	μ [−]	θ_{outlet} [°]	q [m³/s/m]
Chatou (2003)	Model 1	.12	.08	.1	.1	.4	.2	2 (steel)	.3 (21.8° inclined upstream face)	.667	−33.7	[.01;.25]
	Model 2	.08	.12									
	Model 3	.12	.08				.15				−26.6	
	Model 4	.10	.065									
Biskra (2006)		.09	.075	.103	.103	.412	.155	2 (steel)	0	.667	−26.6	[.01;.18]
Liege (2008)	Model 1	.08	.10	.25	0	.5	.2	10 (PVC)	.2	.429	−21.8	[.01;.4]
	Model 2	.11	.07									
	Model 3	.06	.12									
	Model 4	.13	.05									
Liege (2009)		.18	.18	.185	.185	.63	.525	10 (PVC)	.2	.429	−49.7	[.01;.5]

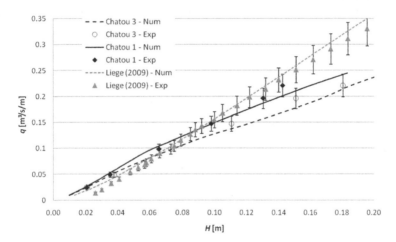

Figure 3. Comparison between experimental (points) and numerical (lines) head/discharge curves for the physical models Chatou 1, Chatou 3 and Liege (2009). +/−10% variation bars on the experimental discharges.

experimental ones. The numerical results are often around 10 percents of the experimental observations, and very similar values are obtained with some geometries, such as Chatou 4, Biskra and Liege (2009) (Fig. 4).

For low head ratios and thick crest (Liege (2009), Fig. 4), the non monotonous variation of the release capacity because of the weir thickness is not accounted for in the numerical model, and the numerical curve doesn't show the inflexion of the experimental data curve. For thin crest (Chatou 3 and Biskra for instance, Fig. 4), the numerical model assumptions seem in agreement with the real weir behavior for decreasing head ratios (constant lateral discharge coefficient).

The numerical model convergence is limited for high head ratios, depending on the PK-Weir geometry. When the water depth over the weir is high, the assumption of distinct flows in the inlet and the outlet is no more valid and no stable solution is reached by the solver.

Tests have been conducted with the Biskra model geometry to assess the influence of the discretization step and the α coefficient value in the momentum exchange terms (Fig. 5). A simulation has been performed with α equal to 1 in the inlet (full loss of momentum) and 0 in the outlet (no gain in momentum from the discharged water). When the free surface in the outlet is under the lateral crest level, the numerical results are not affected by the α value in the outlet. When the free surface level becomes higher, the weir efficiency decreases more when α is equal to 0 in the outlet than when it is equal to 1. The comparison with the experimental results is better when α is equal to 1

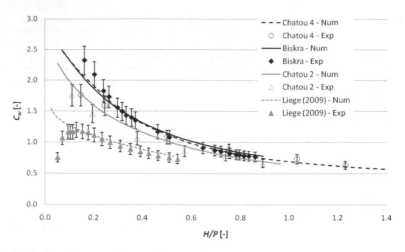

Figure 4. Comparison between experimental and numerical non-dimensional head/discharge curves for the physical models Chatou 2, Chatou 4, Biskra and Liege (2009). +/−10% variation bars on the experimental discharge coefficients.

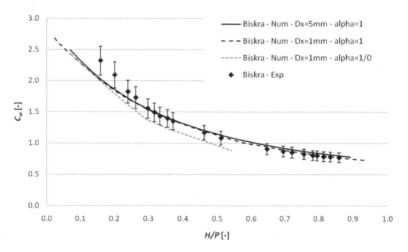

Figure 5. Biskra model geometry – Effect of the space discretization step and α coefficient value on the numerical results. +/−10% variation bars on the experimental discharge coefficients.

in both the inlet and the outlet. This suggests an exchange of momentum between the two alveoli, at least for moderate and high heads, i.e. when the water level in the outlet is higher than the lateral crest height on a significant length.

The results of the simulations with the Liege (2008) model geometry enable interesting observations. With the numerical model configuration depicted in chapter 3, the curves of figure 6 – config A show the poor comparison with experimental data for models 2 and 4 with very small outlets width compared to the width of the inlet. Tests have been performed considering an increased inlet and outlet width, taking into account the thickness of the lateral crest (figure 6 – config. B). The width of half the inlet is $(W_i + T_s)/2$ and the width of the half outlet $(W_o + T_s)/2$. Calculations have also been done with a lateral discharge coefficient equal to 0.5 instead of 0.429, in order to increase the efficiency of the 1 cm thick lateral crest (figure 6 – config. C).

With increased widths, the numerical results are closer to the experimental ones for all heads. With an increased lateral discharge coefficient, the numerical results are more satisfactory for lower heads, when only the inlet flow governs the weir efficiency, and are unchanged for higher heads, when the weir is flooded (few exchange of discharge between the inlet and the outlet. The

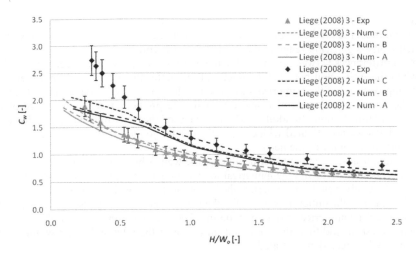

Figure 6. ULg (2008) model geometry – Config. A: inlet width is $W_i/2$, outlet width is $W_o/2$, $\mu = .429$, config. B: inlet width is $(W_i + T_s)/2$, outlet width is $(W_o + T_s)/2$, $\mu = .429$, config. C: inlet width is $W_i/2$, outlet width is $W_o/2$, $\mu = .5$. $+/-10\%$ variation bars on the experimental discharge coefficients.

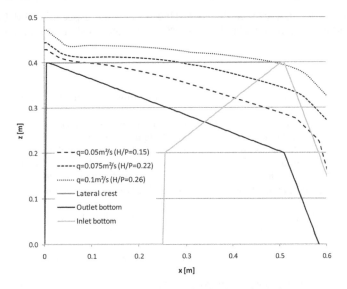

Figure 7. ULg (2008) model 2 – Computed free surface levels along the outlet.

inflexion point in the non dimensional head/discharge curve of figure 6 for the ULg (2008) model 2 geometry ($H/W_0 = 0.7$) can be explained by an analysis of the free surface elevation curves along the outlet (Fig. 7). When the discharge becomes higher than 0.05 m²/s, the water level in the outlet gets quickly over the lateral crest level along a significant length, and thus strongly decreases the water exchange between the inlet and the outlet.

These tests pave the way for further improvements of the numerical model.

5 CONCLUSIONS

A simplified numerical model has been set up to compute the flow over a PKW. It is based on a 1D modeling of the inlet and the outlet separately, with a single upstream reservoir and interactions between both flows by exchange of mass and momentum along the lateral crest.

The comparison of the numerical results with several experimental data has shown the ability of the numerical model to predict with reasonable accuracy the release capacity of a PKW, whatever its geometry, on a significant range of the head in the reservoir.

The main differences between numerical and experimental results have been observed for low heads, when the interaction between the inlet and the outlet flows takes only place at the outlet entrance. Improvement of the numerical model would thus need a better modeling of the transition between the reservoir and the outlet, and especially the prediction of the curvature of the free surface at this critical point. In addition, the accuracy of the experimental results for these low heads should also be investigated with care, as the discharge coefficient has a high sensitivity to the measurement accuracy of the head.

For very high heads, when the water depth over the weir is high, the assumption of two distinct water lines is no more valid and the numerical model is unable to reach a stable solution.

The tests of the solver enable to underline the importance to consider the exchange of momentum between the inlet and the outlet flow to predict the weir efficiency. They also show that significant improvements of the numerical model may lie in a better evaluation of the discharge coefficient of the lateral crest. In addition, the important influence of the inlet and outlet width on the weir release capacity has been demonstrated.

ACKNOWLEDGMENTS

The authors gratefully acknowledge the NGO Hydrocoop, and especially MM Lempérière and Vigny, for providing the experimental data from "Chatou" and "Biskra" scale models.

REFERENCES

Blanc, P. & Lempérière, F. 2001. Labyrinth spillways have a promising future. *Hydropower & Dams* 8(4): 129–131.

Erpicum, S. 2006. *Optimisation objective de paramètres en écoulements turbulents à surface libre sur maillage multibloc*. Ph.D. Thesis, HACH, University of Liège, Belgium. In French.

Erpicum, S., Dewals, B.J., Archambeau, P., & Pirotton, M. 2010a. Dam-break flow computation based on an efficient flux-vector splitting. *Journal of Computational and Applied Mathematics* 234: 2143–2151.

Erpicum, S., Dewals, B.J., Archambeau, P., Detrembleur, S., & Pirotton, M. 2010b. Detailed inundation modeling using high resolution DEMs. *Engineering Applications of Computational Fluid Mechanics* 4(1): 196–208.

Erpicum, S., Dewals, B.J., Archambeau, P. & Pirotton M. 2010c. Reliable hydraulic numerical modeling with multiblock grids and linked models. *Proc. of SimHydro: Hydraulic modeling and uncertainty*, Sophia Antipolis, France.

Laugier, F. 2007. Design and construction of the first Piano Key Weir spillway at Goulours dam. *Hydropower & Dams* 14(5): 94–100.

Laugier, F., Lochu, A., Gille, C., Leite Ribeiro, M. & Boillat, J-L. 2009. Design and construction of a labyrinth PK-WEIR spillway at Saint-Marc dam, France. *Hydropower & Dams* 16(5): 100–107.

Le Doucen, O., Leite Ribeiro, M., Boillat, J-L. & Schleiss, A. 2009. Etude paramétrique de la capacité des PK-Weirs. *Colloque SHF "Modèles physiques hydrauliques"*, Lyon, France.

Luck, M., Lee, E-S., Mechitoua, N., Violeau, D., Laugier, F., Blancher, B. & Guyot, G. 2009. Modélisations physique et numérique 3D pour l'évaluation de la débitance et le design des évacuateurs de crue. *Colloque SHF "Modèles physiques hydrauliques"*, Lyon, France.

Machiels, O., Erpicum, S., Archambeau, P., Dewals, B.J. & Pirotton, M. 2009. Large scale experimental study of piano key weirs. *Proc. of 33rd IAHR Congress*, Vancouver, Canada.

Nujic, M. 1995. Efficient implementation of non-oscillatory schemes for the computation of free-surface flows. *Journal of Hydraulic Research* 33(1):101–111.

Ouamane, A. & Lempérière, F. 2003. The piano keys weir: a new cost-effective solution for spillways. *Hydropower & Dams* 10(5): 144–149.

Ouamane, A. & Lempérière, F. 2006. Design of a new economic shape of weir. Berga et al. (eds), *Dams and Reservoirs, Society and Environment in the 21st Century*: 463–470. Taylor & Francis: London.

Truong Chi, H., Huynh Thanh, S., Ho Ta Khanh, M. 2006. Results of some piano keys weir hydraulic model tests in Vietnam. *22ème congrès des grands barrages – Barcelona*: 581–596. ICOLD: Paris.

Labyrinth and Piano Key Weirs – PKW 2011 – Erpicum et al. (eds)
© 2011 Taylor & Francis Group, London, ISBN 978-0-415-68282-4

Influence of structural thickness of sidewalls on PKW spillway discharge capacity

F. Laugier, J. Pralong & B. Blancher
EDF – Hydro Engineering Center, France

ABSTRACT: Piano Key Weir (PKW) labyrinth spillways are a technico-economical optimization of traditional labyrinth weirs and an interesting option to increase the discharge capacity of many dams. The concept was first introduced by Lemperiere and Ouamane in 2003. Compared to traditional labyrinth spillways, their main advantage is that they can be installed at the top of main existing concrete dams (especially concrete gravity dams). PKW discharge capacity can be 2 to 5 times higher than classical Creager weir one. Sidewall thickness is a significant factor impacting PKW discharge capacity with potential loss larger than 15%. Based on validated 3D numerical models, parametric studies were carried out at EDF Hydro Engineering Center to try and quantify the impact of labyrinth wall thickness. Conclusions raise the issue of hydraulico-structural optimizations of labyrinth spillways.

1 INTRODUCTION

PKW labyrinth spillways are innovative solutions which increases discharge capacity from 2 to 5 times higher than usual free flow Creager spillway. Their main advantage when compared to traditional vertical wall labyrinths spillways, is that they can be installed on the crest of main concrete dams, especially concrete gravity ones.

PKW sidewall thickness T_s is a fundamental parameter. Indeed, the crest width available at the top of the dam usually limits the PKW dimensions. Thinner sidewalls would enable to increase the inlets and outlets width, and therefore enhance the PKW discharge capacity.

The improvements in structure engineering enable new alternatives to traditional concrete such as steel or composite materials that could reduce sidewall thickness.

As no such studies have been yet described in the literature, a sensitivity analysis has been carried out using validated CFD models (Pralong et al., 2011), to quantify the gains that could occur by optimizing (or decreasing) the sidewall thickness. A range of thickness from 10 to 50 cm has been tested on the PKW type A (Lemperiere & Ouamane, 2003), keeping the total width of the PKW constant (figure 4).

2 REFERENCE NUMERICAL MODEL

3D numerical modeling turned out to be an interesting opportunity as it could enable to quickly simulate various geomtries. Therefore, a *Flow-3D®* model has been calibrated on experimental data from Lemperiere & Ouamane (2003) before being validated on data collected from physical model studies. The model has demonstrated a good correlation with the experimental results. Comparisons with data available on EDF PKW physical models have also led to encouraging results, with a mean relative deviation in specific flowrates not exceeding 5%. The model has been validated, and is considered reliable since for further sensitivity analysis on PKW geometrical parameters.

A sensitivity analysis has been carried out based on this model to assess optima for several geometrical parameters of PKW. The initial configuration is the PKW type A (Lemperiere & Ouamane, 2003) defined by equal upstream and downstream overhangs and equal inlets and outlets width.

Table 1. Reference PKW type A characteristics.

B	W_i	W_o	P	T_s	W_u	L_u	$N = L_u/W_u$
m	m	m	m	m	m	m	
12	2,4	2,4	4	0,02~0	4,8	28,8	6

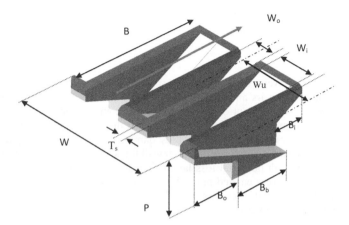

Figure 1. PKW main parameters (Pralong et al., 2011).

Figure 2. Steel model PKW used by Lemperiere.

Initial type A characteristics are given by table 1 below with figure 1 reminding PKW main parameters.

PKW models are made with steel sheets panels whose thickness is very small with $T_s << W_i$.

The numerical model considers 2 cm thick sidewall with regard to mesh issues and calculation time. Figure 2 provides a view of steel physical models used by Lemperiere & Ouamane (2003) corresponding to PKW type A.

3 SENSITIVITY STUDY PRINCIPLE

A range of thickness from 10 to 50 cm has been tested on the PKW type A reference configuration (Lemperiere & Ouamane, 2003) keeping the width of the PKW unit W_u constant (figure 3) and the ratio $W_i/W_o = 1$. Sidewall thickness is equally increased by 10 cm steps in both inlet and outlet keys.

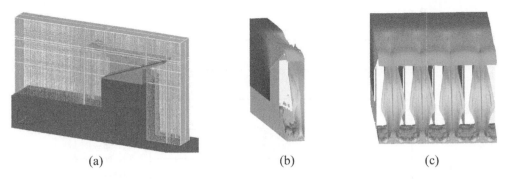

| (a) | (b) | (c) |

Figure 3. 3D view of the mesh (a), the modeled (b) and reconstituted (c) flows (PKW type A, 1.5 m head).

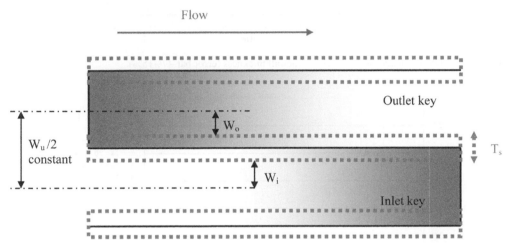

Figure 4. Parametrical sidewall thickness T_s with constant W_u.

Four water heads are tested for each sidewall thickness configuration: $0,5 - 1 - 1,5 - 4$ m. This range of water heads cover different types of flows patterns in the PKW: Configuration with limited downstream effect in the outlet key (0,5 m) and configurations which gradually increase the saturation in the outlet key.

4 NUMERICAL RESULTS

Numerical results of 3D simulations are given below in terms of PKW specific discharge capacity q_s and PKW discharge coefficient C_{dw} being:

$$q_s = Q/W \tag{1}$$

with Q: PKW discharge capacity W: PKW witdh

$$Q = Cd_w.(2.g)^{0,5}.H^{1,5} \tag{2}$$

PKW unit width Wu being constant, it is logical that the PKW unit discharge capacity decreases when the sidewall thickness increases. Figure 5 shows that the loss of discharge capacity is larger than 20% between 10 and 50 cm thick sidewalls configurations for lower heads.

Figure 7 below shows the same results in terms of specific discharge loss (dq_s/q_s) for each thickness configuration compared to reference configuration ($T_s = 0,02$ m)

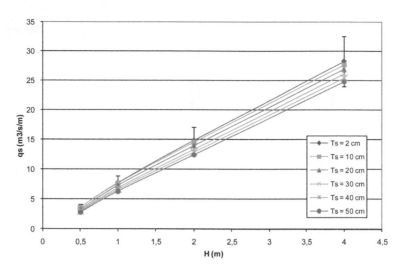

Figure 5. PKW discharge capacity versus sidewall thickness (error bar +/−15%).

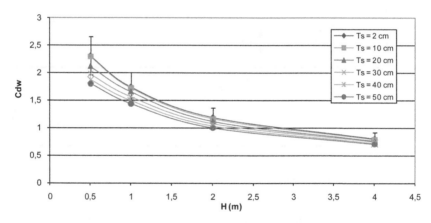

Figure 6. PKW discharge capacity coefficient versus sidewall thickness (error bar +/−15%).

Figure 7 also shows that:

- PKW unit discharge capacity decreases when the sidewall thickness increases. Loss of discharge capacity is larger than 20% between 2 cm and 50 cm thick sidewalls configurations.
- Apart from configuration $T_s = 10$ cm, the discharge capacity loss decreases for higher heads. This might be explained for different reasons:
  - The "effective" length of PKW developed crest is smaller for thicker sidewalls. This is more sensitive for low heads when the flow really "sees" the geometrical developed crest length.
  - For small heads and thick sidewalls, the sidewall nearly becomes a broad crested weir whose discharge capacity coefficient is significantly smaller than the one of sharp crested weir (typically 0,35 versus 0,43).
  - For small heads, it is likely that the flow gets detached on the downstream edge of the sidewall. On the opposite, for high heads, the detachment point is located on the sidewall upstream head as shown in figure 8. This upstream detachment flow virtually enlarges the outlet key width by T_s.

Figure 9 shows the loss of discharge capacity dq_s/q_s versus sidewall thickness for identical heads.

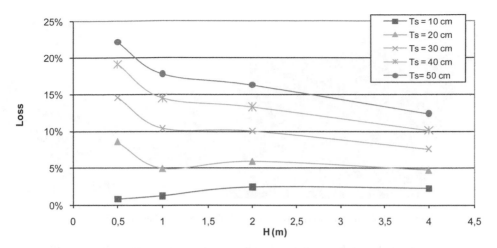

Figure 7. PKW discharge capacity loss versus sidewall thickness (error bar +/−15%).

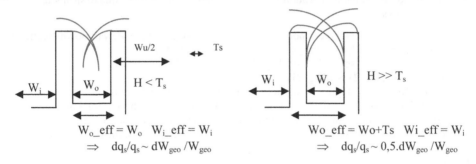

Figure 8. PKW Flow on sidewalls for small and high heads.

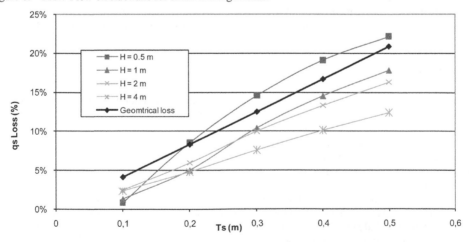

Figure 9. PKW discharge capacity loss versus sidewall thickness T_s compared to reference PKW ($T_s = 2$ cm).

It also shows the "geometrical loss" dW_{geo}/W_{geo} defined as the relative loss of geometrical crest width called "W_{geo}" seen by the flow. For a PKW unit, that means:

- $W_{geo}(T_s) = (W_i(T_s) + W_o(T_s))/2$ $W_{geo}(T_s = 0) = (W_i(T_s = 0) + W_o(T_s = 0))/2$
- $W_{geo}(T_s = 0) - W_{geo}(Ts) = ((W_i(T_s = 0) - W_i(Ts)) + (W_o(T_s = 0) - W_o(T_s))/2$
  $= (T_s/2 + T_s/2)/2 = T_s$
- $\Rightarrow dW_{geo} = T_s$ $dW_{geo}/W_{geo} = T_s/(W_u/2)$

163

Figure 10. Gloriette PKW construction (2010) – Sidewall thickness varying from 25 to 35 cm.

Figures 7 and 9 highlight the effect of sidewall thickness on the capacity loss with PKW when increasing the sidewall thickness in addition to simple geometrical effects due to shrinkage of PKW inlet and outlet keys:

– For small heads, the detachment point of the flow is located on the sidewall downstream edge.
  • The outcoming flow "sees" the outlet key efficient width W_o.
  • The incoming flow "sees" the inlet key efficient width W_i.
  ⇨ **dqs /qs ~ Ts/(Wu/2) ~ dW$_{geo}$/W$_{geo}$**
  This is confirmed by Figure 9 whereby diamond dot curve is close to square dot curve corresponding to smallest calculated heads (0,5 m), especially for higher thicknesses.
– For high heads, the detachment point of the flow is located on the sidewall upstream edge.
  • The outcoming flow "sees" the outlet key efficient width $W_o + Ts$, thus $W_o/2 + T_s/2$ for half an outlet key.
  • The incoming flow "sees" the inlet key efficient width W_i
  ⇨ **dqs /qs ~ 0,5.Ts/(Wu/2) ~ 0,5.dW$_{geo}$/W$_{geo}$**
  This analyses is confirmed Figure 9 whereby diamond dot curve "geometrical loss" values are close to half the values of ring dot curve corresponding to highest calculated heads (4 m).

5 DISCUSSION

From a structural point of view, actual thickness of sidewall will depend on construction material (concrete, steel, composite material ...) and on the applied loads (water head, ice, seismic load etc.). Other issues such as the chemical nature of water (more or less pure, containing acid or basic ions etc.) might also impact structural design.

Practically, most PKW built until now were fabricated in concrete. Goulours PKW walls (Laugier, 2007) are 20 cm thick while Saint-Marc PKW walls are 30 cm thick and Malarce PKW walls would be thicker than 40 cm.

The aforementioned results show that 20 cm thickening sidewall can lead to a discharge capacity loss of 10% for small heads (1 m) and 5% for higher heads (4 m). Structural design of sidewall should therefore include hydraulics issues in order to get a technico-economical optimization.

Three technical solutions might appear to be able to reduce PKW sidewall thickness:

– Construction of PKW with steel elements. A detailed cost analysis should be performed. It is likely that this option would be slightly more expensive as it was not proposed as an option by contractors which built the four first PKW in France. In addition, one can see at least two crucial issues:
  • Steel corrosion in water environment. This issue might be addressed either by inox steel (more expensive) or by painting coatings (maintenance issues on regular basis).
  • Treatment of flow vibration is not an easy issue and is difficult to address even with a physical model.

– Construction of PKW with composite concrete / steel sheets. It consists of using steels panels as concrete formwork and taking advantage of concrete mass for vibrations issues. This technical seems to be very promising. However, it has not been much developed yet and includes the same corrosion issue than pure steel solutions.
– A third solution was used for Gloriettes PKW (Vermeulen et al., 2011) consisting in having a variable thickness from 25 cm at the top of the PKW top and 35 cm at the bottom where more loads apply. Actual construction on site do not highlight particular difficulty or overcost because of a non vertical sidewall.

We recommend the Gloriettes configuration for the moment.

6 CONCLUSION

The model implemented with *Flow-3D®* has proven its efficiency and relevance for sensitivity analysis of PKW geometrical parameters. It has proven effective and relevant for performing sensitivity analysis to some parameters, like the sidewall thickness.

This study has shown that this parameter can have an impact larger than 20% of PKW discharge capacity between a PKW with very thin sidewalls (steel plates of a physical model) and 50 cm thick sidewall PKW. These results should significantly change according to the actual geometrical parameters of the PKW.

Therefore an optimization in sidewall structural design could significantly improve the PKW discharge capacity by reducing their thickness. This parameter should also be taken into account when operating physical models with fixed thickness. Comparing PKW discharge capacity with different sidewall thicknesses is a real issue and it should be envisaged to propose a correction factor to take into account the actual sidewall thickness.

REFERENCES

Lempérière, F. & Ouamane, A. 2003. The piano keys weir: a new cost-effective solutions for spillways. *Hydropower & Dams* (5): 144–149
Laugier. F. 2007. Design and construction of the first Piano Key Weir spillway at Goulours dam. *Hydropower and dams* (issue 5)
Pralong, J., Blancher, B., Laugier, F., Machiels, O., Erpicum, S., Pirotton, M., Leite Ribeiro, M., Boillat, J-L., & Schleiss, A.J. 2011. Proposal of a naming convention for the Piano Key Weir geometrical parameters. *International Workshop on Labyrinth and Piano Key weirs,* Liège, Belgium
Pralong J., Montarros F., Blancher B. & Laugier. F. 2011. A sensitivity analysis of Piano Key Weirs geometrical parameters based on 3D numerical modeling. *International Workshop on Labyrinth and Piano Key Weir,* Liège. Belgium
Vermeulen, Laugier, F., Faramond, L. & Gilles, C. 2011. Lessons learnt from design and construction of EDF first Piano Key Weirs. *International Workshop on Labyrinth and Piano Key Weir,* Liège. Belgium

Labyrinth and Piano Key Weirs – PKW 2011 – Erpicum et al. (eds)
© 2011 Taylor & Francis Group, London, ISBN 978-0-415-68282-4

Experimental study of side and scale effects on hydraulic performances of a Piano Key Weir

G.M. Cicero, J.M. Menon & M. Luck
National Hydraulic and Environmental Laboratory, Chatou, France

T. Pinchard
EDF CIH, Le Bourget, France

ABSTRACT: The Malarce dam was initially equipped with 3 gated spillways designed to discharge $4100\,m^3/s$ at the Maximum Water Level. To increase the spillway capacity to $4600\,m^3/s$, the addition of a Piano Key Weir on the right bank wing of the dam has been studied on physical models. To optimize the PKW and the upstream and downstream civil structures, four alternatives were tested on a 1/60 scaled model representing the whole geometry. To examine side effects and scale effects on hydraulic performances, the second alternative was tested in a straight channel. This model, at scale 1/30 represented the PKW with a part of the dam only. The PKW discharge characteristics were measured on both physical models, for various crest levels and various side geometries (size and shape of the river bank). This article is devoted to the analyses of the common tests performed on both physical models.

1 INTRODUCTION

Located on the Chassezac river (France), the Malarce dam is currently equipped with 3 gated spillways with a Creager type crests, designed to discharge $4100\,m^3/s$ at the Maximum Water Level (221 NGF). The update evaluation of the design flood (time return 1000 years) needs to increase the spillway capacity to $4600\,m^3/s$.

EDF CIH proposed to set up a Piano Key Weir between the existing spillways and the right bank of the dam. Many tests were carried out on a physical model at scale 1/60 (Cicero et al. 2010) to optimize the PKW design, the crest level and the civil engineering structures upstream and downstream of the dam. The scale of this model allowed to represent the dam body, the existing spillways and the PKW with an important part of the reservoir. The upstream limit was set at 900 m from the dam body to represent the flow behaviour after 2 tight curves of the river. Four alternatives were tested at various crest levels and with various topographies of the right bank.

The second alternative was tested on a complementary model to check possible scale effects on the discharge characteristics. This model at scale 1/30, represents only the PKW with a part of the dam in a straight channel. Tests were carried out with various shapes of the right bank to study the effects on the release capacity of the PKW.

This article presents the tests results of the 1/30 scale model and the comparisons with the tests performed on the 1/60 scale model, for rather similar geometries.

2 EXPERIMENTAL SETUP AND PROGRAM

The experimental setup at scale 1/30 (Fig. 1) is a part of the Malarce dam with the PKW and a pier of the closest existing spillway. The model was tested in a straight channel (70 m long, 2 m wide and 1 m deep), with measurements of the discharge (1% accuracy) and of the upstream level ($+/-0.2\,mm$ accuracy) at about 2.5 m from the PKW.

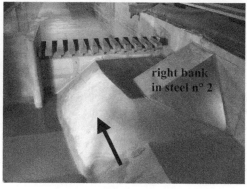

Figure 1. Upstream views of the 1/30 scale model – Without right bank (high left) – With right bank n°1 in concrete and deflector (high right) – With right bank n°1 in concrete and n°2 in steel (low left) – Saturated flow over the PKW (low right).

Table 1. Characteristics of the tests performed on the scale model 1/30.

Tests	Test 1	Test 2	Test 3	Test 4	Test 5	Test 6
Crest level	219	219	219	219.5	219.75	219.75
Right bank	no	n°1	n°1	n°1	n°1	n°1 & 2
Deflector	no	no	yes	yes	yes	no

The PKW geometry corresponds to the PKW type B1 tested on the 1/60 scale model. A complete description of the model is found in Cicero et al. (2010). The main parameters of the PKW are the spillway width ($W = 42.72$ m), the widths of inlet ($W_i = 1.95$ m) and of outlet ($W_o = 1.68$ m) keys, the lateral length ($B = 11.03$ m),the sidewall thickness ($T_s = 0.33$ m) and the weir height ($P = 4.4$ m).

Many tests were carried out on the 1/30 scale model to study the influence of the right bank topography, of a deflector and of the crest level on the release capacity of the PKW (Fig. 1 and Table 1).

The first tests (Test 1) were performed with the sole PKW in the channel (Fig. 1 high left), at crest level 219 NGF corresponding to a dam height of 14.6 m. For Test 2, a right bank in concrete (Fig. 1 high right) was set up to represent schematically the initial topography of the right bank on the 1/60 scale model (Fig. 2 left). For Test 3, a steel deflector (Fig. 1 high right) was implemented above the right bank in concrete (n°1) to simulate the cross-flows effects that may occur on the 1/60 scale model since the river reaches the dam after two tight curves (Cicero et al, 2010). For Test 4 and Test 5, the crest level were increased by increasing the dam height. For Test 6 (Fig. 1 low left), the deflector was replaced by a right bank in steel (n°2), to simulate the final topography of the right bank (Fig. 2 right) on the 1/60 scale model. The results of the two scaled models were compared for the tests performed at the same crest level and for similar geometries.

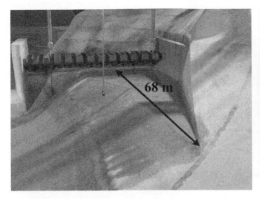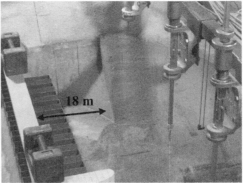

Figure 2. Upstream views of the 1/60 scale model – Initial topography of the right bank (left) – Final topography reducing the material extraction of the right bank (right).

For all tests, the discharge Q and the upstream (piezometric) head H were measured and the discharge coefficient C_{dW} related to the spillway width W was calculated by:

$$C_{dW} = \frac{Q}{W\sqrt{2g}H^{1.5}}$$ (1)

The uncertainity on C_{dW} was estimated with the measurements accuracy of Q and H and with the root mean square values (σ_Q and σ_H) of the time fluctuations by:

$$\frac{\Delta C_{dW}}{C_{dW}} = \frac{\Delta Q}{Q} + \frac{\sigma_Q}{Q} + 1{,}5\left(\frac{\Delta H + \sigma_H}{H}\right)$$ (2)

$\Delta Q/Q = 0.01$ on both models. $\Delta H = 6\,\text{mm}$ on the 1/30 scale model and $\Delta H = 30\,\text{mm}$ on the 1/60 scale model.

3 EXPERIMENTAL RESULTS

3.1 Side effects

We compare (Fig. 3 left) the non-dimensional curves $C_{dW}(H/P)$ measured at the same crest level (219 NGF) for Test 1 to Test 3. The right bank seems to reduce the release capacity of the PKW, but the differences on the discharge coefficient, which decrease with the head from 4% to 2%, are lower than the measurements uncertainities (shown by error bars). The effects of the deflector are quite negligible.

At crest level 219.75 NGF (Fig. 3 right), the comparison of the results of Test 5 and Test 6, shows that the implementation of the right bank in steel has negligible effect on the discharge coefficient.

The PKW is less sensitive to side effects than other classical weirs probably because most of the discharge does not come from the surface level but from the bottom part (Laugier 2007).

3.2 Rating curves analysis

The first tests (Fig. 3 left) showed a peak of the discharge coefficient around $H/P = 0.1$. Thus, for Tests 5 and 6, the low heads domain was investigated with more measurements to check the increase of the discharge coefficient with the upstream head for $H/P < 0.1$.

We observed that the peak on the discharge coefficient was linked to the transition between two flow regimes. At very low heads, the flow is adhering to the walls of the structure (Fig. 4 left) and the PKW behaves like a linear weir with the total crested length (Leite Ribeiro et al. 2009). The discharge coefficient increases with the head until the jets coming from the outlet keys and

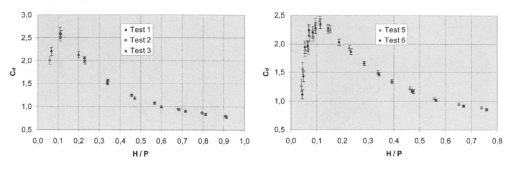

Figure 3. Scale model 1/30 – Effects of the right bank and of the deflector on the discharge coefficient at crest level 219 NGF (left) and 219.75 NGF (right).

Figure 4. Scale model 1/30 – Flow regimes at low heads during test 4 – Clinging flow for H/P = 0.07 (left) and springing flow for H/P = 0.1 (right).

from the sidewalls begins to detach from the walls (Fig. 4 right). Then the discharge coefficient decreases because the jets coming from the oulet keys and from the sidewalls are over-crossing. The discharge coefficient tends to an asymptotic value when the PKW operates saturated (Fig. 1 low right).

The flow regime transitions are due to a wall thickness effect as described by Johnson (2000) who observed three flow regimes on flat-topped weirs. "Clinging flow" occurs at very small heads, when surface tension and gravity forces cause the water to adhere to the downstream face of the weir. "Leaping flow" begins when the flow energy is sufficient to detach the flow from the downstream face of the weir. "Springing flow" occurs when the energy is sufficient to detach the flow from the upstream edge of the flat-topped crest. Johnson (2000) identified that the transition between leaping and springing flow depends on the ratio H/T_s and on the weir geometry, and it provides a change in the slope of the discharge coefficient for $H/T_s > 1.8$. In our experiments on the 1/30 scale model, the peak of the discharge coefficient is observed for $1.5 < H/T_s < 2$.

3.3 Crest level effects

The Tests 3 to 5 performed at various crest levels confirmed the shape of the rating curves (Fig. 5 left) with a peak of the discharge coefficient located around $H/P = 0.1$.

For very low heads ($H/P < 0.1$), we observed differences up to 20% on the discharge coefficient. Those differences are greater than the measurements accuracy. On the 1/30 scale model, the heads over the crest were lower than 1.5 cm. Thus the model might not be representative of prototypes since the surface tension effects become significant. The Weber number was calculated by equation 3 to characterize the surface tension effect ($\sigma = 7.3 \; 10^{-2}$ N/m). Figure 5 right shows that the peak

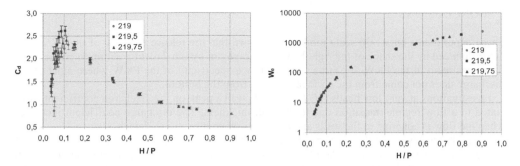

Figure 5.   Scale model 1/30 – Crest level effect on the discharge coefficient (left) and on the Weber number (right).

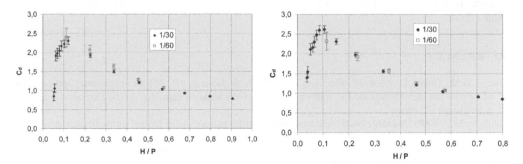

Figure 6.   Scale effects on the discharge coefficient for PKW at crest level 219 NGF (Left) and 219.5 NGF (right).

location of the discharge coefficient ($H/P = 0.1$) corresponds to $We = 30$.

$$We = \frac{\rho g H^2}{\sigma} \qquad (3)$$

For $H/P > 0.1$, the crest level has no influence on the rating curves $C_{dW}(H/P)$, meaning that the release capacity of the PKW does not depend on the dam height. Nevertheless, in these experiments the variations of the crest level were lower than 0.75 m, i.e. 4% of the dam height. These results should be confirmed for greater variations of the crest level.

3.4   Scale effects

We could compare the curves of the discharge coefficient measured on both models for rather similar geometries of the right bank. At crest level 219 NGF (Fig. 6 left), we compare Test 3 performed on the model 1/30 with the right bank in concrete and the deflector (Fig. 1 high left) to the tests performed on the 1/60 scale model with the first topography (Fig. 2 left). At crest level 219.5 NGF (Fig. 6 right), we compare Test 6 performed on the 1/60 scale model with the right banks in concrete and in steel (Fig. 1 low left) to the tests performed on the 1/60 scale model with the final topography (Fig. 2 right).

The discharge coefficients measured on both models show differences less than 8%, lower than the measurements uncertainties. At very low heads, the uncertainities are much higher on the 1/60 scale model and it was not possible to make measurements. The results of both models are rather close for $H/P > 0.2$, i.e. $H > 1.5$ cm or $We > 30$ on the 1/60 scale model.

Considering the significant differences of geometries (dam representations, lay-out, topography), the hydraulic performances of the PKW are rather similar on both scale models. Indeed, the scale model 1/60 is a complete representation of the prototype area, while the scale model 1/30 is a partial and schematic representation in a straight channel.

171

4 CONCLUSIONS

These experiments on a 1/30 scale model have shown that the release capacity of the PKW was not influenced by the shape of the right bank topography or by the deflector set up to create cross flows upstream of the PKW. It can be explained by a specificity of the PKW where most of the discharge does not come from the surface level but from the bottom.

Investigations at very low heads ($H/P < 0.1$) showed a peak on the discharge coefficient associated to a flow regime transition due to the wall thickness. The discharge coefficient increases when the flow is adhering to the walls and decreases when it begins to detach.

For higher heads where the surface tension effects become negligible, the rating curves are not influenced by the crest level (or dam height). Nevertheless, in these experiments the crest level variations represented only 4% of the dam height and it should be confirmed for higher variations.

The discharge coefficient measured on both models at 1/30 and 1/60 scales showed differences lower than the measurements uncertainties. The results of both models are rather close if the Weber number (calculated by Equation 3) is greater than 30, or if the head on the models is greater than 1.5 cm.

REFERENCES

Cicero, G.M., Guene, C., Luck, M., Pinchard, T., Lochu, A. & Brousse, P.H. 2010. Experimental optimization of a piano key weir to increase the spillway capacity of the Malarce dam. *1rst IAHR European congress, Edinbourgh, 04–06 mai.*

Johnson, M.C. 2000. Discharge coefficient analysis for flat-topped and sharp crested weirs. *Irrigation science,* 19: 133–137.

Laugier, F. 2007. Design and construction of the first Piano Kew Weir (PKW) spillway at the Goulours dam. *Hydropower & Dams,* issue 5: 94–101.

Le Doucen, O., Leite Ribero, M., Boillat, J.L., Schleiss, A., Laugier, F., Lochu, A., Delorme & Villard, J.F. 2009. Réhabilitation de la capacité d'évacuation des crues- Intégration de pk-weirs sur des barrages existants. *Colloque CFBR-SHF: Dimensionnement et fonctionnement des évacuateurs de crues, Paris, 20–21 janvier.*

Leite Ribero, M., Bieri M., Boillat, J.L., Schleiss, A., Delorme, F. & Laugier, F. 2009. Hydraulic capacity improvment of existing spillways design of piano key weirs. *23ème congrès des grands barrages, Brasilia.*

Machiels, O., Erpicum, S., Archambeau, P., Dewals, B. & Pirotton, M. Analyse expérimentale du fonctionnement des déversoirs en touche de piano. *Colloque CFBR-SHF: Dimensionnement et fonctionnement des évacuateurs de crues, Paris, 20–21 janvier.*

Ouamane, A. & Lemperiere, F. 2006. Design of a new economic shape of weir. *Proceedings of the International Symposium on Dams in the Societies of the 21st Century, Barcelona.*

Chi Hien, T., Thanh Son, H. & Ho Ta Khanh, M. 2006. Results of some piano key weir hydraulic model tests in Vietnam. *22ème congrès des grands barrages, Barcelona.*

Hydraulic design

Labyrinth and Piano Key Weirs – PKW 2011 – Erpicum et al. (eds)
© 2011 Taylor & Francis Group, London, ISBN 978-0-415-68282-4

Study of optimization of the Piano Key Weir

A. Noui
Hydraulic Laboratory planning and Environment, Biskra University, Algeria
Centre for Scientific and Technical Research on Arid Regions CRSTRA, Biskra, Algeria

A. Ouamane
Hydraulic Laboratory planning and Environment, Biskra University, Algeria

ABSTRACT: Labyrinth weirs are adequate solution in dam rehabilitation projects when the storage and\or discharge capacity has to be increased. Piano key weirs (PKW) are particularly interesting due to their structure as well as their upstream and downstream flow conditions. PKW have a different geometrical shape than the classic labyrinth weirs. The keys are rectangular shaped with inclined key bottoms allowing the use of overhangs. Two different types of PKW were defined: Type A shows two overhangs, one upstream and one downstream. Type B does not have the downstream overhangs. Physical modelling tests have showed that the efficiency of Type B is higher than of Type A. The study reveals that flow on the PKW is influenced by different geometrical parameters. The dimensional analysis allowed the development of relations between discharge capacity and the shape of the PKW.

1 INTRODUCTION

Labyrinth weirs are interesting alternatives to reduce the risk of dam failure. They are often applied in dam rehabilitation projects when the storage and/or spillway capacity has to be increased. The Piano Key Weir (PKW) is a particularly interesting solution due to its rigid structure and its upstream and downstream flow conditions. The PKW was developed in 2003 (Lempérière & Ouamane 2003). Its geometry is different from the classical labyrinth weir. The keys are rectangular shaped and have an inclined bottom to favor the use of key overhangs. Two different types of PKW are distinguished (Ouamane & Lempérière 2006a, b): Type A is characterized by two overhang keys, one upstream and one downstream. Type B differs from Type A by the lack of the downstream overhangs. The physical modeling tests results obtained on Type B showed that discharge capacity is higher compared to Type A. The study focused on the optimization of PKW. It shows that flow depends on various geometric parameters, which characterize this particular type of spillway. The dimensional analysis of flow allowed the development of mathematical relationships, helpful for PKW design.

2 SYSTEM OF GEOMETRIC PIANO KEY WEIR

Lempérière and Ouamane (2003) mention the following main PKW characteristics:

- A rectangular arrangement of keys is similar to the keys of a piano;
- The alternatively placed inlet and outlet keys have downstream respectively upstream overhangs;
- Even on small platforms, considerable crest lengths can be achieved;
- The overhangs reduces the base width of the PKW;
- The surface of the side walls is reduced due to inclined keys.

3 EXPERIMENTAL SETUP

An upstream channel of a section of 0.75 m to 0.75 m and a length of 3.5 m supplies the squarely shaped basin. It is 3 m wide, 3 m long and 1.1 m deep. The upstream inlet of the basin is equipped

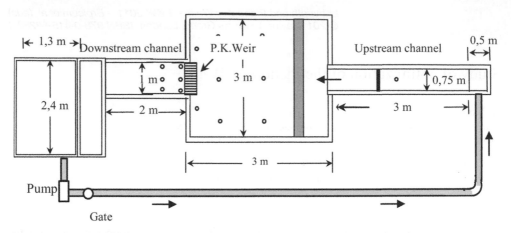

Figure 1. Plan view of the experimental device.

with a metal gate and a brick wall that provide uniform upstream flow conditions. A series of pressure taps is placed at different locations in the basin, to define the water column. The PKW models are installed at the outlet of basin. Downstream of the spillway, a 2 m long and 1 m wide channel is located.

4 RESULTS

The flow coefficient is derived from the universal equation that expresses the flow passes through a weir:

$$C_W = \frac{Q}{W \sqrt{2g} H^{3/2}} \quad (1)$$

where C_w = flow coefficient; Q = discharge; W = weir width; and H = upstream head.

The flow coefficient C_w can be defined by the measured flow Q and the upstream head H. It is suitable to represent the flow coefficient as a function of dimensionless parameters. The π theorem of Buckingham allows a dimensional analysis on the discharge capacity of Type B PKW. Following relationships could be developed:

$$\Pi_1 = Q / \sqrt{g} W_o^{5/2} \quad (2)$$

$$\Pi_2 = H / W_o \quad (3)$$

$$\Pi_3 = L / W_o \quad (4)$$

$$\Pi_4 = W / W_0 \quad (5)$$

$$\Pi_5 = P / W_o \quad (6)$$

$$\Pi_6 = W_i / W_o \quad (7)$$

where W_o = outlet key width; W_i = inlet key width; L = crest length; and P = total weir height.

By combining these parameters, the relationship is obtained:

$$C_W = f(H / P, L / W, W / P, W_i / W_o) \quad (8)$$

To better understand the effect of these dimensionless parameters on the flow of the Type B PKW, ten models were tested.

4.1 Comparison between PKW model A and B

A Type A (Fig. 2) and a Type B PKW (Fig. 3) have been tested in the test flume.

Table 1. Geometrical parameters of the tested PKW (W_u width of PKW unit, $n = L/W$).

Model	L [cm]	W [cm]	W_u [cm]	P [cm]	n [–]	W_i [cm]	W_o [cm]	B [cm]	L/W [–]	W_u/P [–]	W_i/W_o [–]
A	600	100	16.67	15	6	9	7.5	41	6	1.11	1.2
B01	600	100	16.67	15	6	9	7.5	41	6	1.11	1.2
B02	600	100	16.67	15	6	10	6.5	41	6	1.11	1.5
B03	400	100	25	15	4	15	10	41	4	1.67	1.5
B04	800	100	12.5	15	8	7.6	4.8	41	8	0.83	1.5
B05	600	100	16.67	15	6	6.6	9.9	41	6	1.11	0.7
B06	600	100	25	15	4	15	10	62.5	6	1.67	1.5
B07	400	100	25	20	4	15	10	41	4	1.25	1.5
B08	600	100	16.67	20	6	9	7.5	41	6	0.83	1.2
B09	400	100	25	25	4	15	10	41.2	4	1	1.5

Figure 2. Model of PKW model A.

Figure 3. Model of PKW model B.

The test results show that Type B without downstream overhangs is an appropriate solution for high flow rates. Type A is a symmetric and therefore a very economical solution. Prefabricated elements can be used for construction. It is recommended to design the PKW without downstream overhangs if it is structurally stable during the periods when the level in the reservoir is lower than the threshold weir level.

4.2 Influence of L/W ratio

Two PKW with different L/W ratios were tested. Figure 5 shows that the discharge capacity can be significantly increased by increasing L/W ratio from 4 to 6. This increase is about 15% for $H/P = 0.2$ and 8% for medium heads of $H/P = 0.4$.

4.3 Influence of W/P ratio

Three models of $L/W = 4$ have been tested by models B03, B07 and B09 (Fig. 6). The difference between the curves of model B03 ($W/P = 1.67$) and B07 ($W/P = 1.25$) is about 7% for low heads

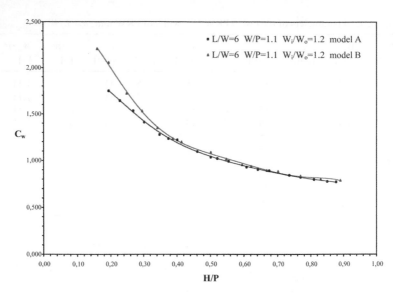

Figure 4.   Comparison of flow coefficients of PKW model A and B.

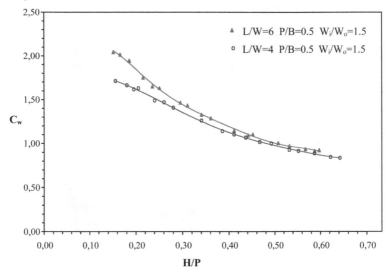

Figure 5.   Flow coefficient depending on L/W ratio.

of $H/P = 0.25$ and 4% for medium heads of $H/P = 0.5$. The difference between the curves of model B07 ($W/P = 1.25$) and B09 ($W/P = 0.99$) is about 9% for low heads of $H/P = 0.25$ and 6% for $H/P = 0.5$.

4.4   Influence of key widths W_i/W_o

To achieve the hydraulically optimal ratio between the inlet key width W_i and the outlet key width W_o, three PKW models with the same ratio $L/W = 6$ were tested (Fig. 7). The comparison between the flow coefficients of the models of $W_i/W_o = 1.2$ and $W_i/W_o = 1.5$ shows that the lower W_i/W_o ratio increases the discharge capacity of 5% for low heads and of 3% for medium ones.

4.5   Influence of PKW position

The first test concerning the position of model A was performed with a downstream water level at the same level as PKW base. The second test was performed by raising the position of the PKW

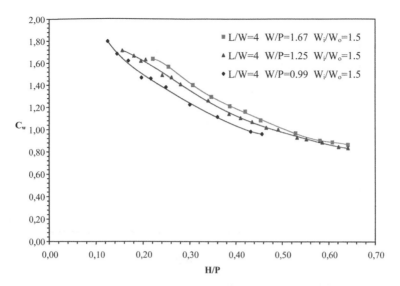

Figure 6. Flow coefficient depending on W/P ratio.

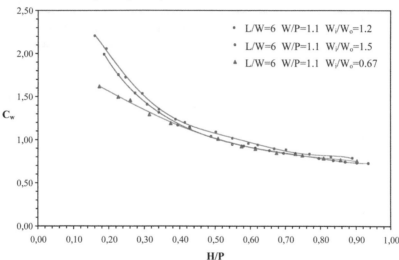

Figure 7. Flow coefficient depending on upstream and downstream key widths.

by 16 cm. Figure 8 shows flow coefficients of the two tested positions of PKW model A. The two curves are almost identical. The same performance for both locations was achieved.

4.6 *Influence of inlet key slope*

To determine the influence of the inclination of key bottom on the flow, two models were tested. The first one had horizontal inlet key slopes and inclined outlet keys. The second model was characterized by inclined inlet and outlet keys. Figure 9 shows that the model with strike angles increased the discharge capacity of the PKW by about 12% for $H/P > 0.6$ compared with the model with zero strike.

4.7 *Influence of the key length*

Side walls with door-to-short overhangs are more stable and easier to implement than door-to-long fakes. Two models of the same ratio $L/W = 6$ were tested to determine their hydraulic efficiency.

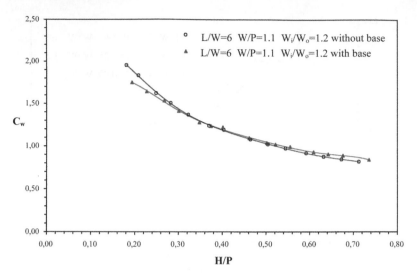

Figure 8.   Flow coefficient based on PKW position.

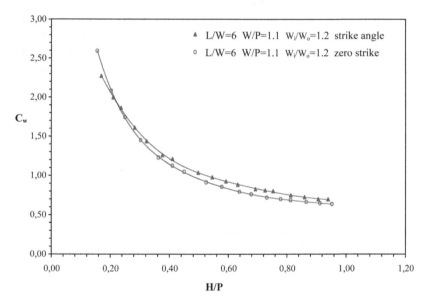

Figure 9.   Flow coefficient based on the inlet key slope.

Figure 10 shows that increasing length of the sidewalls has no impact on the performance of the PKW as long as H/P is lower than 0.7. For important heads of $H/P > 0.7$, the model with $B = 62.5$ cm is by 3% more efficient than the model with $B = 41$ cm.

4.8   *Influence of the outlet key*

To check the impact of a ski jump at the downstream end of the oulet key on the performance of PKW, two tests were performed on a model with $L/W = 4$, $W/P = 1.25$ and $W_i/W_o = 1.5$. The first test was done without obstacle at the outlet key outlet (Fig. 11). For the second test a ski jump at the downstream end of the outlet key was implemented (Fig. 12). Figure 13 shows that the presence of an obstacle at the downstream end of the outlet key of the PKW reduces the performance of PKW significantly. The reduction is about 20% for $H/P < 0.5$.

180

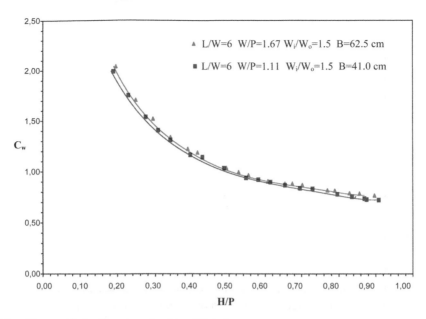

Figure 10. Flow coefficient based on the sidewall length.

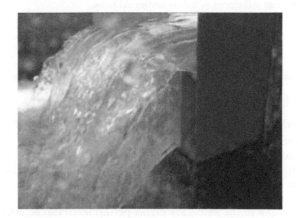

Figure 11. Flow on model without ski jump.

Figure 12. Flow on model with ski jump.

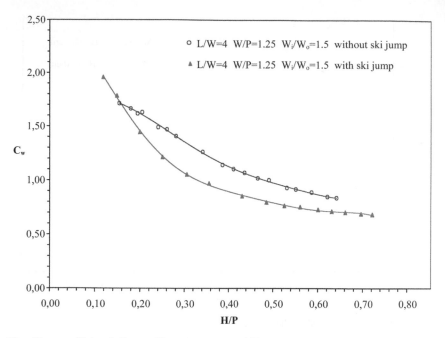

Figure 13.   Flow coefficient influenced by a downstream ski jump.

5   CONCLUSION

This study showed that the flow coefficient is related to several dimensionless parameters:

$$C_W = f(H / P, L / W, W / P, W_i / W_o) \qquad (9)$$

The results of the physical modeling tests with fourteen different PKW models reveal:

- The PKW is a more efficient solution than the classic Creager weir;
- The performance of PKW Type B is higher than Type A;
- The PKW has a higher performance when the H/P ratio is lower. Therefore a PKW should be operated for heads lower than half of the weir height P;
- For steep key slopes the effect of L/W is very apparent;
- By increasing the height of the weir, 25% higher performance can be achieved for low heads, for medium heads only 5%;
- PKW Type B allows a better performance if the width of the inlet key is 1.2 times the width of the outlet one. From a practical and economic point of view this fact is particularly interesting because the concrete volume is the same;
- The PKW is designed to be placed on gravity dams or in spillway channels of earth dams. Tests have shown that the same performance for both cases of location can be achieved;
- A ski jump at the downstream end of the outlet key allows avoiding the impact of the water jet on the dam body; however the performance of the PKW decreases.

REFERENCES

Lempérière, F. & Ouamane, A. 2003. The Piano Keys Weir: a new cost-effective solution for spillways, *The International Journal on Hydropower & Dams* Issue Four.
Ouamane, A. & Lempérière F. 2006. New design of weir for increasing the capacity of dams. *International Symposium on the protection and preservation of water resources, Blida.*
Ouamane, A. & Lempérière, F. 2006. Design of a new economic shape of Weir. *International Symposium on Dams In The Societies Of The XXI Century, Barcelona.*

Experimental parametric study for hydraulic design of PKWs

M. Leite Ribeiro, J.-L. Boillat & A.J. Schleiss
Laboratory of Hydraulic Constructions (LCH), Ecole Polytechnique Fédérale de Lausanne (EPFL), Switzerland

O. Le Doucen
Département du Territoire, Service de l'écologie de l'eau, CH-1219 Aïre, Switzerland

F. Laugier
*Electricité de France (EDF), Centre d'Ingénierie Hydraulique, Savoie Technolac,
Le Bourget-du-Lac, France*

ABSTRACT: The implementation of a new type of labyrinth spillway, called PKW (Piano Key Weir) reveals as a performing alternative for increasing the overflow capacity of existing dams. The optimal hydraulic design of such structure is however not obvious due to the large number of involved parameters. The present work explores experimentally the most relevant geometrical parameters, such as width, length, height and slope of keys and also the upstream and downstream flow conditions. The rating curve of a unit structure is then analysed considering the following dimensionless parameters: (i) the total developed crest length over PKW width (L/W), (ii) the inlet over outlet key widths (W_i/W_o), (iii) the vertical over the horizontal shapes (P/W_i), and (iv) the vertical height of the dam over the PKW height (P_d/P). A non-linear global stepwise regression approach is then applied to fit the most influent parameters in order to provide a mathematical formulation for the hydraulic design of a PKW. The discussion reveals that there is not only one optimal solution from the hydraulic point of view. The ideal solution can then only be selected when considering local and economical constraints related to excavation, materials, construction techniques and particularly to the downstream energy dissipation system.

1   INTRODUCTION

With the increase of hydrological data records and the development of new methodologies for flood discharge estimation, as well as higher requirements of the communities on safety issues, a large number of existing dams require spillway rehabilitation in order to improve their hydraulic capacity. For such projects, the new shape of labyrinth spillways, called Piano Key Weir (PKW) is an interesting alternative (Lempérière and Ouamane, 2003; Leite Ribeiro et al., 2009). As for labyrinth shapes, this structure provides a longer total effective crest length for a given spillway width, with the advantage that it can be placed in the upper part of most existing dams, due to its reduced base surface.

Over the last years, many efforts have been made in order to understand the hydraulic behavior of PKWs but for the moment, only few systematic laboratory tests as well as design basis can be found (Ouamane and Lempérière, 2006). Nowadays, the hydraulic behavior of PKWs is still not complete understood. All the current projects under development to assess the hydraulic capacity are based on physical modeling (Laugier 2007, Laugier et al., 2009, Bieri et al., 2009).

In the present research, an experimental study is conducted in order to evaluate the influence of selected geometrical parameters on the hydraulic efficiency of a PKW. The objective is (i) to analyze the influence of the main geometric parameters on the PKW discharge capacity, using non dimensional ratios and (ii) to propose an empirical formulation for the preliminary design of new PKWs.

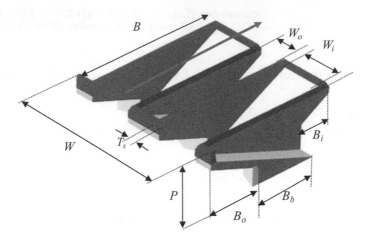

Figure 1.   Fundamental parameters on an entire PKW – 3D-view (Pralong et al., 2011).

2   PARAMETRIC STUDY

The total discharge over a PKW is a function of several parameters which can be summarized as follows:

$$Q = f(\rho, g, \nu, H, L, P, P_d, W, W_i, W_o, B, T_s, R, \alpha) \qquad (1)$$

In Equation 1, the fluid is characterized by its density ρ and the kinematic viscosity ν, g is the acceleration of gravity, and H is the total upstream hydraulic head. The other parameters are related to the geometry of the PKW. L is the total developed crest length, P is the total height of the PKW, P_d is the dam height, W is the total width of the PKW, W_i and W_o are the widths of the inlet and outlet keys, B is the length of the side weir, α is the angle between the inlet/outlet key crest and the side weir of the PKW, T_s is the sidewall thickness and R is the radius of crest curvature. With exception of α, P_d and R, the other geometric parameters are shown in Figure 1.

In order to analyze the PKW efficiency, a comparison to a sharp-crested weir with crest length W is made. As discussed by Leite Ribeiro et al. (2007), a PKW tends to behave like a linear weir as the upstream head increases. Consequently, a discharge enhancement ratio r between the PKW discharge (Q_{PKW}) and the corresponding rectangular sharp-crested weir discharge (Q_W) has been adopted for the analysis.

$$r = \frac{Q_{PKW}}{Q_W} = \frac{C_d L_{eff} \sqrt{2g} H^{\frac{3}{2}}}{C_d W \sqrt{2g} H^{\frac{3}{2}}} \qquad (2)$$

In Equation 2, the discharge coefficient C_d is assumed as constant to 0.42, characteristic average value of sharp-crested weirs (Hager and Schleiss 2009), and L_{eff} is the effective crest length of the PKW that theoretically contributes to the overflow. L_{eff} decreases with increase of head, due to the interference of the overflow layers (Falvey 2003). For the analysis, the discharge capacity of the PKW was measured in the laboratory, while the theoretical discharge of a sharp-crested weir was calculated with reference to the measured hydraulic head.

The present investigation is only based on geometrical parameters of the PKW, thus the terms ρ, g and ν are not considered. The tested PKWs are all rectangular shaped ($\alpha = 90°$) and therefore α is nor included in the analysis. The length of the sidewall B can be omitted in the analysis because it is a function of L/W. Therefore, Equation 1 can be replaced by the following dimensionless relation:

$$r = \frac{L_{eff}}{W} = f(\frac{L}{W}, \frac{W_i}{W_o}, \frac{P}{W_i}, \frac{P_d}{P}, \frac{T_s}{P}, \frac{T_s}{R}, \frac{H}{P}) \qquad (3)$$

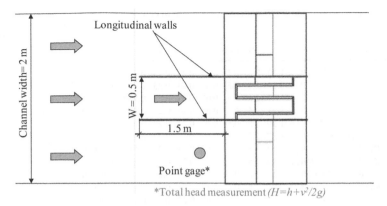

Figure 2.   Schematic overview of the experimental set-up.

In the present study, T_s and R were maintained constant and therefore, only the parameters L/W, W_i/W_o, P_d/P, P/W_i and H/P will be discussed.

3   EXPERIMENTAL SET-UP

The experimental set-up is installed in a 2 m wide straight flume, on a platform placing the bottom of the PKW 0.5 m over the ground of the channel ($P_d = 0.5$ m). A one and half unit configuration (1.5 inlet key + 1.5 outlet key) is constructed over a constant width ($W = 0.5$ m). Two longitudinal guide walls (1.5 m long) allow the water supply to be uniform when approaching the weir. A "reference PKW configuration" is defined, corresponding to the following geometrical values:

- ratio $W_i/W_o = 1.25$, with $W_i = 0.163$ m and $W_o = 0.130$ m;
- ratio $L/W = 5$, with a total developed crest $L = 2.5$ m;
- height of the sidewall $P = 0.217$ m;

Sidewalls are 0.02 m thick ($Ts = 0.02$ m) and have semi-circular crests. Moreover, semicircular noses are installed under the outlet keys of the PKW. A schematic view of the experimental set-up is shown in Figure 2.

The experimental program was organized in order to allow a first analysis of the individual influence of the main dimensionless parameters. All the tested PKWs are described in Table 1. The analysis is divided into four parts.

- Influence of the dam height (P_d) on the discharge capacity of the reference PKW: For these experiments, a movable bottom was installed inside the longitudinal walls and different ratios P_d/P were tested.
- Influence of P/W_i: Different values of P were tested for the configurations with $L/W = 3$ and 5.
- Influence of W_i/W_o: Different W_i/W_o ratios were tested with constant values of $L/W = 5$, $P = 0.217$ m and $P_d = 0.50$ m. For all tests, the sum $W_i + W_o$ was maintained constant and equal to 0.293 m.
- Influence of L/W: Three values of $L/W = 3$, 5 and 7 were tested for constant values of $W_i/W_o = 1.25$, $P = 0.217$ m and $P_d = 0.50$ m.

For each PKW configuration, rating curves (Q versus H) are established. Measurement of the total hydraulic head is performed outside of the guide walls by a point-gage whereas the discharge is measured by an electromagnetic flowmeter.

185

Table 1. Summary of the tested PKW configurations with main geometric and dimensionless parameters.

[°] Configuration	L [m]	W [m]	B [m]	B_i [m]	B_o [m]	W_i [m]	W_o [m]	P [m]	P_d [m]	L/W [–]	W_i/W_o [–]	P/W_i [–]
1 $L/W=5; W_i/W_o=1.25; P/W_i=1.33$	2.5	0.5	0.67	0.23	0.23	0.163	0.130	0.217	0.50	5	1.25	1.33
2 $L/W=5; W_i/W_o=1.25; P/W_i=0.96$	2.5	0.5	0.67	0.23	0.23	0.163	0.130	0.157	0.50	5	1.25	0.96
3 $L/W=5; W_i/W_o=0.80; P/W_i=1.67$	2.5	0.5	0.67	0.23	0.23	0.130	0.163	0.217	0.50	5	0.80	1.67
4 $L/W=5; W_i/W_o=0.80; P/W_i=1.21$	2.5	0.5	0.67	0.23	0.23	0.130	0.163	0.157	0.50	5	0.80	1.21
5 $L/W=5; W_i/W_o=1.60; P/W_i=1.20$	2.5	0.5	0.67	0.23	0.23	0.181	0.113	0.217	0.50	5	1.60	1.20
6 $L/W=5; W_i/W_o=1.60; P/W_i=1.87$	2.5	0.5	0.67	0.23	0.23	0.181	0.113	0.157	0.50	5	1.60	0.87
7 $L/W=5; W_i/W_o=0.63; P/W_i=1.92$	2.5	0.5	0.67	0.23	0.23	0.113	0.181	0.217	0.50	5	0.63	1.92
8 $L/W=5; W_i/W_o=0.63; P/W_i=1.39$	2.5	0.5	0.67	0.23	0.23	0.113	0.181	0.157	0.50	5	0.63	1.39
9 $L/W=5; W_i/W_o=2.00; P/W_i=1.11$	2.5	0.5	0.67	0.23	0.23	0.195	0.098	0.217	0.50	5	2.00	1.11
10 $L/W=5; W_i/W_o=2.00; P/W_i=0.81$	2.5	0.5	0.67	0.23	0.23	0.195	0.098	0.157	0.50	5	2.00	0.81
11 $L/W=5; W_i/W_o=0.50; P/W_i=2.21$	2.5	0.5	0.67	0.23	0.23	0.098	0.195	0.217	0.50	5	0.50	2.21
12 $L/W=5; W_i/W_o=0.50; P/W_i=1.60$	2.5	0.5	0.67	0.23	0.23	0.098	0.195	0.157	0.50	5	0.50	1.60
13 $L/W=7; W_i/W_o=1.25; P/W_i=1.33$	3.5	0.5	1.00	0.40	0.40	0.163	0.130	0.217	0.50	7	1.25	1.33
14 $L/W=7; W_i/W_o=0.80; P/W_i=1.67$	3.5	0.5	1.00	0.40	0.40	0.130	0.163	0.217	0.50	7	0.80	1.67
15 $L/W=7; W_i/W_o=2.00; P/W_i=1.11$	3.5	0.5	1.00	0.40	0.40	0.195	0.098	0.217	0.50	7	2.00	1.11
16 $L/W=3; W_i/W_o=1.25; P/W_i=1.33$	1.5	0.5	0.33	0.07	0.07	0.163	0.130	0.217	0.50	3	1.25	1.33
17 $L/W=3; W_i/W_o=1.25; P/W_i=0.82$	1.5	0.5	0.33	0.07	0.07	0.163	0.130	0.134	0.50	3	1.25	0.82
18 $L/W=3; W_i/W_o=1.25; P/W_i=0.59$	1.5	0.5	0.33	0.07	0.07	0.163	0.130	0.096	0.50	3	1.25	0.59
19 $L/W=3; W_i/W_o=0.80; P/W_i=1.67$	1.5	0.5	0.33	0.07	0.07	0.130	0.163	0.217	0.50	3	0.80	1.67
20 $L/W=3; W_i/W_o=0.80; P/W_i=1.03$	1.5	0.5	0.33	0.07	0.07	0.130	0.163	0.134	0.50	3	0.80	1.03
21 $L/W=3; W_i/W_o=0.80; P/W_i=0.74$	1.5	0.5	0.33	0.07	0.07	0.130	0.163	0.096	0.50	3	0.80	0.74

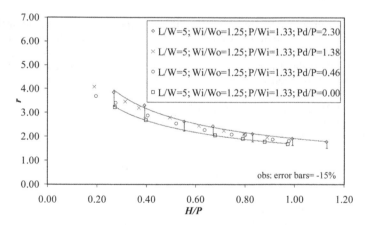

Figure 3. Discharge enhancement ratio r as function of H/P for different ratios of P_d/P.

4 EXPERIMENTAL RESULTS

The analysis is made by comparison of the discharge enhancement ratio (r) versus H/P. Only values of H/P higher than 0.2 are considered in order to avoid capillarity effects, responsible for aspiration due to negative pressure under the overflowing jet.

4.1 Influence of the height of the dam (P_d)

Experiments revealed that the flow approach conditions, characterized by the height of the dam (P_d), can have a significant influence on the discharge capacity. As illustrated in Figure 3, the influence of P_d tends to diminish with the increase of P_d/P. However, it can reduce the PKW efficiency to about 15% for $P_d/P = 0$. This effect can be related to a head loss increase associated to the approach velocity in front of the outlet keys.

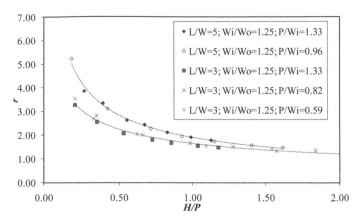

Figure 4. Discharge enhancement ratio as function of H/P for different values of P/W_i.

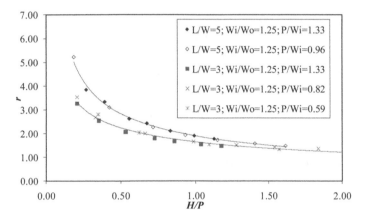

Figure 5. Discharge enhancement ratio r as function of H/P for different values of W_i/W_o.

4.2 *Influence of P/W_i*

The ratio of the vertical to horizontal dimensions of the inlet key (P/W_i) does not affect the discharge capacity of a PKW. As illustrated in Figure 4, for same values of H/P, the PKWs with identical ratios L/W and W_i/W_o have approximately the same efficiency for different P/W_i.

4.3 *Influence of W_i/W_o*

The ratio between the inlet (W_i) and outlet widths (W_o) reveals a higher efficiency when $W_i/W_o > 1$ than for $W_i/W_o < 1$ (Fig. 5). This suggests that the most efficient part of the PKW is the inlet key that combines lateral and frontal overflows. However, there are practically no differences between the measurements with $W_i/W_o = 1.25$, 1.60 and 2.00.

4.4 *Influence of L/W*

As expected, the most important parameter influencing the efficiency of a PKW is the developed length ratio L/W. Experimental values let appear differences of about 50% between tests with L/W = 7 and L/W = 3 for low ratios H/P. However, with the increase of H/P, the differences tend to decrease (Fig. 6).

187

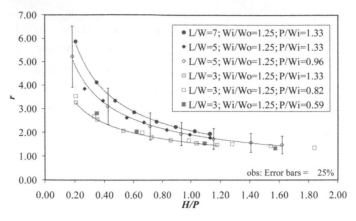

Figure 6. Discharge enhancement ratio r as function of H/P for different values of L/W.

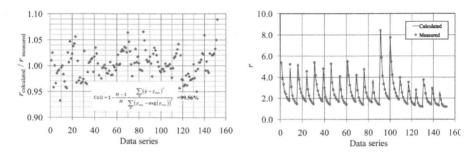

Figure 7. Ratio between the measured and calculated r.

5 HYDRAULIC DESIGN

In order to provide a mathematical formulation for the hydraulic design of a PKW, a non-linear global stepwise regression approach was applied to fit the most influent parameters. The Evolutionary Polynomial Regression (EPR) toolbox (Laucelli et al., 2009) was applied to all the measurements (r versus H/P) performed on the models with ratio $P_d/P = 2.3$ ($P_d = 0.50$ m).

5.1 Universal formula

The finally adopted mathematical form, for the preliminary design of a PKW consists of an exponential function, where the main geometrical mentioned parameters are considered.

$$r = e \left[\begin{array}{l} -0.25945 \left(\dfrac{P}{W_i}\right)^{1.4} \left(\dfrac{H}{P}\right)^{0.15} + 1.0056 \left(\dfrac{L}{W_i}\right)^{0.1} \left(\dfrac{P}{W_i}\right)^{0.5} \left(\dfrac{H}{P}\right)^{0.7} \\ + 0.067404 \left(\dfrac{L}{W_i}\right)^{0.3} \left(\dfrac{P}{W_i}\right)^{0.1} \left(\dfrac{W_i}{W_o}\right)^{0.25} \left(\dfrac{H}{P}\right)^{0.2} + 13.9156 \left(\dfrac{L}{W}\right)^{0.35} \left(\dfrac{H}{P}\right)^{0.15} \\ - 14.0239 \left(\dfrac{L}{W}\right)^{0.35} \left(\dfrac{H}{P}\right)^{0.2} + 0.094 \end{array} \right] - 1 \qquad (4)$$

The application of Equation 4 to the entire data set shows that the accuracy of the calculated r-values ranges between $\pm 5\%$ with respect to the measured ones (Fig. 7). The coefficient of determination CoD is equal to 99.56%. In the CoD equation, expressed in Figure 7 (left), N is the number of

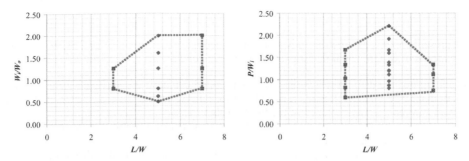

Figure 8.   Application domain of the universal formula.

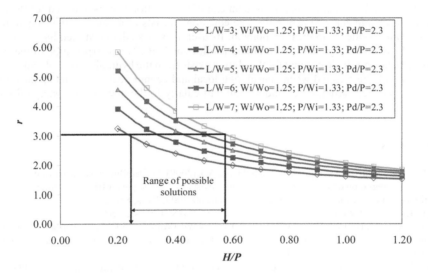

Figure 9.   Generated efficiency curves for PKWs with different developed length ratios L/W and $P/W_i = 1.33$, $W_i/W_o = 1.25$ and $P_d/P = 2.3$ and example of solutions for a discharge enhancement ratio $r = 3$.

data; $avg(yexp)$ is the average value of observations; $\hat{y}$ is the value predicted by the mathematical equation and $yexp$ is the corresponding observation.

5.2   Application domain

It is important to notice that the proposed empirical equation is based on a limited number of configurations and parameters. Therefore, its application is rigorously valid inside the corresponding application domain. Figure 8 illustrates the curves W_i/W_o versus L/W (left) and P/W_i versus L/W (right) where the coefficient of determination (CoD) shown in Figure 7 is valid. Moreover, it is applicable for $0.2 < H/P < 1.2$ and for $P_d/P \geq 2.3$ (no significant influence of the dam height).

5.3   Generation of efficiency curves

Figure 9 illustrates a series of efficiency curves (r versus H/P) generated with Equation 4. The curves represent PKWs with developed length ratios L/W of 3, 4, 5, 6 and 7 and constant values of $P/W_i = 1.33$, $W_i/W_o = 1.25$ and $P_d/P = 2.3$. The interpretation of Figure 9 shows that there is not only one optimal solution of a PKW, from the hydraulic point of view. PKWs with different developed length ratios L/W can evacuate the same discharge for different values of H/P.

6 CONCLUSIONS

In the present paper, a parametric study on the main geometrical parameters of a PKW was carried out by means of experimental investigations. The main conclusions to be highlighted are:

- The height of the dam on which the PKW is installed can have a significant influence on the PKW discharge capacity. Results revealed that for a PKW with $L/W = 5$, values of P_d/P near to zero can reduce the efficiency of the PKW in about 15%.
- The vertical-to-horizontal dimensions ratio (P/W_i) does not influence considerably the efficiency of the PKW for a same value of H/P.
- PKW with ratios $W_i/W_o > 1$ are more efficient. However, no remarkable differences could be pointed out between values of 1.25, 1.60 and 2.0.
- The developed length ratio L/W is the most influent parameter on PKW capacity.

From the available experimental dataset, an empirical equation was established for the calculation of the discharge enhancement ratio. The equation consists of an exponential function that is rigorously valid inside an application domain. Its usefulness was demonstrated by the generation of several PKW efficiency curves with different values of L/W. The interpretation of these curves demonstrates that no single optimal PKW solution exists from the hydraulic point of view. The ideal solution can only be selected when considering local and economical constraints related to excavation, materials, construction techniques and particularly to the downstream energy dissipation system.

REFERENCES

Bieri, M., Leite Ribeiro, M., Boillat, J.-L. and Schleiss, A.J. (2009). "Réhabilitation de la capacité d'évacuation des crues – intégration de PK-Weir sur des barrages existants". Colloque CFBR-SHF, Dimensionnement et fonctionnement des évacuateurs de crues. 20–21 January 2009. Paris, France.

Falvey, H.T. (2003). "Hydraulic Design of Labyrinth Weirs." ASCE Press, USA.

Hager, W. and Schleiss, A.J. (2009). "Traité de Génie Civil, Volume 15 – Constructions Hydrauliques – Ecoulement Stationnaires, EPFL". Presses polytechniques et universitaires romandes. Lausanne, Switzerland. ISBN 978-2-88074-746-6.

Laucelli, D., Berardi, L. and Doglioni, D. (2009). Evolutionary Polynomial Regression (EPR) Toolbox Version 2.SA. Technical University of Bari, Italy. Department of Civil and Environmental Engineering. January, 2009.

Laugier, F. (2007). "Design and construction of the first Piano Key Weir (PKW) spillway at the Goulours dam". Hydropower & Dams. Issue 5, 2007. pp. 94–101.

Laugier, F., Lochu, A., Gille, C., Leite Ribeiro, M. and Boillat, J-L. (2009). "Design and construction a labyrinth PKW spillway at St-Marc dam, France". Hydropower & Dams. Issue 5, 2009. pp. 100–107.

Lempérière, F. and Ouamane, A. (2003). "The Piano Keys weir: a new cost-effective solution for spillways". Hydropower & Dams. Issue 5, 2003. pp. 144–149

Leite Ribeiro, M., Boillat, J.-L., Schleiss, A., Laugier, F. and Albalat, C. (2007). "Rehabilitation of St-Marc Dam – Experimental Optimization of a Piano Key Weir". Proceedings of the 32nd Congress of IAHR. 01–06 July 2007. Venice, Italy.

Leite Ribeiro, M., Bieri, M., Boillat, J-L., Schleiss, A., Delorme, F. and Laugier, F. (2009). "Hydraulic capacity improvement of existing spillways – Design of piano key weirs". 23rd Congress of Large Dams. Question 90, Response 43. 25–29 May 2009. Brasilia, Brazil.

Machiels, O., Erpicum, S., Archambeau, P., Dewals, B. J. and Pirotton, M. (2009). "Large scale experimental study of piano key weirs". Proceedings of the 33rd Congress of IAHR. 09–14 August 2009. Vancouver, Canada.

Ouamane, A. and Lempérière, F. (2006). "Design of a new economic shape of weir". Proceedings of the International Symposium on Dams in the Societies of the 21st Century. 18 June 2006. Barcelona, Spain. pp. 463–470.

Pralong, J., Blancher, B., Laugier, F., Machiels, O., Erpicum, S., Pirotton, M., Leite Ribeiro, M., Boillat. J.-L., and Schleiss, A.J. 2011. Proposal of a naming convention for the Piano Key Weir geometrical parameters. International Workshop on Labyrinth and Piano Key weirs, Liège, Belgium.

Tullis, J. P., Amanian, N. and Waldron, D. (1995) "Design of Labyrinth Spillways". Journal of Hydraulic Engineering, Vol. 121, N°3, March 1995. pp. 247–255.

Labyrinth and Piano Key Weirs – PKW 2011 – Erpicum et al. (eds)
© 2011 Taylor & Francis Group, London, ISBN 978-0-415-68282-4

Main results of the P.K weir model tests in Vietnam (2004 to 2010)

M. Ho Ta Khanh
VNCOLD and CFBR, France

Truong Chi Hien
Head of the Hydraulic Laboratory of HCMUT, Vietnam

Nguyen Thanh Hai
Director of HYCONMECH, SIWRR, Vietnam

ABSTRACT: Model tests were carried out since 2004 in Vietnam to determine some important parameters and functioning conditions of the P.K weirs. They concerned in particular the following topics: determination of the discharges versus the nappe depths for different shapes and sizes of labyrinth and P.K weirs, in case of free overflow, for P.K weirs placed in a channel, determination of the head versus the discharge and comparison with the labyrinth weirs, in case of submerged flow and finally observation of the aeration of the flow on the downstream face of a gravity dam with P.K weir placed on the crest, the downstream face being smooth or with 2D and 3D steps and measurement of the scour at the dam toe. This paper presents the main results of these tests to allow a comparison with the results of other hydraulic laboratories, in particular in France and Algeria and provides also useful information for the projects of P.K weirs under design and construction in Vietnam.

1 DETERMINATION OF THE DISCHARGES VERSUS THE NAPPE DEPTHS FOR DIFFERENT SHAPES AND SIZES OF LABYRINTH AND P.K WEIRS

1.1 *First series of tests*

They were carried out by the Hydraulic Laboratory of HCMUT in a channel, $W = 60$ cm, at the scale 1/25. The P.K weirs are of type A with symmetrical upstream and downstream overhangs (Fig. 1), very similar to the basic shape defined by HydroCoop, with the parameters $P_m \approx 4.5$ m and for $n = 4$, $n = 5$, $n = 7$ (normalized notation). The results of the labyrinth weir of Song Mong ($n = 3.6$), performed by the SIWRR Laboratory, were also used for the comparison. Due to the limited capacity of the pump, the tests were only performed for the low values of the head H, with the specific discharges <16 m^3/s.m and for free overflow (Truong Chi Hien, 2006).

The figure 2 shows the tests of a P.K weir and of the labyrinth weir. On the figure 3 are indicated the specific discharges q versus the nappe depths H for different P.K weirs, for the labyrinth weir and for a straight Creager weir.

We provide here under the best correlation for the specific discharge q versus the nappe depth H (H < 2.2 m) for respectively the P.K weirs, the labyrinth weir and a straight Creager weir:

- For P.K weir with $n = 4$ (PKA3) $\rightarrow q_4 = 5.6\ H^{1.22}$
- For P.K weir with $n = 5$ (PKA1) $\rightarrow q_5 = 6.4\ H^{1.28}$
- For P.K weir with $n = 7$ (PKA2) $\rightarrow q_7 = 8.7\ H^{1.15}$
- For the labyrinth weir $\rightarrow q_L = 5.1\ H^{1.11}$
- For a straight Creager weir $\rightarrow q_c = 2.15\ H^{1.50}$

It can be observed that for the low nappe depths, the discharges of the labyrinth and P.K weirs depend mainly on the values of n ($q = \alpha n\ H^{1.5\beta}$, with $\alpha > 1$ and $\beta < 1$).

Figure 1.   Upstream view of a PKW type A (left) and downstream view of a PKW type B (right).

Figure 2.   Test of a P.K weir type A (left) and test of a labyrinth weir (Song Mong dam) (right).

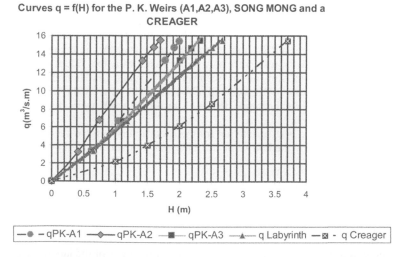

Figure 3.   Discharge efficiency of varied labyrinth and PK Weirs.

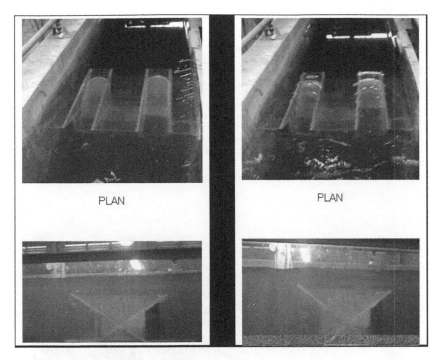

Figure 4.   P.K weir type A. Examples of highly submerged flow.

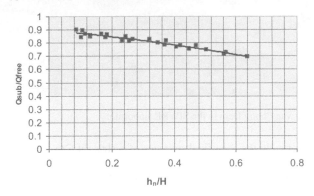

Figure 5.   Ratio between submerged and free overflows.

1.2   *Second series of tests*

The second series of tests (Truong Chi Hien et al., 2006) were carried out in the Hydraulic Labora-tory of HCMUT to measure the coefficients of discharge for the previous P.K weirs with submerged flows (Fig. 4). They showed that for submerged flows – with $h_n/H = 0.6$, h_n being the difference between the downstream water level and the P.K weir crest – the discharge capacity of the P.K weir is reduced about 30% compared with the free overflow case (Fig. 5). This discharge is however much higher than the discharge of the labyrinth weir in the same condition (see the comment about the figures 6 and 8 hereafter and also Tullis (Tullis et al., 1995) and Yildiz et al., 1996). This outcome can be explained by the plunging flows in the outlet keys – which do not exist with the labyrinth weir – these outlet keys acting as inclined shafts. One advantage of a P.K weir installed in a river or a channel is that it provokes a very low rising of the upstream water level when it is submerged.

On the figure 6, the relationship between the specific discharge q and the upstream water level Zt – for the Phuoc Hoa labyrinth weir ($n = 4$, crest at El = 42.9 m then $H = Zt - 42.9$) tested by the SIWRR Laboratory in 2006 – indicates a very low increase of q versus Zt for a submerged overflow

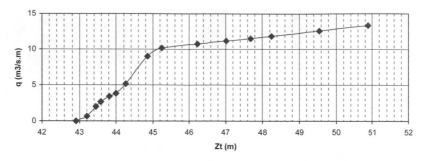

Figure 6.   Phuoc Hoa labyrinth weir. Relationship between q and U/S water level Zt.

Figure 7.   Model of the Van Phong Dam in the SIWRR Laboratory.

(Zt > 44.8 m) and practically the specific discharges of a straight Creager weir of the same width W for Zt > 46 m with a "fully submerged flow", i.e. when the downstream water level is higher than the weir crest. These outcomes are coherent with the results indicated in 1.1 and 1.2 taken into account the location of this labyrinth on a "duck mouth" spillway (with then a bit lower values of q versus H). It is to note that the Phuoc Hoa project was designed in 2002 before the development of the P.K weirs by HydroCoop.

Tests were also carried out in 2009 in the Hydraulic Laboratory of the SIWRR for the Van Phong dam which will be fully submerged during the high floods. These tests were interesting since they used models at different scales and especially a very large model representing the whole scheme (Fig. 7). The main results are provided in the paper "P.K weirs under design and construction in Vietnam" (Ho Ta Khanh et al., 2011) and confirm those of the former HCMUT tests (Truong Chi Hien et al., 2006), with some additional results. The figure 8 shows the specific discharge versus the nappe depth for a free overflow P.K weir (n ≈ 5) [upper curve], for the fully submerged Van Phong P.K weir (H > 2 m) [2 medium curves showing the range of variation of the measures] and for a free flow with a straight Creager weir (and approximately for a fully submerged straight Creager or labyrinth weir) [lower curve].

1.3   *Third series of tests*

The third series of tests were carried out in the Hydraulic Laboratory of HCMUT with the P.K weirs type B without downstream overhangs (Fig. 1) in the same conditions than the type A. The increase

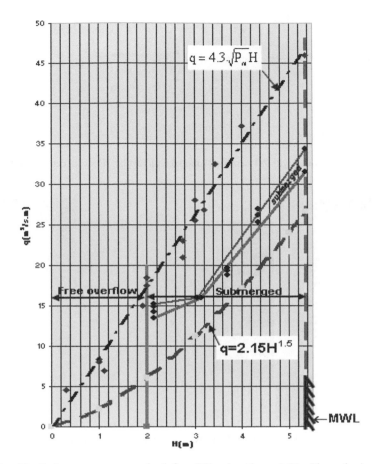

Figure 8.   Specific discharge versus nappe depth for a P.K weir with n ≈ 5 (Van Phong dam).

of the pump capacity has allowed measuring the specific discharges <28 m³/s.m. The capacities of the 2 types of P.K weirs are very similar in our two series of tests (F. Lemperiere and A. Ouamane have noted a little improvement in their type B tests) and we think for the moment that the choice between them must be based more on consideration of implementation of the works, structural analysis and method of construction than on hydraulic performance (Truong Chi Hien et al., 2006).

2   COMBINATION OF P.K WEIRS AND STEPPED SPILLWAY

2.1   *Introduction*

The P.K weirs allow evacuating a relative high discharge, with a short spillway width and a low overflow depth, compared with a traditional Creager weir. They can also be easily placed on the crest of a gravity dam. Stepped spillways are often used in RCC gravity dams. The main advantages are a good adaptation between the structure of the spillway and the method of construction of the RCC dam and, more importantly, the possibility to dissipate on the steps a significant part of the water energy permitting a reduction of the size of the stilling basin. It seems therefore interesting to study in laboratory a combination of these two devices to appraise its advantages and drawbacks and eventually to optimize its use. The tests (Ho Ta Khanh et al., 2010) were carried out in the Hydraulic Laboratory of the HCMUT with the measurement of the scour at the toe of the dam for the following cases: straight Creager weir and P.K weirs type B combined with smooth, 2D-steps and 3D-steps spillway, with different values of discharges and downstream water levels.

Figure 9. Model PKW type B with 2D steps (left), model with regular 3D steps (center) and model with pier baffles (right).

Figure 10. PKW type B + 2D steps (left), PKW type B + 2D steps (center) and Creager weir + 3D steps (right).

The figure 9, left photo, shows the model for a combination of a P.K weir type B and a 2D-stepped spillway. The central and right pictures of figure 9 show the models for two different types of 3D-stepped spillway.

The purposes of the tests, scale 1/75 with a sand bed model, were:

- to establish the specific discharge versus H of a P.K weir type B, for high values of q (but with q < 45 m³/s.m) thanks to the scale reduction,
- to observe the aeration of the overflow on the downstream face of the dam (smooth, 2D-stepped and 3D-stepped),
- to measure directly the scour at the toe of the dam, since the calculation of the residual energy in the stilling basin is not easy for a mix of air and water.

The pictures of figure 10 show the overflow for a P.K weir with a 2D-stepped spillway and a high specific discharge: the aeration of the nappe appears just downstream the weir. The picture on the right of figure 10 shows the overflow for a Creager weir with a 3D-stepped spillway and a high specific discharge: the aeration of the nappe appears only after several steps (the limit is sometimes called the point of inception).

2.2 Main results

The main results of these tests are:

− For medium and high specific discharges, $q \approx K\,H\,\sqrt{Pm}$, for $n \approx 5$. K varies between 4 and 4.3. This empirical simple equation, provided by Lempérière (Lempérière and al., 2003) following the results of the Chatou Laboratory, may be a first approximation with K = 4.15 (our results

196

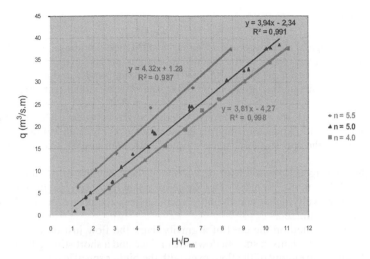

Figure 11.   q (m³/s.m) versus $H\sqrt{P_m}$ (H and P_m in m) for different values of n.

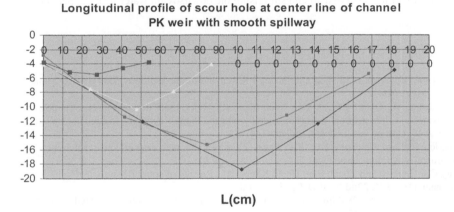

Figure 12.

are a bit lower, K = 4, but probably due to the use of a narrow channel for the model with the side effect, figure 11).

– The flows just downstream the P.K weirs are always well aerated (Fig. 10). For a low and medium dam height (maybe lower than approximately 50 m), the aeration is maintained up to the toe of the dam, even with a smooth downstream face, and the scour is well reduced compared with a Creager weir (an example of scour measurement for different values of q is given in the figure 12). If the dam is high, the aeration by the P.K weirs disappears at the bottom of the dam if the downstream face is smooth. With a stepped spillway – on the entire height or only on the lower part – this aeration can be maintained up to the dam toe. With a good aeration of the outflow, the length of the stilling basin can be reduced and its efficiency is furthermore improved with an end sill in order to highen the downstream water level: the tests show that the scour in the stilling basin decreases in proportion of the water level rising.

– Compared with the 2D-stepped, the 3D-stepped spillway improves a little the aeration but the results concerning the scour are not very modified. Taken into account the more complicated shapes of the steps and also the risk of damaging the baffles by the possible floating debris and logs, it seems that this kind of 3D-stepped spillway is not to be recommended for the moment. However some tests carried out in South Africa with a Creager weir (Wright, 2010) suggest that the addition of triangular protrusions on the steps can be effective to increase the energy dissipation of the overflow.

3 CONCLUSION

For a preliminary design of a P.K weir and with the standard model proposed by HydroCoop (with $n = 5$) (ICOLD, 2010), the results of the specific discharges versus H ($q = 6.4\ H^{1.28}$ for $H < 2\ m$ and $q = 4.15H\sqrt{P_m}$ for $H > 2\ m$) allows a first sizing of the structure and the calculation of the flood routing. Further detailed studies (shape optimization) and model tests may improve a bit these results.

For low nappe depths H, the ratio of the specific discharges between a P.K weir and a Creager weir is superior to 3 and depends mainly on the parameter n. For high values of H and a free overflow, the ratio of the specific discharges between a P.K weir and a Creager weir is generally superior to 2.

For a highly submerged flow in a river or a channel, the reduction of the discharge for a P.K weir is about 30%, but its use, in place of a Creager or a labyrinth weir, is still interesting as this discharge is comparatively higher and the difference between the upstream and the downstream water levels is then minimized.

With a P.K weir installed on the crest of a gravity dam, the flow just downstream the weir is very well aerated. For low dams, a smooth downstream face and a short stilling basin are generally sufficient to dissipate the energy of the flow, even with the higher specific discharges. For medium and high dams, a stepped downstream face, on the entire height or only on the lower part, combined with a short stilling basin and an end sill raising the downstream water level, appears to be a good solution.

REFERENCES

Ho Ta Khanh, M., Truong Chi Hien, Combining P.K weir and stepped spillway, 78th ICOLD Annual Meeting, Hanoi, May 2010.
Ho Ta Khanh, M., Dinh Sy Quat, Dau Xuan Thuy, P.K weirs under design and construction in Vietnam, February 2011, International Workshop on Labyrinth and Piano Key Weirs, Rotterdam: Balkema.
ICOLD. 2010. Cost savings in dams, Bulletin 144.
Lempérière, F. and Ouamane, A., The Piano Keys weir: a new cost-effective solution for spillways, Hydropower & Dams, Issue Five, 2003.
Truong Chi Hien, Huynh Thanh Son, Ho Ta Khanh, M., Results of some P.K weir hydraulic model tests in Vietnam, Q87, R36, 22nd ICOLD, Brasilia, 2006.
Tullis, J.P., Amanian, N., Waldron, D., Design of Labyrinth Spillways, Journal of Hydraulics Engineering, Vol. 121, N° 3, March 1995.
Wright, H.-J., Improved energy dissipation on stepped spillways with the addition of triangular protrusions, 78th ICOLD Annual Meeting, Hanoi, May 2010.
Yildiz, D. and Uzücek, E., Modeling the performance of labyrinth spillways, Hydropower & Dams, Issue Three, 1996.

Labyrinth and Piano Key Weirs – PKW 2011 – Erpicum et al. (eds)
© 2011 Taylor & Francis Group, London, ISBN 978-0-415-68282-4

Piano Key Weir preliminary design method – Application to a new dam project

O. Machiels*, S. Erpicum, P. Archambeau, B. Dewals & M. Pirotton

Laboratory of Hydrology, Applied Hydrodynamics and Hydraulic Constructions, ArGEnCo Department, Liège University, Liège, Belgium
*Fund for education to Industrial and Agricultural Research, F.R.I.A.

ABSTRACT: Within the framework of development of PKW use for rehabilitation as well as for new dam projects, the development of efficient design methods is essential. Currently, the design of PKW projects is done by mean of scale modeling, on the basis of a geometry initially designed based on the general knowledge issued from idealized experimental studies and former PKW buildings. This paper presents a systematic method for PKW preliminary design. The method is based on extrapolation of existing experimental results to develop one or several available project designs, regarding the project constraints (discharge, reservoir levels, available space, ...). The model permits then the optimization of the design depending on technical and economical interests (increase of the safety, decrease of the maximal head, limiting the weir dimensions, limiting construction costs, ...). The application of the method to a new large dam project is presented, highlighting the interest of the method for the optimization of the design depending on technical and economical interests.

1 INTRODUCTION

The Piano Key Weir (PKW) is a particular form of labyrinth weir, initially developed by Lempérière (Blanc & Lempérière 2001, Lempérière & Ouamane 2003), using up- and/or downstream overhangs to limit its basis length and permit its use directly on dam crest (Fig. 1). The PKW is a cost effective solution for rehabilitation but also for new dam projects with either low space for spillway building or reservoir segment available to release a large design discharge. The first scale model studies showed that this new type of weir can be four times more efficient than a traditional ogee-crested weir at constant head and crest length on the dam (Ouamane & Lempérière 2006).

The geometric specificities of the PKW involve a large set of parameters increasing the difficulty of a systematic optimization of the design. The "PKW-unit" can be defined as the basic structure of a PKW, composed of two transversal walls, an inlet and two half-outlets. The main geometric parameters of a PKW are the weir height P, the weir width W, the number of PKW-units N_u, the basis length B_b, the inlet and the outlet widths W_i and W_o, the up- and downstream overhang lengths B_o and B_i, and the wall thickness T (Fig. 1).

Even if the first prototype size PKW were built by "Electricité de France (EDF)" in France since 2006 (Laugier 2007, Bieri, et al. 2009, Laugier, et al. 2009, Leite Ribeiro, et al. 2009), the definition of the optimal geometry of the structure has been still poorly approached. Until now, the hydraulic design of a PKW is mainly performed on the basis of experimental knowledge and scale model studies, modifying step by step an initial geometry following the ideas of the project engineers (Leite Ribeiro, et al. 2007, Cicero, et al. 2010), without systematic study of the efficiency of the initially designed geometry.

In order to improve the efficiency of the PKW design, a systematic method is presented in this paper (Section 2). The method is based on an extrapolation of existing experimental results to develop one or several available project designs, regarding the project limitations (discharge, reservoir levels, available space, ...). The optimization of the design is then realized depending on technical and economical interests.

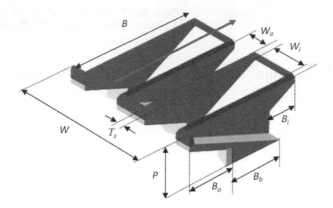

Figure 1. 3D sketch of two PKW-units and main geometric parameters (Pralong, et al. 2011).

To illustrate the method, its application to a PKW design for a new large dam project is presented in section 3. This application highlights the main interest of the model, permitting an optimization of the PKW geometry based on various project constraints.

2 PRELIMINARY DESIGN METHOD

The proposed design method is based on an extrapolation of existing experimental results from a reference scale model to respond to the project constraints (discharge, reservoir levels, available space, ...). The final design is after chosen among the different hydraulically and geometrically acceptable possibilities, depending on secondary technical and economical interests (increase of the safety, increase of the reservoir capacity, decrease of the structure dimensions, ...).

The basis elements of the method are categorized in project elements (single notations) and reference model elements (notations with *). The hydraulic and geometric specificities of the project are necessary to define the constraints, and the release capacity curve of the reference model, as well as its geometric characteristics, are necessary to define the scaling coefficient and calculate the performance of the project design.

The first step of the method consists in defining the different acceptable PKW solutions as a function of the number of PKW-units in the structure, by scaling up the geometric and hydraulic parameters of a selected reference model.

The PKW-unit width W_u is defined as a function of the number of PKW-unit N_u and the available width for the project W:

$$W_u = \frac{W}{N_u} \tag{1}$$

The scale of the project solution x is than defined as the ratio between the widths of the PKW-units on the project design W_u and on the reference model W_u^*:

$$x = \frac{W_u}{W_u^*} \tag{2}$$

Applying this scaling coefficient to the design head H, the corresponding head on the reference model H^* is:

$$H^* = \frac{H}{x} \tag{3}$$

The discharge coefficient C_{dW} of the Poleni equation related to PKW total width (Eqn (4)) is a non-dimensional number, assumed to suffer no scaling effect. The C_{dW} value of the project design

Table 1. Project dam and spillway characteristics.

Type	Gravity dam
Available crest length	100 m
Maximal head	4.5 m
Maximal overhangs length	5 m

for the design head H is thus equal to the C_{dW}^* value of the reference model at the corresponding head H^*:

$$Q = C_{dW} W \sqrt{2gH^3} \qquad (4)$$

$$C_{dW}(H) = C_{dW}^*(H^*) \qquad (5)$$

where Q is the flow discharge on the complete project structure and g is the gravitational acceleration.

Inserting equations (1) to (3) and (5) in the Poleni equation (4), a head/discharge relation for the design project, depending on the hydraulic and geometric characteristics of the reference model (W_u^* and C_{dW}^*) and the project constraint (W), is obtained for each value of N_u:

$$Q = C_{dW}^* W \sqrt{2gH^3} \quad \text{with} \quad C_{dW}^* = f\left(\frac{HW_u^* N_u}{W}\right) \qquad (6)$$

This relation enables to compute the head/discharge curve of the PKW for each value of N_u.

If the discharge coefficient of the reference model is not known for the reference model initial heads, its value is calculated by interpolation between existing values or extrapolation of the reference model curve. The accuracy of the design method is thus directly linked to the experimental tests accuracy and to the number of available results for the reference model.

Limiting the project head/discharge curves defined by equation (6) under the design head and over the design discharge, zero, one or several acceptable designs may exist. If more than one design respond to the project constraints, the second step of the proposed method is to optimize remaining parameters depending on secondary technical and economical interests of the project engineer to identify the optimum solution.

By application of the scaling coefficient to the reference model dimensions X^*, the project model dimensions X are completely defined for each acceptable solution, permitting the optimization of the final design including structural, economical or hydraulic criteria:

$$X = xX^* \qquad (7)$$

The last step of a complete design is to validate the preliminary design by exploitation of a scale model representing the complete project integration. The scale model so enables to taking account project specificities and local effects (piers, abutments, . . .).

3 APPLICATION

To illustrate the proposed design method, it has been applied to a new PKW project on a large gravity dam, whose characteristics are summarized in Table 1. Considering preliminary flood mitigation calculations in the reservoir, the design discharge should be around 4000 m³/s, what represents a specific discharge of 40 m³/s/m to be evacuated for a maximal head of 4.5 m. Traditional Creager weirs are unable to achieve this goal. Due to the restricted place and the necessary placement of the weir on the dam crest, the PKW seems to be a good solution to ensure the dam safety.

For building considerations, the overhangs length is limited to 5 m by the project engineers. The overhangs length is a common constraint for PKW projects. Building considerations, as the limit length of crane, structural considerations, as the quantity of steel necessary to the concrete

Table 2. Geometric characteristics of the reference models.

	Ho Chi Minh	Biskra	Liege
W_u^* [m]	0.20	0.165	0.2
P^* [m]	0.22	0.155	0.2
W_i^* [m]	0.12	0.09	0.12
W_o^* [m]	0.08	0.075	0.08
B_i^* [m]	0.15	0.103	0
B_o^* [m]	0.15	0.103	0.25
B^* [m]	0.60	0.412	0.5

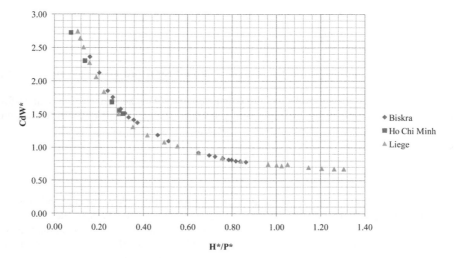

Figure 2. Adimensional head/discharge curves of reference models.

reinforcement, and stability considerations, as the response of the structure to seism, may limit the overhangs length.

The first step of the PKW design is to choose a reference model. Regarding the existing experimental results in the literature, 3 models seem to be suitable for an efficient design: the model B from Ho Chi Minh (Hien, et al. 2006), a model from Biskra (Ouamane 2006) and the model 2 from Liege (Machiels, et al. 2010). The geometric characteristics of these 3 models are given in Table 2 and the experimental results are summarized on Figure 2 in terms of discharge coefficient C_{dW}^* function of the adimensional ratio between the water head H^* and the weir height P^*.

The limited number of results for the Ho Chi Minh model should decrease the accuracy of the design method. The two other models show similar efficiency, but the one from Liege uses only upstream overhangs longer than those on the Biskra model. Due to the constraint on the overhangs length, the model from Biskra is chosen as the reference model. The influence of the choice of the reference model is discussed in section 4, confirming the performance of the Biskra model compared to Liege and Ho Chi Minh models.

Considering a 10 units PKW, the width of each unit is 10 m (Eqn (1)), what represents a scale ratio of 60.6 between the project and the reference model (Eqn (2)). The corresponding design head on the reference model is equal to 0.074 m (Eqn (3)). By interpolation in-between the experimental results, the discharge coefficient of the PKW for the design head is equal to 1.16 (Eqn (5)). The project characteristics, calculated considering 10 to 17 PKW-units, are represented in Table 3. Applying the Eqn (3) to (6) for varying values of the upstream head H, the head discharge curves could be drawn for each value of N_u (Fig. 3).

A PKW with 10 units placed on the crest of the projected dam enables to discharge 4900 m³/s (Eqn (6)). However, the overhangs length of this PKW is 6.24 m (Eqn (7)). A 10 units PKW is unable to respect the project constraints (not enough discharge capacity).

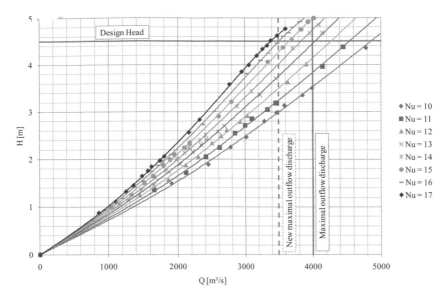

Figure 3. Project head/discharge curves for various N_u values with the Biskra model as reference.

Table 3. Project characteristics for various N_u values.

	10	11	12	13	14	15	16	17
W_u [m]	10.00	9.09	8.33	7.69	7.14	6.67	6.25	5.88
P [m]	9.39	8.54	7.83	7.23	6.71	6.26	5.87	5.53
W_i [m]	5.45	4.96	4.55	4.20	3.90	3.64	3.41	3.21
W_o [m]	4.55	4.13	3.79	3.50	3.25	3.03	2.84	2.67
$B_i = B_o$ [m]	6.24	5.67	5.20	4.80	4.46	4.16	3.90	3.67
B [m]	24.97	22.70	20.81	19.21	17.84	16.65	15.61	14.69
C_{dW}^1 [–]	1.16	1.08	1.02	0.96	0.90	0.87	0.83	0.80
Q^1 [m³/s]	4900	4562	4300	4038	3820	3673	3508	3371
V_c [m³]	966	878	805	743	690	644	604	568

^{1} for $H = 4.5$ m

Regarding project constraints, the single acceptable solution is the one considering 13 PKW-units (For $N_u > 13$, the volume capacity is not enough, and for $N_u < 13$, the overhangs are too long). However, the design discharge of 4000 m³/s has been calculated considering the flood mitigation provided with a weir working as an ogee crested weir. The PKW discharge for low heads is significantly higher than ogee-crested weir discharge, what improves the flood mitigation. As the proposed method provides the complete head/discharge curves for all studied geometries, the spillway design discharge could be reevaluated considering the new expected flood mitigation. Regarding the curve provided for the 16 units PKW, the design discharge is reduced to 3500 m³/s.

Many solutions exist to ensure the discharge of 3500 m³/s under a head of 4.5 m ($N_u \leq 16$). However, solutions with less than 13 PKW units involve overhangs longer than 5 m, what is over the technical constraint of the project. Four solutions, with N_u between 13 and 16, enable thus to ensure the safety of the dam, respecting the constraints of reservoir level and overhangs length.

The final optimization of the design depends on the project engineer's interests. Considering the direct link between the weir cost and the material quantities necessary to its building, the economical optimum can be defined based on the volume of concrete in the structure V_c (Table 3). Regarding construction costs, the solution with 16 PKW-units seems thus to be the best design.

However, economical optimum is not only dependent of the construction costs. The limitation of the weir height, also involving a 16 units PKW design, enables the reservoir filling earlier during

203

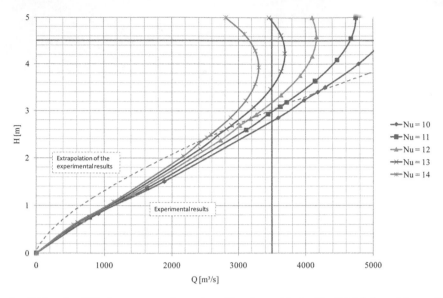

Figure 4. Project head/discharge curves for various N_u values with the Ho Chi Minh model as reference (Dotted line – separation between interpolation in-between the available reference results and extrapolation over the available reference results).

the weir construction, what limit the total building time and thus the project costs. Increasing the normal reservoir level may also involve economical interests (increasing the head for hydroelectric plan by example). Increasing the reservoir level of 0.5 m limits the acceptable designs to those with 13 or 14 PKW-units. Increasing the weir discharge capacity, by reducing the number of PKW-unit, enables mitigation of floods with higher occurrence what may reduce damage costs of extreme floods.

The optimal design could thus be the design responding to technical ($W = 100$ m and $B_{o\,max} = B_{i\,max} = 5$ m) and hydraulic constraints ($Q = 3500$ m^3/s, $H_{max} \leq$ Maximal reservoir level) and assuring economical interests (Minimal V_c value with $P_{max} = 7$ m and normal reservoir level increased by 0.5 m). In this case the final design considers 13 PKW-units.

4 INFLUENCE OF THE REFERENCE MODEL

As mentioned before, the accuracy of the proposed design method lies on the experimental tests accuracy and on the number of available results for the reference model. The choice of the reference model is thus of high importance. In the application presented before, the Biskra model has been favored to Ho Chi minh and Liege ones.

Regarding the head/discharge curves given by application of Eqn (3) to (6) with the Ho Chi Minh model as reference (Figure 4), a non physical phenomenon appears for heads approaching 4.5 m. For this range of head, the weir discharge begins to decrease with increasing heads. This phenomenon results from the extrapolation of the available experimental C_{dW}^* values for heads till 2 times higher than the maximal corresponding head studied on the reference model. These results highlight the importance of choosing a reference model with a wide range of available results near the corresponding design head.

Regarding now the head/discharge curves given by application of Eqn (3) to (6) with the Liege model as reference (Figure 5), more geometries than with using the Biskra model are able to release a discharge of 3500 m^3/s under a head of 4.5 m (Table 4). However, even with the 17 unit PKW, which is the solution with the smallest dimensions, the upstream overhang length is equal to 7.35 m. Using the Liege model as reference provides thus no available solutions. The choice of the reference model must be consistent with the project constraints. Considering a project constraint on the overhangs length favors a reference model with symmetric overhangs.

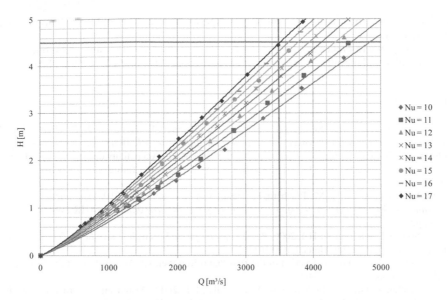

Figure 5. Project head/discharge curves for various N_u values with the Liege model as reference.

Table 4. Project characteristics for various N_u values with the Liege model as reference.

	10	11	12	13	14	15	16	17
W_u [m]	10.00	9.09	8.33	7.69	7.14	6.67	6.25	5.88
P [m]	10.00	9.09	8.33	7.69	7.14	6.67	6.25	5.88
W_i [m]	6.00	5.45	5.00	4.62	4.29	4.00	3.75	3.53
W_o [m]	4.00	3.64	3.33	3.08	2.86	2.67	2.50	2.35
B_o [m]	12.50	11.36	10.42	9.62	8.93	8.33	7.81	7.35
B [m]	25.00	22.73	20.83	19.23	17.86	16.67	15.63	14.71
C_{dW}^1 [–]	1.14	1.08	1.03	0.99	0.94	0.90	0.87	0.84
Q^1 [m³/s]	4808	4548	4366	4168	3965	3802	3666	3536
V_c [m³]	1033	939	860	794	738	688	645	607

[1] for $H = 4.5$ m

5 CONCLUSION

To help project engineers in the research of optimal design solutions for PKW projects, a systematic method has been presented. Based on existing experimental results, the PKW design is realized by scaling up the different geometric and hydraulic parameters of a reference model, to meet as well as possible the project constraints.

The application of the method to a new large dam project highlights its interests for project needing to evacuate important specific discharge under limited head. The project head/discharge curves, calculated based on the reference elements, enable a closer calculation of the exact flood mitigation realized by the PKW use. The solutions ensuring the safety of the dam are thus highlighted.

Furthermore, the interest of the method for the choice in-between different acceptable solutions is also highlighted. The exploitation of the dimensions and head/discharge curves of the different acceptable designs, given as a result of the method, permits a quick and efficient optimization of the final design based on secondary technical and economical criteria defined by the project engineers.

The accuracy of the method is mainly function of the accuracy of the reference model data. The choice of the reference model should be made regarding available experimental results and project constraints. The insertion in this method of new experimental results would enable to improve the design efficiency.

In a next step, the reference model analysis could be completed with numerical modeling, more and more reliable and permitting a separate optimization of each parameters of the PKW geometry.

REFERENCES

Bieri, M., Leite Ribeiro, M., Boillat, J.-L., Schleiss, A., Laugier, F., Delorme, F. & Villard, J.-F. 2009. *Réhabilitation de la capacité d'évacuation des crues: Intégration de "PK-Weirs" sur des barrages existants.* Colloque CFBR-SHF: "Dimensionnement et fonctionnement des évacuateurs de crues". Paris, France.

Blanc, P. & Lempérière, F. 2001. Labyrinth spillways have a promising future. *Int. J. Hydro. Dams* 8 (4): 129–131.

Cicero, G.-M., Guene, C., Luck, M., Pinchard, T., Lochu, A. & Brousse, P.-H. 2010. *Experimental optimization of a Piano Key Weir to increase the spillway capacity of the Malarce dam.* 1st IAHR European Congress, Edinburgh.

Hien, T.C., Son, H.T. & Khanh, M.H.T. 2006. *Results of some piano keys weir hydraulic model tests in Vietnam.* 22nd ICOLD congress, Barcelona: CIGB/ICOLD.

Laugier, F. 2007. Design and construction of the first Piano Key Weir (PKW) spillway at the Goulours dam. *Int. J. Hydro. Dams* 14 (5): 94–101.

Laugier, F., Lochu, A., Gille, C., Leite Ribeiro, M. & Boillat, J.-L. 2009. Design and construction of a labyrinth PKW spillway at Saint-Marc dam, France. *Hydropower & Dams* 16 (5): 100–107.

Leite Ribeiro, M., Albalat, C., Boillat, J.-L., Schleiss, A.J. & Laugier, F. 2007. *Rehabilitation of St-Marc dam. Experimental optimization of a piano key weir.* 32th IAHR Congress, Venice, Italy.

Leite Ribeiro, M., Bieri, M., Boillat, J.-L., Schleiss, A.J., Delorme, F. & Laugier, F. 2009. *Hydraulic capacity improvement of existing spillways – Design of Piano Key Weirs.* 23rd congress of CIGB/ICOLD, Brasilia.

Lempérière, F. & Ouamane, A. 2003. The piano keys weir: a new cost-effective solution for spillways. *Int. J. Hydro. Dams* 10 (5): 144–149.

Machiels, O., Erpicum, S., Archambeau, P., Dewals, B. & Pirotton, M. 2010. Analyse expérimentale de l'influence des largeurs d'alvéoles sur la débitance des déversoirs en touches de piano. *La Houille Blanche* (2): 22–28.

Ouamane, A. 2006. *Hydraulic and costs data for various Labyrinth Weirs.* 22ème congrès des grands barrages. Barcelona: CIGB/ICOLD.

Ouamane, A. & Lempérière, F. 2006. *Design of a new economic shape of weir.* International Symposium on Dams in the Societies of the 21st Century, Barcelona, Spain: 463–470.

Pralong, J., Blancher, B., Laugier, F., Machiels, O., Erpicum, S., Pirotton, M., Leite Ribeiro, M., Boillat. J-L., & Schleiss, A.J. 2011. Proposal of a naming convention for the Piano Key Weir geometrical parameters. *International Workshop on Labyrinth and Piano Key weirs,* Liège, Belgium.

Labyrinth and Piano Key Weirs – PKW 2011 – Erpicum et al. (eds)
© 2011 Taylor & Francis Group, London, ISBN 978-0-415-68282-4

Method to design a PK-Weir with a shape and hydraulic performances

G.M. Cicero

National Hydraulic and Environmental Laboratory, Chatou, France

ABSTRACT: PK-Weirs are a new shape of labyrinth spillway often studied by laboratories since 2000. Until now, there is no general correlations to estimate the hydraulic performances of a PK-Weir according to its numerous geometrical parameters but many experimental results have been published for different shapes of PK-Weir. This article presents a simple method to design a PK-Weir for a given shape and hydraulic performances. Characterizing the PK-Weir geometry by six main parameters (inlet and outlet widths, lateral length, sidewall thickness, spillway width and weir height), the method shows how to calculate the scale factor and the number of PKW units (½ inlet + ½ outlet) to fit the project data (discharge at maximum water head and spillway width). Then we applied this method to design three different shapes of PK-Weir with the same project data.

1 INTRODUCTION

Pk-Weirs are a new shape of labyrinth spillway (Fig.1) often studied by laboratories since 2000. It has been used on various EDF Rehabilitation projects such as Goulours (Laugier 2007), St-Marc and Gloriettes (Bieri et al. 2009), and Malarce (Cicero et al. 2010) to increase the spillway flow capacity of these dams. All these projects needed to design a pk-weir fitting the original project data (discharge at maximum water level, spillway width and crest level).

The hydraulic performances of a PK-Weir depends on a large set of geometrical parameters, and experimental studies have been or are currently carried out in different laboratories to characterize their influence on its discharge capacity (Ouamane & Lemperiere 2006, Chi Hien et al. 2006, Le Doucen et al. 2009, Machiels et al. 2009). These experiments highlighted important qualitative results. The discharge coefficient increases with the relative length (crest length to spillway width ratio) and is homogenized by the head to weir height ratio. The optimal value of the width ratio of inlet and outlet keys depends on the upstream head.

Until now, there is no general correlation to estimate the discharge capacity according to the main geometrical parameters. Nevertheless, many experimental results of hydraulic performances have been published for different shapes of PK-Weir.

This article presents a simple method to design a PK-Weir for a given shape and hydraulic performances. The method uses six main geometrical parameters (widths of inlet and outlet keys, lateral length, weir height, sidewall thickness and spillway width) to calculate the scale factor, the number of PKW units and the discharge capacity. This method is applied to design three different shapes of PK-Weir with the same project data.

2 DESCRIPTION OF THE METHOD

2.1 *Definitions and needed data*

The objective is to design a PK-Weir (so called project) fitting with the following data: the discharge Q_P at the maximum water head H_P and the spillway width W_P. We propose to use a shape of PK-Weir (so called reference) of which the hydraulic performances have been measured experimentally.

The PK-Weir geometry (Fig. 1) is characterized by six main parameters: the spillway width (W), the widths of inlet (W_i) and of outlet (W_o) keys, the lateral length (B), the sidewall thickness (T_s)

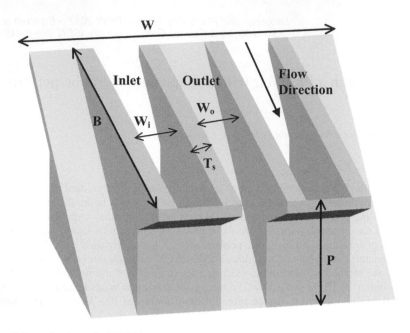

Figure 1. Schematic view of a PK-Weir.

and the weir height (P). A PKW unit is composed of a sidewall between half an inlet and half an outlet so that the number of units is:

$$N_u = \frac{2W}{W_i + W_o + 2T_s} \tag{1}$$

The scale factor e between the project and the reference PK-Weir is defined by:

$$e = \frac{W_{iP}}{W_{iR}} = \frac{W_{oP}}{W_{iR}} = \frac{P_P}{P_R} = \frac{B_P}{B_R} \tag{2}$$

Note that the sidewall thickness of the project PK-Weir T_{sP} is a needed data designed with respect to material resistance criteria.

The rating curves $Q(H)$ or $C_{dW}(H)$ measured on the reference PK-Weir are also needed data, where C_{dW} is the classical discharge coefficient related to the spillway width W:

$$C_{dW} = \frac{Q}{W\sqrt{2g}H^{1.5}} \tag{3}$$

where H is the upstream head and g the gravity acceleration.

2.2 *Methodology*

First the discharge coefficient C_P is calculated with the project data by:

$$C_P = \frac{Q_P}{W_P\sqrt{2g}H_P^{1.5}} \tag{4}$$

The curve $C_{dW}(H)$ of the reference PK-Weir gives the head H_R corresponding to C_R, the closest value of the project discharge coefficient C_P. The scale factor (e^*) and the number of PKW units

$(N_u{}^*)$ of the project PK-Weir can be estimated by:

$$e^* = \frac{H_P}{H_R}$$ (5)

$$N_u{}^* = \frac{2W_P}{W_{iP} + W_{oP} + 2T_{sP}} = \frac{2W_P}{e^*(W_{iR} + W_{oR}) + 2T_{sP}}$$ (6)

Then an arithmetic value of N_u is chosen to correct the scale factor e by:

$$e = \frac{2(W_P - N_u T_{sP})}{N_u(W_{iR} + W_{oR})}$$ (7)

And the discharge curve $Q_P(H)$ of the project PK-Weir is calculated by:

$$Q_P(H) = C_{dW}(H/e)W_P\sqrt{2g}H^{1.5}$$ (8)

The discharge can also be calculated with the non-dimensional curve $C_{dW}(H/P)$, which is identical on the project and the reference PK-Weirs:

$$Q_P(H) = C_{dW}(H/P_P)W_P\sqrt{2g}H^{1.5}$$ (9)

Equations 8 and 9 are valid within the head range ($eHmin, eHmax$), where ($Hmin, Hmax$) is the head range measured on the reference PK-Weir. The curves $Q_P(H)$ are calculated for various N_u values to select the optimal number of PKW units. Then the scale factor is obtained by Equation 7 and the dimensions of the project PK-Weir by Equation 2.

3 APPLICATIONS

3.1 Data of the reference PK-Weirs

The methodology is applied to design 3 different shapes of PK-Weirs with the following project data: $Q_P = 400 \text{ m}^3/\text{s} - H_P = 2 \text{ m}$ and $W_P = 35 \text{ m}$. The 3 chosen shapes are PK-Weir of type A (Ouamane & Lemperiere 2006), with upstream and downstream overhangs.

PK-Weir R1 is the model of "St-Marc I" tested in Switzerland (Bieri et al. 2009). Pk-weir R2 is the model of "Malarce type A" tested in France (Cicero et al. 2010). Pk-weir R3 is the "model B" tested in Vietnam (Chi Hien et al. 2006).

For the sidewall thickness of the project PK-Weir, the same value as the reference PK-Weir has been kept ($T_{sP} = T_{sR}$).

See the main dimensions of the PK-Weirs (Table 1) and the operation domains of the discharge coefficient (Fig. 2).

The discharge coefficient C_{dW} versus the relative head H/P were calculated by regression equations adjusted on the measurements.

For PK-Weir R1 ($P_{R1} = 5.35 \text{ m}$), within the range $0.08 < H/P_{R1} < 0.36$:

$$C_{dW}(\frac{H}{P_{R1}}) = 3.49\left(\frac{H}{P_{R1}}\right)^2 - 4.04\left(\frac{H}{P_{R1}}\right) + 2$$ (10)

For PK-Weir R2 ($P_{R2} = 4.40 \text{ m}$), within the range $0.17 < H/P_{R2} < 0.61$:

$$C_{dW}(\frac{H}{P_{R2}}) = 5.5\left(\frac{H}{P_{R2}}\right)^2 - 7.24\left(\frac{H}{P_{R2}}\right) + 3.46$$ (11)

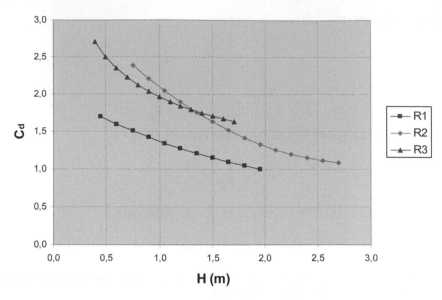

Figure 2.   Discharge coefficient of the reference PK-Weirs.

Table 1.   Dimensions of the reference PK-Weirs.

Reference PK-Weir	Unit	R1	R2	R3
Spillway width	m	15.6	42.5	15
Inlet width	m	3.10	1.25	2.60
Outlet width	m	2.45	1.58	1.60
Lateral length	m	12.1	13.0	15.0
Sidewall thickness	m	0.35	0.40	0.40
Weir height	m	5.35	4.40	5.50

For PK-Weir R3 ($P_{R3} = 5.50$ m), within the range $0.07 < H/P_{R3} < 0.31$:

$$C_{dW}(\frac{H}{P_{R3}}) = 1.0815\left(\frac{H}{P_{R3}}\right)^{-0.35} \tag{12}$$

3.2 Results of applications

With the project data, the discharge coefficient calculated by Equation 4 is $C_P = 0.91$.

For PK-Weir R1, the rating curve (Fig. 2) gives $H_R = 1.95$ m for $C_R = 1$ which is the closest value of $C_P = 0.91$. Then, Equation 5 allows to estimate the scale factor $e^* = 1.03$, and Equation 6 the number of PK-Weir units $N_u^* = 11.60$. For $N_u = 10$, 11 and 12, the scale factor e is calculated by Equation 7, the weir height P_P by Equation 2 and the discharge curve $Q(H)$ of the project PK-Weir by Equations 9 and 10. For $N_u > 11$ (Fig. 3), the head project ($H_P = 2$ m) is outside the operation domain. The optimal value is $N_u = 11$ with $Q(H_P) = 455$ m³/s.

For PK-Weir R2, Figure 2 gives $H_R = 2.7$ m for $C_R = 1.09$, then $e^* = 0.74$ and $N_u^* = 24.17$. For $N_u = 23$, 24 and 25, the discharge curves $Q(H)$ calculated by Equations 9 and 11 gives the optimal value $N_u = 24$ with $Q(H_P) = 479$ m³/s.

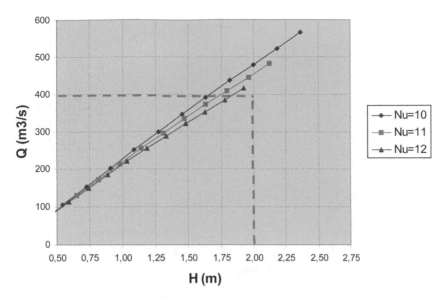

Figure 3.   Effect of the number of PK-Weir units on the discharge capacity of the project PK-Weir R1.

Table 2.   Dimensions of the project PK-Weirs.

Project PK-Weir	Unit	R1	R2	R3
Units		11	24	12
Spillway width	m	35	35	35
Inlet width	m	3.38	0.93	3.12
Outlet width	m	2.28	1.18	1.92
Lateral length	m	13.18	9.72	17.98
Sidewall thickness	m	0.35	0.40	0.40
Weir height	m	5.83	3.29	6.59

For PK-Weir R3, Figure 2 gives $H_R = 1.7$ m for $C_R = 1.63$ which is much higher than the project coefficient $C_P = 0,91$. Then $e^* = 1.18$ and $N_u^* = 12.19$. For $n = 11$, 12 and 13, the discharge curves $Q(H)$ calculated by Equations 9 and 12 gives the optimal value $N_u = 12$ with $Q(H_P) = 720$ m^3/s.

Table 2 presents the dimensions of the project PK-Weirs designed by this method with the chosen reference PK-Weirs.

Figure 4 shows the discharge curves of the project PK-Weir predicted with each reference. With PK-Weir R3, the discharge at the head project ($H_P = 2$ m) is strongly overestimated because the range of C_{dW} measurements (Fig. 2) are much greater than the project discharge coefficient ($C_P = 0.91$).

Since the PK-Weir dimensions of Table 2 depend on the data of the sidewall thickness, it can be necessary to adjust the thickness value to these dimensions, and apply again the method to correct the PK-Weir dimensions. The sidewall thickness must be calculated independently with respect to material resistance criteria.

4   CONCLUSIONS

A method has been proposed to design a PK-Weir fitting with the project data (discharge at Maximum Water Head and spillway width) with a shape of PK-Weir when the rating curve has been measured. This method has been applied with 3 different shapes of PK-Weirs with the same project data.

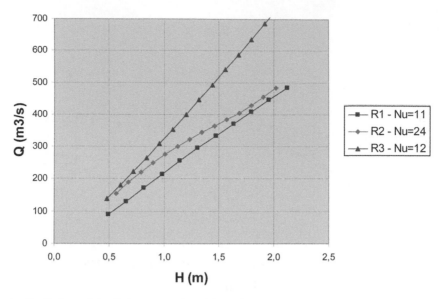

Figure 4.   Predictions of the discharge capacity of the project PK-Weirs.

The project data and the measurements of the discharge coefficient of the reference PK-Weir allows to estimate the scale factor and the number of PK-weir units. Then the number of PK-Weir units is optimized to correct the scale factor and calculate the dimensions and the discharge capacity of the project PK-Weir.

The results of the method depend on the data of the sidewall thickness, which must be designed separately with respect to material resistance criteria. Thus, it could be necessary to correct the thickness value and the PK-Weir dimensions by an iterative process.

It is recommended to choose a shape of PK-Weir where the measurements of the discharge coefficient are close to the project discharge coefficient.

REFERENCES

Bieri, M., Leite Ribero, M., Boillat, J.L. & Schleiss, A. 2009. Etude paramétrique de la capacité des pk-weirs. *Colloque SHF: Modèles physiques hydrauliques, Lyon, 24–25 novembre.*

Cicero, G.M., Guene, C., Luck, M., Pinchard, T., Lochu, A. & Brousse, P.H. 2010. Experimental optimization of a piano key weir to increase the spillway capacity of the Malarce dam. *1rst IAHR European congress, Edinbourgh, 04–06 mai.*

Laugier, F. 2007. Design and construction of the first Piano Kew Weir (PKW) spillway at the Goulours dam. *Hydropower & Dams,* issue 5: 94–101.

Le Doucen, O., Leite Ribero, M., Boillat, J.L., Schleiss, A., Laugier, F., Lochu, A., Delorme & Villard, J.F. 2009. Réhabilitation de la capacité d'évacuation des crues- Intégration de pk-weirs sur des barrages existants. *Colloque CFBR-SHF: Dimensionnement et fonctionnement des évacuateurs de crues, Paris, 20–21 janvier.*

Machiels, O., Erpicum, S., Archambeau, P., Dewals, B. & Pirotton, M. Analyse expérimentale du fonctionnement des déversoirs en touche de piano. *Colloque CFBR-SHF: Dimensionnement et fonctionnement des évacuateurs de crues, Paris, 20–21 janvier.*

Ouamane, A. & Lemperiere, F. 2006. Design of a new economic shape of weir. *Proceedings of the International Symposium on Dams in the Societies of the 21st Century, Barcelona.*

Chi Hien, T., Thanh Son, H. & Ho Ta Khanh, M. 2006. Results of some piano key weir hydraulic model tests in Vietnam. *22ème congrès des grands barrages, Barcelona.*

Planned and existing projects

Labyrinth and Piano Key Weirs – PKW 2011 – Erpicum et al. (eds)
© 2011 Taylor & Francis Group, London, ISBN 978-0-415-68282-4

Lessons learnt from design and construction of EDF first Piano Key Weirs

J. Vermeulen, F. Laugier, L. Faramond & C. Gille
EDF Hydro Engineering Center, Le Bourget du Lac, France

ABSTRACT: Four piano key weirs (PKW) have already been built by EDF on its existing dams from 2006 to 2010: Goulours dam on the Lauze river (East Pyrenees), Saint Marc dam and Etroit dam on the Taurion river (Centre of France) and Gloriettes dam on the Gave d'Estaubé (West Pyrenees). EDF Hydro Engineering center has designed these PKW from pre feasibility stages to the Concept Designs and has subsequently supervised their constructions. Feedbacks have been performed on varied topics related to design and construction in order to enhance the development of PKW. These feedbacks will offer EDF strong savings in design and construction steps in future and ongoing projects.

1 INTRODUCTION

Lempérière & Ouamane (2003) have designed the piano key weirs (PKW) as a "cost effective solution for spillways". The objectives were to design an efficient spillway structure, cheap and easy to build even in developing countries. PKW have effectively proven to be very efficient and cost effective solutions as spillways which is why EDF Hydro Engineering center have taken an interest in PKW very early. With more than 400 dams, 55 years old in average, EDF has to ensure the safety of its installations against floods, which are the main cause for dam's breakdown in the world along with first fulfilling (CIGB Bulletin n°99). Growing hydrological data base, evolution of hydrological prediction methods (Gradex, Schadex), and the will to improve the dams safety lead EDF to increase the discharge capacity of some of its dams. PKW enable discharge coefficients of 0.7 to 2 (related to spillway total width Cdw), low risks with floating trees or debris and without operation constraints during floods. These are parts of the reasons why EDF has already retained PKW solution on four of its recently rehabilitated dams and a dozen to-be-rehabilitated dams. Many PKW will be implemented in the near future. This is the reason why the feedbacks of the firsts designed and built PKW turn out to be essential.

2 DAMS AND PKW DESCRIPTIONS

The first PKW has been built in the Pyrenees in 2006 at Goulours dam. From 2008 to 2010, three other PKW have been built, two in the center of France (St Marc and l'Etroit dams) and one in Pyrenees (Gloriettes dam). All are concrete dams but three are gravity curved dams and one is a gravity straight dam. Their heights vary from 20 to 43 m. In one case, the PKW has been built on a concrete gravity dam beyond the abutment of a slim arch dam.

Table 1 compares the main geometric and hydraulic parameters between the four PKW built. This table is related to the proposal for a naming convention for the PKW (Pralong et al., 2011a).

3 TECHNICAL CONSIDERATIONS

As shown in figure 1, significant improvements have been achieved during the last years regarding PKW hydraulic efficiency partly due to an improvement of the understanding of the hydraulic behaviour.

Table 1. Dams and PKW descriptions.

	Goulours	St Marc	Etroit	Gloriettes
Dam type	Concrete gravity curved	Concrete straight gravity	Concrete gravity curved	Slim arched
Height on ground level (m)	20	40	25	43
Dam's date of construction	1946	1932	1933	1951
PKW's date of construction	2006	2008	2009	2010
PKW localisation	On the lower part of dam near the bank	On the crest dam	On the crest dam	On a gravity wall extending the abutment dam
Dam design flowed (m^3/s)	162	660	501	150
PKW Discharge capacity at MWL (m^3/s)	68	138	82	90
Total project cost (k€)	400	1800	1500	1600
PKW Cost (k€)	300	400	400	400
Head on PKW at MWL (m)	0.95	1.35	0.95	0.8
Time construction (month)	3	6	6	3
Upstream-downstream length B (m)	9.3	12.7	12.2	10
Upstream overhang crest length B_o (m)	3.35	4	3.2	3.5
Downstream overhang crest length B_i (m)	1.5	4	2	2.6
Base length B_b (m)	4.4	4.7	7	3.6
Total height P (m)	3.1	4.2	5.3	3
Total width W (m)	12	18	18.7	18.5
Inlet key width W_i (m)	2.7	3.1	2.45 to 2.75	2.3 to 2.5
Outlet key width W_o (m)	1.5 and 1.8	2.2	1.5	1.5
Sidewall thickness T_s (m)	0.2	0.30	0.35	0.3 to 0.2
Crest thickness at inlet key extremity T_i (m)	0.2	0.25	0.30	0.25
Crest thickness at outlet key extremity T_o (m)	0.2	0.25	0.30	0.25
Total developed length along the overflowing crest axis L (m)	59	77	78	86.8
Height of the outlet parapet wall P_{po} (m)	0	0	0.5	0
Length of the nose B_n (m)	1	1.5	1	1.5
Nose form	Triangular	Triangular	Triangular	Rounded

Total width of the PKW (W), difference between maximum water level (MWL) and normal water level (NWL) or base length (B_b) are major parameters in the PKW efficiency. But in case of rehabilitation of existing dams, these parameters are constrained, which explains the various PKW solutions retained for the four PKW already built. This must be taken into account when comparing the spillways.

3.1 Partial demolition of the existing dam to establish the PKWeir

To limit shockwaves in the existing dams due to explosives, it seems essential to first saw the existing dam with a diamond cable before using blasting devices. Before the blast, the protection of existing structures can be done with simple straw balls.

3.2 Upstream and downstream overhangs

Several authors (Lempérière & Ouamane 2003, Truong Chi Hien et al. 2006, Pralong et al. 2011b) have considered the importance of upstream and downstream overhang crest length (B_o and B_i).

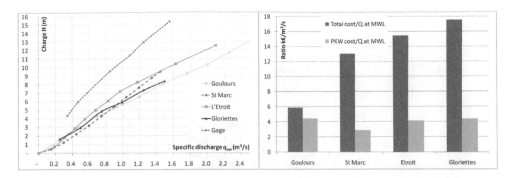

Figure 1. Hydraulic performance (left) and economical ratios (right).

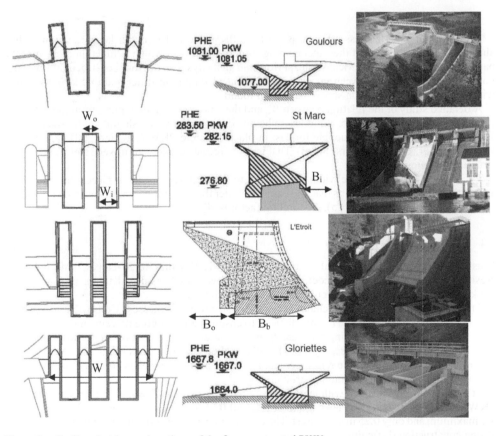

Figure 2. Outline sketches and sections of the four constructed PKW.

It has been rapidly measured that for a given upstream-downstream length (B), PKW with high B_o are slightly more efficient than symmetrical PKW. The better inlet key alimentation with high B_o can explain these results. If PKW with high B_o is more efficient, they also require costly scaffoldings during construction. On the four PKW built by EDF, only Goulours PKW is strictly symmetrical.

3.3 Outlet and inlet keys width (W_o)

There is no study only dedicated to the outlet width. Authors prefer to analyse the ratio Wi/Wo. Le Doucen et al. (2009) advises $W_i/W_o > 0.8$ while Pralong et al. (2011b) advises $W_i/W_o > 1$. Machiels et al. (2009) has elaborated a relation between the ratio W_i/W_o and the ratio (H/P).

Figure 3. Available space between sidewalls and scaffoldings for upstream overhangs (Etroit dam).

The outlet crest is the less efficient part of the PKW crest. The outlet key's function is then almost limited to the spill of the water coming from the sidewalls which is function of each of the parameters P, W_i, B and H. Meanwhile, performing an analysis on the four constructed PKW does not show a correlation between the assessed amount of incoming water and the designed outlet width though an optimization should have been found.

A detailed study of the relation between W_o and the amount of water coming from the sidewalls can lead to efficiency increase. But it has to be noted that outlet width should not be smaller than 1.5 m to keep it easily constructible for cast in place concrete PKW (Laugier, 2007) and to deal with the starter reinforcing bars (Fig. 3). That is why precast solutions or mixed steel and concrete structures are now explored.

The minimum constructible width also applies to the inlet key in cast in place concrete (about 1.5 m). Nevertheless, as it seems that W_i should be wider than Wo for hydraulic reasons (LeDoucen et al. 2009, Pralong, et al. 2011b), the limiting factor is W_o.

3.4 Crest thicknesses (T_s, T_i and T_o)

Crest thicknesses are fundamental parameters for the PKW hydraulic efficiency. Indeed narrow crests can lead to better discharge coefficient Cdl. Moreover, smaller sidewall thickness T_s increases the key widths (W_i and W_o) for a given total PKW width (W).

The Goulours PKW design includes 0.2 m sidewall crest thicknesses. Construction feedback reveals that this thickness made it more difficult to ensure a good aggregate/cement mixing. Moreover, the rather high reinforcing bar density made the concrete vibration more difficult which increases the risk of concrete segregation. Finally, as there are numerous reinforcing bar layers, it was not possible to increase the reinforcement bar cover beyond French normalized rules (BAEL) as it is usually applied for hydraulics structures.

St Marc PKW was adapted with larger sidewall crest thickness ($T_s = 0.35$ m). For the two other PKW a good compromise has been found: a sidewall 0.35 m thick at the bottom where the efforts are maximum and only 0.25 m at the crest. The construction of non vertical sidewall has not induced strong construction difficulties.

PKW structure has many 90° angles reinforcing bars. The better option seems to fold reinforcing bars in the sidewalls (Fig. 4). After removing the forms and picking the concrete, these reinforcing bars are simply unfold one time (Fig. 3) which is authorized for reinforcing bars diameter smaller than 16 mm.

3.5 Crest shape

Depending on the crest thickness and the water head, the PKW crest might be considered either as sharp or broad crest weirs from a hydraulic point of view. Ribeiro et al. (2006) have tested three different crest shapes. If the discharge capacity was not influenced, they advise the use of a quart-rounded crest with the rounded face on the upstream part of the wall for a higher efficiency in terms of energy dissipation. The influence of crest shape has also been studied on labyrinth weirs

Figure 4.   Reinforcing bars in sidewalls of Goulours PKW and starter reinforcing bars on Gloriettes.

Figure 5.   Crest shapes for inlet key PKW of Gloriettes and parapet walls (0.5 m high) on L'Etroit PKW.

(Tullis et al., 1995) and the conclusions can be transferred to PKW: crest shape does not have a significant effect on the performance. Quarter round and half round shaped are commonly found and efficient even if ogee shapes are slightly better at small heads. Different crest shapes have been tested on EDF PKW: Quarter round (Fig. 5), half round, chamfered, triangular. Finally, quarter round or chamfered shapes should be preferred for both efficiency and lifetime considerations.

3.6   *Parapet walls*

The use of vertical parapet walls on the outlet key crest increases the discharge capacity (Pralong et al., 2011b). Two physical explanations can be found on that topic: the outlet key parapet wall improves stream lines shape upstream the PKW and increases the outlet key volume and so its spilling capacity.

The two first PKW (Goulours, St Marc) were not equipped with parapet walls. L'Etroit PKW was designed with a 0.5 m high parapet wall all along the crest of the PKW (Fig. 5). Hydraulic model has shown that the discharge capacity has actually increased by more than 10% because of the parapet. Latter simulations have shown that parapet walls are not quite favorable on the inlet key crest. Present PKW designs include parapet wall only on the outlet key crest.

3.7   *Nose*

The noses (Fig. 7), built under the upstream overhangs, improve inlet key feeding and consequently the PKW discharge capacity. Triangular noses have been designed in Goulours, St Marc and l'Etroit PKW, while rounded one have been built on Gloriettes PKW. These designs choices do not rely on a proper sensitivity analysis. There is a need for hydraulic studies aiming in defining optimum nose forms and dimensions.

Figure 6.   Counterweight under construction in L'Etroit and L'Etroit and St Marc cranes.

3.8   PKW Boundaries

If the PKW unit dimensions and its impact on the discharge capacity have been well evaluated, the question of boundaries has not been studied. How should a PKW begin and how should it finish on its extremities?

Given that the discharge capacity is more important in the inlet key than in the outlet key, the PKW should be surrounded by inlet keys. The number of outlet keys will then be one less than the number of inlet keys. But with such a configuration, some water will splatter downstream on the left and on the right of the PKW. If this is not acceptable (non protected downstream dam face) the possibility is to build semi-inlet keys on both boundaries. These semi-inlet keys will not overhang downstream of the dam face. This configuration has been used for all the four PKW built (Fig. 7 and 9).

3.9   PKW overall stability and anchorage

Symmetrical PKW are naturally self-balanced structures. Structural analysis shows that PKW overall stability is nearly achieved by the mean of gravity and frictions forces for usual loadings such as *NWL* or moderate water head. For extreme cases (strong earthquakes, probable maximum flood), it might be required to anchor the PKW on the dam structure.

Swelling concrete behaviour is suspected for St Marc and L'Etroit dams. In this condition PKW anchorage on the supporting structure was excluded to minimize the constraint and designs were based on a free standing structure stabilized with gravity and frictions forces. Calculations showed that the adjunction of a concrete counterweight in abutment upstream the dam was necessary (Fig. 6). This upstream concrete brings two features: Firstly it increases PKW overall weight and its gravity stability by adding additional weight on an adequate upstream position. Secondly and as an ultimate safety barrier, it can be considered as an abutment preventing the structure from sliding in case friction forces would not be overwhelmed by destabilizing loads.

This counterweight should be built before the PKW with heavy scaffolding anchored in the dam. If this method is expensive, it is essential to decrease the risk of cracking in the structure. This counterweight could be reduced or even suppressed by designing a PKW with a nose of bigger dimensions and/or by increasing upstream wall thickness.

3.10   Waterproofing and drainage

PKW ultimate stability calculations are made with a conservative hypothesis of loss of all waterproofing barriers, which means triangular uplift pressure under the PKW. In general, three waterproofing barriers have been used on the boundaries: upstream a Waterstop, then a hydroswelling joint (Fig. 7) and downstream grouting pipes for later injection in case of water-proofing loss. A forth barriers has even been created for St Marc and L'Etroit PKW thanks to an upstream membrane.

Figure 7.   Waterstop and hydroswelling joint at the boundaries of L'Etroit PKW (left). Nose and construction joint protection on Gloriettes PKW (right).

Figure 8.   Aeration on St Marc (left) and Goulours (right) PKW.

To limit structural loads when the PKW is built on different blocks of the existing dams, it has been decided to provide in the continuity of the existing dam joint, a structural joint in the PKW. This joint has been placed as far as possible from the sidewalls to assure a good implementation. In L'Etroit PKW, the joints have been setup in the inlet or outlet key without any further considerations. For currently design project the position in the outlet key has been avoided to limit the constraint on the joint when the PKW is overflowing. A waterstop ensures the waterproofing between the two PKW blocks but hydroswelling joint and grouting pipes have also been put up. The hard task is to assure a good connection between the horizontal and vertical waterproofing barriers.

Downstream the waterproofing barriers and under the PKW, drain pipes have been installed at the interface between the existing dam and the new PKW structure to release unlikely uplift pressures. Trenches are digged in the existing dam concrete, filled with gravels and surrounded by geotextile layer.

To reduce steel corrosion, the protection of all construction joint and concrete patching with a waterproofing resin is now generalized.

3.11   *Aeration under the downstream overhangs*

The flow over a PKW is well aerated. Ribeiro et al. (2006) have measured vibration in the structure when the flow is not artificially aerated. Measurements on a real PKW overflowing each year (on Malarce dam) are planned by EDF to study this question. Meanwhile, it is advice to provide aeration under the downstream overhangs. This is easily done by putting a pipe in the base of the PKW (diameter of 200 to 500 mm), sealed in concrete and linked to the free air, over the crest of the dam, on the boundaries of the PKW (Fig. 8).

Figure 9.   Different construction constraints between Gloriettes (left) and L'Etroit (right) PKW.

4   ECONOMICAL CONSIDERATIONS

4.1   *Other spillway type solutions*

Several other solutions were studied at the first stages of the discussed projects. General crests overflowing were forbidden and solutions maintaining the *MWL* were promoted for stability reasons and to save the gateway freeboard if existing. On the other hand, free overflowing spillways are preferred because of many reasons: general system reliability (no inappropriate opening), less floating debris blocking risks, less operating constraints, little maintenance cost. Free overflowing spillways are also quite less sensible in case of design flood reevaluations.

The remaining solutions are frequently the construction of a new channel next to the dam or the construction of a new spillway at the top of the crest. Several possibilities can be envisaged for each configuration: flap or radial gates on ogee spillway, fuse gates or PKW. Flap gates are usually more expensive (40% more) and operators are reluctant to the use of fuse gates because of cost and negative public image in case of gate loss. PKW were often the cheapest solutions.

4.2   *Economical comparison of the four PKW built*

A strict comparison between the four PKW is not possible because each project is specific. L'Etroit dam is situated in a very narrow and deep valley. More than half of the total cost was due to on site logistic (access and crane equipment). For Goulours PKW, on site logistic represents only 25% of the total cost. However, removing the on-site logistic costs does not allow a fair comparison: a different solution can lead to different on site logistic costs. The fairest economical analysis should therefore compare different solutions for the same project. As showed earlier, this work was done in the first stages of each projects and the PKW solution was often in the cheapest ones. The objectives for EDF are now to compare the cost of different PKW dimensions for the same project.

At this time, the only and obvious conclusion is that PKW built on natural ground are cheaper than the ones built on top of the dam crests (Fig. 9). This gives an extended foundation area on the natural ground and makes the general works and structural stability quite easier. But important savings can also be made thanks to smaller lift-on installations. For instance, for Goulours and Gloriettes PKW, the use of a mobile crane was sufficient, while for St Marc and L'Etroit PKW, the installation of a very high fixed crane was necessary (Fig. 6). This crane needed a proper platform and a mobile crane for its assembly and disassembly.

The counterpart of all these advantages is the necessity to create a downstream channel. For Gloriettes dam, the channel of 250 meters long, represents 2/3 of the total project cost.

Economical ratios have been estimated for the four PKW built (fig. 1) by calculating the total project cost on total PKW discharge capacity at *MWL* and the PKW cost only on total PKW discharge capacity at *MWL*. The first ratio is very dependant of the constraints sites (access, channel construction, downstream protection slab necessity, difference between *MWL* and *NWL*, . . .) and varies from 5 to 20. The second is focused on PKW construction cost (demolition, concrete,

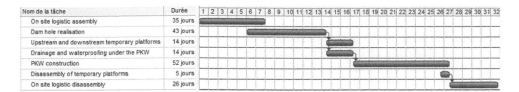

Nom de la tâche	Durée	1	2	3	4	5	6	7	8	9	10	11	12	13	14	15	16	17	18	19	20	21	22	23	24	25	26	27	28	29	30	31	32
On site logistic assembly	35 jours																																
Dam hole realisation	43 jours																																
Upstream and downstream temporary platforms	14 jours																																
Drainage and waterproofing under the PKW	14 jours																																
PKW construction	52 jours																																
Disassembly of temporary platforms	5 jours																																
On site logistic disassembly	26 jours																																

Figure 10.   St Marc PKW construction schedule (weeks).

reinforcing bars, formworks, drainage and waterproofing, . . .) and is quite constant: 3 to 4. This shows that PKW costs are a small part of the total project costs and that different structural options on PKW are not economically significant.

4.3   *Operating losses during works*

The dam partial demolition induces a water level lowering. It is advised to fix the demolition level higher than the minimum operating level in order to minimize operating losses. The PKW height (P) can then be another fixed parameter, like W or B_b.

5   METHODS AND TOOLS FOR STUDIES

Heavy means have been displayed to the design and verification of the PKW projects. For each one, a specific physical model was built. These physical models also allowed estimating existing spillway discharge capacity. Leite Ribeiro et al. (2009) have already described the hydraulic capacity improvement of the four PKW designed with physical model.

For current projects, 3D numerical studies are proceeded before physical ones with Flow3D® software. Numerous tests have been carried out to validate numerical models for discharge estimation on PKW structure thanks to numerical and physical results comparison based on the first PKW projects.

PKW stability was studied thanks to a spreadsheet program and the dam stability was verified with the EDF software Stabet®, taking into account the new PKW. Stability calculations were based on several loading cases: empty dam, dam at *NWL*, design flow and seism. The estimation of foundation and existing concrete parameters was made through several geotechnical campaigns. Aster® and Robot® codes were also used for structural calculations on PKW and thermal effects.

6   SCHEDULE AND CONSTRUCTION STAGES

The simplicity of the PKW structure goes hand in hand with very short works duration. It is then another advantage of the PKW to be able to avoid, during the construction, high flow risk period or winter period for mountain dams. For the four PKW already built, the total construction has lasted from 3 to 6 months (Fig. 10). Moreover, during the construction stages, the discharge capacity of the dams was not lowered, even temporarily. As the construction of the PKW requires performing at first the dam partial demolition, the total discharge capacity is increased from the beginning of the work. Of course the downstream face must be protected against scouring. For Gloriettes dam, the whole project has been built in 2 summers: the first one was dedicated to the downstream channel and the second one to the PKW. When the demolition has been performed in the dam, the channel was already operational in case of flood. For l'Etroit PKW, the downstream protection slab has been built before the dam demolition.

The different construction stages of the four PKW built were chronologically: first the pouring of the upstream counterweight for a free standing PKW, secundly the concreting of the PKW base (anchored or not) with the starter reinforcing bars for the sidewalls, then the pouring of the sidewalls and at last the concreting of the inlet and outlet key slabs. Reservations for the later waterproofing injections and for the aeration under the downstream overhangs are not to be forgotten.

7 CONCLUSIONS

PKW have revealed being the best solutions for discharge capacity improvement on four EDF dams. Easy to build in a very short time, they induce low operating losses and ensure maintaining a sufficient total discharge capacity for the dams during works.

Designed to be build on the crest of a dam, it is even cheaper to implement PKW on the bank, close to the dam. Upstream overhangs are hydraulically advantageous, but they require heavy scaffoldings and thus should not measure more than 5 m long. Hydraulic studies should be performed concerning the impact of key width on the PKW hydraulic efficiency. However, it must be kept in mind that for construction limitations with cast in place concrete, a width of 1.5 m seems to be a minimum. Quart-rounded or chamfered crests are to be preferred for lifetime duration and downstream dissipation efficiency. A crest thickness of 0.35 to 0.4 m at the bottom and 0.25 to 0.3 m at the top appears as a minimum for cast in place concrete. Improvements on the nose hydraulic shape are possible and this structure can participate to the PKW stability. In case of swelling concrete behavior in the dam, free standing PKW can be implemented. It seems clever to design several different waterproofing barriers as their implementation is easy and cheap regarding the total cost. Drainage and aeration under the downstream overhangs should not be forgotten.

As the hydraulic behavior of the PKW is not yet perfectly acknowledged, hydraulic improvements are possible. However, it seems that breakthroughs will also come from structural achievements.

REFERENCES

Laugier. F, 2007. Design and construction of the first Piano Key Weir spillway at Goulours dam. *Hydropower & dams* (issue 5).

Le Doulcen, O. Leite Ribeiro, M., Boillat, J.L., Schleiss, A.J. & Laugier, F. 2009. *PK Weir capacity parametric study*. Proceedings of the SHF Congress on hydraulic physical models. Lyon, France.

Leite Ribeiro, M. Boillat, J.L. Schleiss, A.J. Laugier, F. & Albalat, C. 2006. Rehabilitation of St-Marc dam experimental optimization of a piano key weir, *IAHR*

Leite Ribeiro, M. Bieri, M. Boillat, J.L. Schleiss, A.J., Delorme, F. & Laugier, F. 2009. Hydraulic capacity improvement of existing spillways, piano key weir design. 23rd International Congress on Large Dams. Brasilia, Brazil.

Lempérière, F. & Ouamane, A. 2003. The piano keys weir: a new cost-effective solutions for spillway. *Hydropower & Dams* (5).

Machiels, O. Erpicum, S., Archambeau, P. Dewals, B. & Pirotton, M. 2010. *Experimental study of the alveoli widths influence on the release capacity of Piano Key Weirs*. Proceedings of the SHF Congress "Dimensionement des évacuateurs de crues", Paris, France.

Pralong, J., Blancher, B., Laugier, F., Machiels, O., Erpicum, S., Pirotton, M., Leite Ribeiro, M., Boillat, J.-L., and Schleiss, A.J. 2011a. Proposal of a naming convention for the Piano Key Weir geometrical parameters. *International Workshop on Labyrinth and Piano Key weirs*, Liège, Belgium.

Pralong, J. Montarros, F. Blancher, B. & Laugier, F. 2011b. A sensitivity analysis of Piano Key Weirs geometrical parameters based on 3D numerical modeling. *International Workshop on Labyrinth and Piano Key weirs*, Liège, Belgium.

Truong Chi Hien, Huynh Thanh son & Ho Ta Khanh, M. 2006. Results of some "Piano keys" weir hydraulic model tests in Vietnam. in 22ème congrès des grands barrages – Barcelona. CIGB-ICOLD, Paris, 581–596.

Tullis, J. P., Armanian, N. & Waldron, D. 1995. Design of labyrinth spillways. *Journal of Hydraulic Engineering* 121(3): 247–255. ASCE Press.

Labyrinth and Piano Key Weirs – PKW 2011 – Erpicum et al. (eds)
© 2011 Taylor & Francis Group, London, ISBN 978-0-415-68282-4

P.K weirs under design and construction in Vietnam (2010)

M. Ho Ta Khanh
VNCOLD and CFBR, France

Dinh Sy Quat
Director Consulting Center, HEC, Vietnam

Dau Xuan Thuy
Project Manager PECC2, Vietnam

ABSTRACT: The paper presents some P.K weirs under design and construction in Vietnam. These new projects include a P.K weir, combined or not with a gated spillway, this combination being probably the optimal solution for future dams on sites with large floods. The main characteristics of the dams and the results of the hydraulic laboratory tests used for the projects are indicated. The paper exposes some consideration about the design and construction of these first P.K weirs with high discharges.

1   INTRODUCTION

The paper presents some P.K weirs under design and construction in Vietnam. These new projects include a P.K weir, combined or not with a gated spillway, this combination being probably the optimal solution for future dams on sites with large floods. The schemes described in this paper are:

- The Dakmi 2 gravity dam under construction in the Quang Nam province with a 2 gates-surface spillway in the central part and two P.K weirs installed on the sides of the dam with a stepped downstream face. The peak discharge of the design flood is 6 500 m^3/s. The maximum discharge by the P.K weir is 3 440 m^3/s.
- The Van Phong diversion dam under construction in the Binh Dinh province. The barrage has a 10 gates-surface spillway in the central part and two long P.K weirs on the sides. The peak discharge of the design flood is 12 420 m^3/s, the peak discharge of the check flood is 14 440 m^3/s and the maximum discharge by the P.K weirs is 8 700 m^3/s. The P.K weirs are completely submerged during the high floods.
- The Ngan Truoi irrigation scheme under final design in the Ha Tinh province with an embankment dam and a spillway located on a saddle. The spillway includes 5 radial gates on the left bank and a P.K weir on the right bank. The peak discharge of the design flood (inflow) being 4 840 m^3/s, the maximum discharge by the P.K weir is equal to 1 560 m^3/s to limit the total peak discharge of the outflow to 2 280 m^3/s in normal condition and equal to 1 900 m^3/s if all gates are blocked. There is then an important routing by the reservoir.
- The Vinh Son 3 gravity dam under final design in the Binh Dinh province. An alternative with a P.K weir, 100 m long, installed on the crest, in place of the 3 gates-surface spillway of the initial design, was finally selected due to its numerous advantages. The peak discharge of the check flood is 4 000 m^3/s.

Several other Vietnamese projects including P.K weirs, with or without surface gates, are at a feasibility stage or under final design and will benefit the return of experience of these first large projects.

Figure 1.   Model of the dam (left) and outflow for the design flood (right).

2   THE DAKMI 2 HPP

2.1   General description

The main purpose of the Dakmi 2 RCC gravity dam (H = 38 m, L = 144 m) under construction in the Quang Nam province is hydroelectricity. The dam is situated in a narrow site with a steep left bank. The peak discharge of the check flood (1 000-year) is 6 500 m^3/s.

2.2   The design

The spillway, installed on the dam crest, includes a 2 gates-surface spillway in the central part (for each gate H = 12 m, L = 11 m) and two P.K weirs (W = 2 × 37.5 m) on the two sides, with steps (3 m high) on the downstream face and on the left bank. The main parameters of the P.K weirs are n = 5 and P$_m$ = 5 m. This alternative was selected as it allows:

– With the gates, to evacuate 3 000 m^3/s during the design flood and to control the water level during the first filling of the reservoir and in operation. The gates will be opened only after the P.K weirs are overflowed in order to maintain the jump in the stilling basin with an end sill (crest at El. 602.50).
– With the P.K weirs, to evacuate 3 500 m^3/s during the design flood with a short spillway length (W = 75 m) minimizing the excavation in the left abutment and increasing the safety of the dam in case of blocked gates. The combination of the P.K weirs and the stepped spillway allows reducing the length of the stilling basin due to a very good aeration of the overflow (Ho Ta Khanh et al., 2010) (Fig. 1).

The laboratory tests were carried out with a detail model (1/30) for a group of 4 units corresponding to the P.K weir on each side and with a general model (1/60) of the dam and abutments (Fig. 1).
The NWL is 630.00 and the tests confirmed the calculated MWL = 635.20 for the design flood (all gates opened) with a sufficient freeboard in case of blocked gates for the usual floods. The dam crest is at El. 637.50.

3   THE VAN PHONG DAM

3.1   General description

The Van Phong barrage is a diversion dam under construction on the Côn River in the province of Binh Dinh. It is located downstream the multipurpose scheme of Dinh Binh (flood mitigation, irrigation and hydroelectricity) with a RCC gravity dam constructed some years ago.

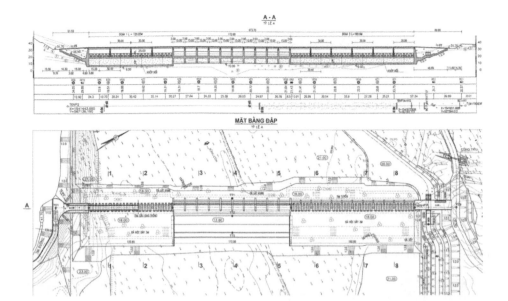

Figure 2. The Van Phong dam. Longitudinal section and plan view.

The purpose of the Van Phong barrage is to re-regulate the flows coming from the upstream HPPs and to maintain the required water level in the main canal supplying an irrigated area of 10 800 ha and the fresh water for the population downstream. The dam crest is at El. 25.00 and the bottom of the river is at El. 22.00. The dam is in general more than 10 m high because the bed rock is covered by a thick layer of alluvium. A consequence of the small difference between the two indicated levels is that the dam is completely submerged during the high floods. Moreover an important requirement of this project is that the raising of the upstream water levels during the floods must be minimized as the river banks between the Dinh Binh and the Van Phong dams are densely inhabited and cultivated. At the site, the river bed is about 475 m broad and a mini power station (6.6 MW) will be installed on the right abutment. The peak discharge of the design flood (200-year flood) is 12 420 m³/s and the peak discharge of the check flood (1 000-year flood) is 14 440 m³/s.

3.2 The design

Several alternatives of dam were contemplated with fixed weir and flap gates or radial gates, but it appears that a satisfying solution may be a combination of 10 radial gates in the central part and a P.K weir on each side (Fig. 2 and 3).

As the floods are high and the size of the P.K weir much larger than all the previous projects, it was decided to carry out thorough studies with several hydraulic models at different scales to complete the previous results (Truong Chi Hien et al., 2006) and (Ho Ta Khanh et al., 2011):

– Two 1 unit-PK weir models, scale 1/15, placed in a flume (Fig. 4). The clear width of the keys is 2.4 m for the first model and 3.6 m for the second model.
– A 6 units-P.K weir model, scale 1/25, placed in a flume (Fig. 4). The clear width of the keys is 2.4 m.
– A general model, scale 1/75 (Fig. 4) with the gated portion (150 m long excluding the thickness of the piers) and the two P.K weirs (total width Wt including the piers = 301.75 m and net width Wn = 289.75 m) with a stretch of river 1 500 m upstream and a length of 750 m between the two banks.

The tests (Fig. 5) were focused on the following topics:

– Optimization of the P.K sizes in particular the width of the keys.
– Measurement of the upstream water levels for different floods, the downstream water levels being fixed by the rating curve of the site. These outcomes are the most important for the project

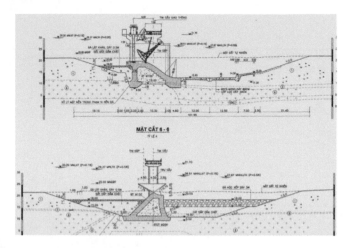

Figure 3.   The Van Phong dam. Cross sections through the gate and the P.K weir.

Figure 4.   1 unit model (left), 6 units model (center) and general model (right).

Figure 5.   Free overflow (left) and submerged flow (right).

and the tests were consequently performed with many hypotheses concerning the operation of the gates (variable numbers of gates completely closed or completed opened).
– Determination of the water velocities and the water pressures at different points of the models.
– Reduction of the discharges through the P.K weirs in case of sediment deposits upstream the inlet keys.

The most interesting results are summarized here under:

– With the 2.4 m width model, the discharges versus the nappe depths are a bit higher for the low nappe depths and free overflow (due evidently to the higher value of n). This difference

Figure 6.   Van Phong dam construction – The base of the P.K weir on the right side.

is however reduced and almost cancelled for the higher nappe depths and with the submerged flows, since the influence of n decreases with the increases of discharge and nappe depth.
– With the 10 gates fully opened and the designed P.K weirs, the maximum upstream water levels are always below the required limits during the design and the check floods. In the worst situation, with all the gates blocked in the closed position during the extreme floods, the P.K weirs alone can maintain the upstream water levels in acceptable limits (if the P.K weir were replaced by a straight Creager weir, the raising of this level would be 1 m more, with heavy consequences upstream).
– The maximum values of the water velocities and the water pressures on the different parts of the concrete works (gate sluices and the P.K weirs) are always in the acceptable ranges.
– Even with a raising of the river bed by the sediment deposits upstream the P.K weirs inlet keys, the discharges of the P.K weirs will not be significantly reduced. Moreover, it can be noted that the existence of a large reservoir upstream (Dinh Binh) and the existence of the 10 gates permitting a frequent flushing will normally limit the sedimentation at the heel of the Van Phong barrage.
– It would be possible to adopt the 3.6 m width for the keys and to reduce the number of gates to 8 (in place of 10), for cost savings and facility of maintenance, but it was decided to keep the initial design since the upstream water levels are a bit lower for the more frequent floods.
– In conclusion, the present design, with a combination of gates and P.K weirs, seems optimal as, by taking advantage of each device, it responds economically and safely to all the requirements of the project.

3.3   The construction

The Van Phong dam is now at the commencement of the construction. The main concerns are the method of construction of the P.K weirs such as for example the possibility to use precast units and some details of construction such as the position of the joints. It is not possible to indicate at present the adopted solutions but these issues will be addressed soon for a rapid and economic construction.

The pictures of figure 6 show the works under construction at the end of September 2010 with the river in its bed and the banks protected by a levee and sheetpiles.

4   THE NGAN TRUOI SPILLWAY

4.1   General description

The Ngan Truoi irrigation scheme under design in the Ha Tinh province includes an embankment dam and a spillway located on a saddle distant from the dam. A particularity of this spillway is that it leads to a natural channel with a normal maximum capacity of 1 200 m³/s. It is admitted however

Figure 7.   Plan view of the Ngan Truoi spillway.

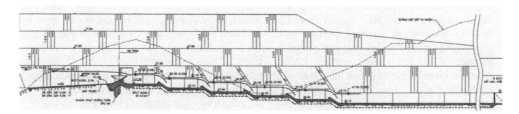

Figure 8.   Longitudinal section of the spillway.

that this discharge can be exceeded during the exceptional floods, but the design must limit it to the max value of 1 850 m³/s.

The peak discharge of the check flood (1 000-year flood) inflow is 6 060 m³/s. The NWL is 52.00 and the MWL is 54.60. The spillway is founded on highly or completely weathered rocks.

4.2   *The design*

To respect the indicated requirement, the only solution is to route the higher floods in the reservoir and to be able to lower the reservoir water level before their arrivals. A safe solution, permitting to minimize the outflow, consists in a combination of 5 radial gates (H = 2.5 m, L = 12 m) on the left bank and a P.K weir (8 units, 12 m long each) on the right bank (figures 3 and 4). The gates allow lowering the water level of the reservoir and, if necessary, limiting the total outflow (P.K weir plus the gates) when this discharge exceeds the maximum capacity of the channel. Furthermore, an important advantage of the gates is the possibility to control the first reservoir filling. The P.K weir allows evacuating a high discharge with a short spillway length (topographically limited by the saddle), with the paramount advantage to guarantee, for usual floods, the safety of the embankment dam in case of blocked gates. Between the spillway and the channel the energy of the outflow is dissipated by a seven-drop channel.

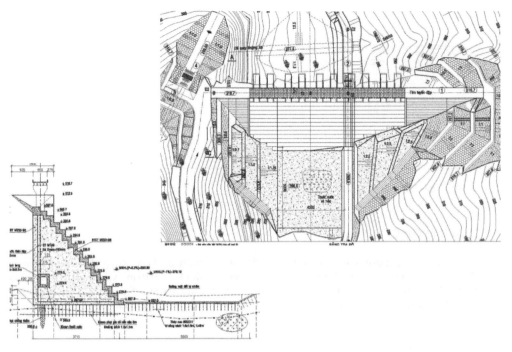

Figure 9. The Vinh Son 3 dam. Plan view (upper right) and cross section (lower left).

5 THE VINH SON 3 HPP

5.1 *General presentation*

The Vinh Son 3 HPP on the Côn River, under final design in the Binh Dinh province, includes a CVC gravity dam ($H_{max} = 51$ m, crest length $= 175$ m) located in a narrow valley. The NWL at El. 312.50 and the MWL at El. 316.45 are fixed by the tailwater levels of the upstream Vinh Son 2 HPP. The peak discharge of the check flood is $4\,000$ m^3/s for the 500-year flood (and $4\,300$ m^3/s for the 1 000-year flood).

5.2 *The design*

Due to the topographical conditions and the existence of the upstream HPP, a traditional alternative with a 3 gates surface spillway ($H = 13$ m, $L = 11$ m) installed on the dam crest, with a ski jump or a flip bucket at the toe of the dam, was contemplated in the initial design. Taken into account the investment and maintenance costs and the risk of blocked gates for this alternative, it was decided to replace these gates by a P.K weir (with the parameters $n = 5$ and $P_m = 5$ m) installed in the central part of the dam crest (Fig. 9). This type of fixed weir has the following advantages: high flood discharge ($4\,000$ m^3/s) with low overflow depth $= 4$ m and short crest length $W = 100$ m (with a Creager weir, the spillway must be extended on the river banks with a total length of 233 m), cost savings and safety improvement in operation. It is foreseen, at this stage of study, to install reinforced concrete steps 3 m high on the entire downstream face of the spillway to maintain the aeration of the overflow produced by the P.K weir, but these steps could be finally installed only on the lower part of the dam if the outcomes of the next model tests confirm this possibility (ICOLD, 2010; Truong Chi Hien and al., 2006). The optimal solution – a combination of gates and P.K weir – is obtained in this project by placing a large bottom outlet in the diversion channel after the dam construction. The role of the bottom outlet is to permit the control of the water level during the first reservoir filling, to lower the reservoir level under the weir crest if required in operation and to increase the dam safety in case of underestimation of the adopted check flood.

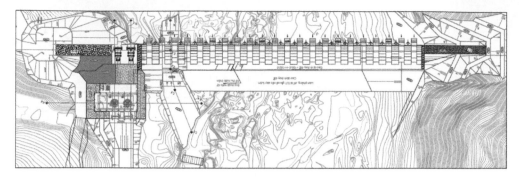

Figure 10.   The Dak Rong 3 scheme. Plan view.

6   OTHER PROJECTS

Several other Vietnamese schemes including P.K weirs such as Dak Rong 3 (Fig. 10) with $Q = 5\,340\,m^3/s$ and a P.K weir 200 m wide, Dak Glun 2 with $Q = 1\,970\,m^3/s$ and a P.K weir 72 m wide, Dak Pru 3 with $Q = 405\,m^3/s$ and a P.K weir 38 m wide are at a feasibility stage or under final design and will take into account the return of experience of these first large projects.

7   CONCLUSION

The probability of totally blocked gates during usual floods may be higher than the probability of very exceptional floods (10 000 year-flood or PMF), as proved by some recent accidents worldwide. The new Vietnamese designs must take into account this risk.

   The dams under design and construction in Vietnam described in this paper show that the combination of gated spillway and P.K weir is often an interesting alternative, in place of an all-gates solution, and is particularly promising when:

– the dam or the spillway has to evacuate high floods with a limited weir length and a minimum overflow depth,
– the spillway is fully submerged during the higher floods and it must minimize the raising of the upstream water level.

   Compared with the traditional solution of all-gates spillways, this combination allows to reduce the investment and maintenance costs and to increase the safety of the dam in case of blocked gates.

   Some optimization and improvement of P.K weirs are certainly possible in the future but their advantages in the present designs are already evident. It remains some issues concerning the economic and rapid construction of the P.K weirs but they will be surely addressed soon.

   Other Vietnamese projects including P.K weirs, with or without surface gates, are at a feasibility stage or under final design and will benefit the return of experience of these first large projects.

REFERENCES

Ho Ta Khanh M., Truong Chi Hien, Combining P.K weir and stepped spillway, 78th ICOLD Annual Meeting, Hanoi May 2010.
Ho Ta Khanh M., Truong Chi Hien, Nguyen Thanh Hai, Main results of the P.K weir tests in Vietnam (2004 to 2010), February 2011, *International Workshop on Labyrinth and Piano Key Weirs*, Liege, Belgium.
ICOLD. 2010. Cost savings in dams, Bulletin 144.
Truong Chi Hien, Huynh Thanh Son, Ho Ta Khanh M. Results of some P.K weir hydraulic model tests in Vietnam, Q87, R36, 22nd ICOLD, Brasilia 2006.

Labyrinth and Piano Key Weirs – PKW 2011 – Erpicum et al. (eds)
© 2011 Taylor & Francis Group, London, ISBN 978-0-415-68282-4

Spillway capacity upgrade at Malarce dam: Design of an additional Piano Key Weir spillway

T. Pinchard & J.-M. Boutet
EDF Centre d'Ingénierie Hydraulique, Le Bourget du Lac, France

G.M. Cicero
EDF Laboratoire National d'Hydraulique et Environnement, Chatou, France

ABSTRACT: The hydrologic study of Malarce dam was updated in 2008 and revealed an insufficiency in flood discharge capacity. The dam shows a deficit of 600 m³/s for the updated design flood of 4600 m³/s. To improve the dam discharge capacity, a complementary free surface spillway will be implemented in the right bank. It is a particular type of labyrinth weir, usually called Piano Key Weir (PKW).

The head was limited to 1.5 m upstream the complementary spillway and its width could not be higher than 42.5 m. Therefore the geometry of the PKW had to be optimized to meet the specific discharge requirement of 14 m³/s/m which is about 4 times higher than an ogee spillway. To do so, an unsymmetrical PKW (type B with minor downstream overhangs, according to Lemperiere et al. – 2003) was chosen. This design allowed lowering the Normal Water Level of only 0.5 m compared to 2.4 m if an ogee spillway had been chosen. Hence, loss of storage and loss of maximum power of the powerhouse are acceptable. Final configuration was tested on two physical models (1/30 and 1/60) at EDF-LNHE laboratory.

Overflow of the PKW is expected to occur more than once a year. Therefore the riverbank rock needs to be protected by a reinforced concrete lining. Energy dissipation structures will re-route the flow of the new weir to the natural river bed, preventing the erosion of the foundation and the banks. The construction of the spillway will require 16.5 m³ of concrete per meter of weir and 56 m³ if one considers every additional structures such as the concrete lining and energy dissipators. Civil works of the new Malarce spillway are due to occur in 2012.

1 INTRODUCTION

In 1968, Malarce dam was constructed on the Chassezac River, in southern France. In this area, climate is Mediterranean, which is very dry but sensitive to stormy weather, with heavy rainfall during less than a day. The catching area, of only 500 km², is made of very dry rocky soil, steep slopes and almost no capacity to retain water. Therefore floods are often very strong with high flowrates during short periods. The design flood in 1968 was 4100 m³/s, but latest hydrologic study revealed an extreme flood (1000 year return period) of 4600 m³/s with a maximum increase gradient of 1400 m³/s/hr, which is very high compared to the 14 m³/s of the natural mean flowrate of the Chassezac.

Today the dam is equipped with 3 tainter gates (14 m × 11.5 m), that include a 2.2 m high flap to ensure progressive flowrate downstream for minor floods and floating debris evacuation. Malarce dam is 31.4 m high above foundations and retains a total volume of water of approx. 3.5 hm³. Total discharge of the dam was measured on hydraulic model (Cicero et al., 2010), and was estimated to 4000 m³/s at the Maximum Water Level, showing that the dam had a discharge capacity deficit of 600 m³/s. Studies have started in 2008 to design a new spillway, which concluded that the adjunction of a Piano Key Weir 42.5 m wide on the right bank side of the dam was the best option.

Figure 1. Downstream view of Malarce dam. New Spillway will be implemented on the right bank side of the dam.

2 DESIGN OF A NEW SPILLWAY

2.1 *Need for an innovative type of spillway*

To improve the discharge capacity of the dam, numerous solutions were studied: additional gate, flap, ogee spillway or tunnel. Because floods are very unpredictable and rough on the Chassezac River, an autonomous solution was preferred. Today, the 3 flaps mounted on top of the gates are the only automatic devices, and they have a discharge capacity of 230 m³/s at Normal Water Level (NWL). The construction of another autonomous spillway increases significantly the safety of the dam if the operators happened to be late in case of trouble to access the dam. Then, a free surface spillway is a very interesting solution.

The problem was complex: Malarce dam needed an additional discharge of 600 m³/s with a reservoir elevation at the Maximum Water Level (MWL) of el. 221 m (1 m above NWL), but only 42.5 m were available on the right bank side of Malarce dam. A specific flowrate of 14 m³/s/m was required. Because MWL could not be raised, every solution which required an upstream head of more than 1 m would imply to lower the NWL, resulting in loss of storage and loss of maximum power at the power plant.

The challenge was to design the best option considering the construction costs and the costs induced by the NWL lowering. The final solution, considered as the optimum, is a Piano Key Weir of acceptable size, requiring only 1.5 m head (NWL lowering of only 0.5 m).

2.2 *Hydraulic design of the Piano Key Weir*

Four different shapes were tested to converge toward Malarce PKW. The first shape was based on Ho Ta Kahn results (Ho Ta Kahn et al. – 2006) and using similitude laws to approach a given discharge capacity. The other three shapes are improved versions of the first design. At the time of the hydraulic design (2008), physical modeling was the only option available. Today, EDF

Table 1. Main characteristics of Malarce PKW project.

Total width of PKW (W)	42.5 m
Number of upstream alveoli	11½
Number of downstream alveoli	11½
Upstream-downstream length of elements (B)	13.46 m
Upstream overhang crest length (B_o)	6.63 m
Downstream overhang crest length (B_i)	2.03 m
Base length (B_b)	4.8 m
Wall height (P)	4.4 m
Width of a unit (W_u)	3.63 m
Inlet key width (W_i)	1.25 m (1.65 m at crest)
Outlet key width (W_o)	1.58 m
Crests & wall thickness at base ($T_s = T_i = T_o$)	0.4 m
Crests & wall thickness at top ($T_s = T_i = T_o$)	0.2 m
Total developed length of a unit (L_u)	29.39 m
Developed length ratio of a unit (n_u)	8.08
Aspect ratio ($2W_u/P$)	1.65
Parapet walls height P_{po}	1.65 m
Parapet walls crest thickness	0.2 m
Nose length B_n (triangular profile)	1.5 m

Table 2. Malarce PKW hydraulic performance.

Head (m)	Specific discharge ($m^3/s/m$)	Total discharge (m^3/s)
0.3	2.2	93.5
0.6	5.6	238
0.9	8.8	374
1.2	11.6	493
1.5	13.8	587
1.8	15.8	672

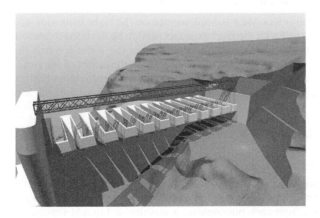

Figure 2. Upstream view of the final design.

knowledge of PKW, EDF Computational Fluids Dynamics experience of PKW (Pralong et al. – 2011) and PKW database of current and completed projects would allow a better and faster design of such spillways. The final PKW is a type B according to Lempérière (Lempérière et al. – 2003) with little downstream overhangs.

This configuration allows a specific flowrate of 14 $m^3/s/m$ for a head of 1.5 m, this configuration was tested on hydraulic model by EDF-LNHE (Cicero et al. – 2010).

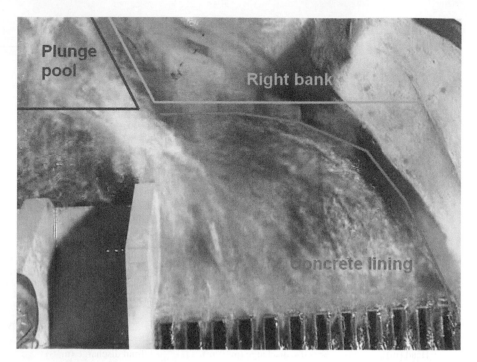

Figure 3.   Downstream view of the PKW flow.

To ensure good feeding of the weir, rock will be ripped off in front of the spillway on a short distance. The model showed that water passing through the PKW comes surprisingly mainly from the bottom of the reservoir. Then, it was unnecessary to take off more rock.

2.3   *Hydraulic design of energy dissipation structures*

Because the new Piano Key Weir is implemented on the right bank side of the dam, PKW flow ($600 \text{ m}^3/\text{s}$) is not aligned with the river axis, and need to be deviated to prevent erosion of the right bank. Consequently, dissipation structures will guide the water towards the natural dissipation basin located 20 to 40 m downstream of the dam and at the center of the river. These structures are composed of two guiding-walls and a flip bucket, allowing redirection of the water with an impact point in the plunge pool area. The design prevents any erosion of the rock located at the dam foot.

3   STRUCTURAL CONSIDERATIONS

3.1   *Dam stability*

Previous studies showed that Malarce dam was sensitive to high water levels, even without the presence of the new spillway. For high upstream water levels, cracking could occur at the base of the dam. To prevent from such phenomena, it was decided to add vertical post-tensioned cables. Theses cables are used to improve dam stability and not PKW stability. However, since it was necessary to keep an access to the anchor heads (for maintenance, force measurements, etc.), it was simpler to have the cables pass through the PKW. Twelve cables of 1300 kN will be installed, one cable by PKW outlet, and their anchor heads will be protected by waterproofed and drained cavities.

3.2   *Stability of the new spillway*

At first, PKW seemed to be very unbalanced structures, and particularly Malarce PKW. Because of its 6.5 m upstream overhangs and its 2 m downstream overhang for a base of only 4.8 m, it

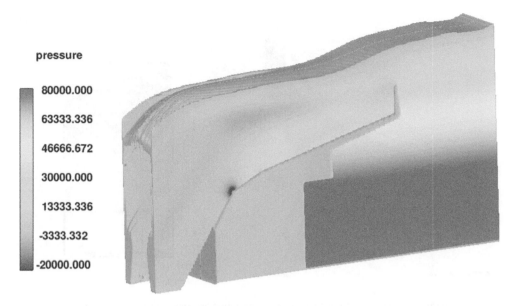

Figure 4.   Pressure field in the Malarce PKW for a 2.5 m head.

seemed at first that it would be hard to ensure natural stability without anchoring the new Malarce spillway. Despite its upstream overhang, the PKW center of mass is located above the base (approx. 0.8 m away from the upstream pavement of the dam) ensuring the stability of the spillway for low water level. Because of its particular shape, it seemed that the buoyancy effect on Malarce PKW (associated with uplift forces) would be too high to ensure stability to a free standing structure for high heads. Flow 3D© (a commercial Computational Fluid Dynamics software used by EDF-CIH) was used to simulate the flow over Malarce PKW and showed quite high pressures on the downstream alveoli sill. Therefore, downstream water exerts significant forces on the sill, which tend to pin the PKW against the dam. If drainage is provided, Malarce PKW stability is ensured even for high heads without any anchor.

Because it was simpler to have the anchor heads in the PKW than in the dam (as discussed in the previous paragraph), the additional force of the cables ensures an even better stability for the new spillway.

The picture below represents the water pressure in Pa exerting on the PKW. The main purpose of this CFD simulation was to provide a better understanding about the PKW stability but showed that the angle connecting the downstream pavement and the outlet sill was too obtuse, which could result in negative pressures. To prevent cavitation phenomena, the shape was then modified with a smoother profile, where no negative pressure is expected.

3.3   Internal stresses in the Piano Key Weir

Because PKW are composed mainly of thin walls, specific studies were undertaken to understand their particular structural behavior. Indeed, PKW happen to be very hyperstatic structures, so usual and unusual loads such as gravity, water pressure, and seismic accelerations are very well supported by the reinforced concrete, with low stresses. As with all thin-walled hyperstatic concrete structures, attention must be paid to the thermal loads and the stresses induced. Thermal calculations showed that high temperature gradients can establish during summer when the sunlight heats up the downstream face and the water colds down the upstream face. Because PKW are very rigid, more steel reinforcement is required for the thermal loads than for classic structural loads.

Because the steel reinforcement shape is complicated and dense, wall-thickness was increased from 33 cm to 40 cm, in order to facilitate the concreting and give better protection to the structure against aggressive elements on a long-term period.

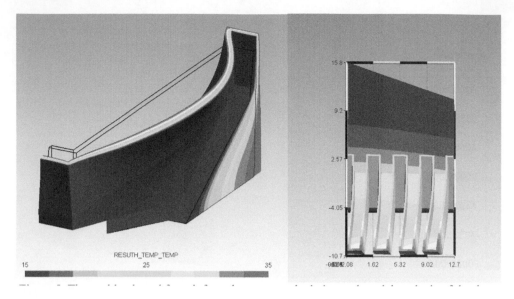

RESUTH_TEMP_TEMP

15   25   35

15.8
9.2
2.57
-4.05
-10.7
-06288802.08   1.62   5.32   9.02   12.7

Figure 5.   Thermal load used for reinforced concrete calculation and modal analysis of the dam and PKW.

In addition, seismic calculation was performed on the structure and showed that the first natural frequencies were high enough to avoid PKW resonance during an earthquake. Nevertheless, because Piano Key Weirs are usually installed at the top of dams, their concrete reinforcements (and stability) must be calculated using the amplified acceleration calculated at the crest of the dam and not using the Peak Ground Acceleration. This would account for dam resonance.

3.4   *Waterproofing and drainage*

3.4.1   *Upstream waterproofing*
The new spillway will be constructed over 3 different blocks which are of different heights. Today, small displacements can be observed between the different blocks, and in order to keep these degrees of freedom, the new spillway will be built in three different parts. New waterstop joints welded on existing ones in the dam will allow displacements and watertightness of the upstream face of the dam.

3.4.2   *Downstream waterproofing*
The downstream face of the dam is in good condition and is expected to withstand the water pressure and erosive power. Since the dam was not designed to be overflowed, there is no waterstop joint on the downstream face of the dam. Hence, new joints will have to be installed on the downstream face of the dam.

3.4.3   *PKW-dam contact waterproofing*
The PKW-dam link is a crucial issue because it will be the contact area between a young and an old concrete. If no attention is paid, cracking can occur easily at the interface causing leakage of the structure and high uplift forces. To prevent this, a triple waterproofing system will be put in place coupled to a drainage system.

– A stripe of geomembrane will be sealed on the upstream face
– A hydro-swelling joint will ensure watertightness if the geomembrane fails
– A network of grouting pipes will be set up all along the interface and will not be injected at construction. If leakage is discovered during the life of the new spillway, these pipes could be injected
– A drainage system will allow capturing the leaks if all the systems above had to fail.

238

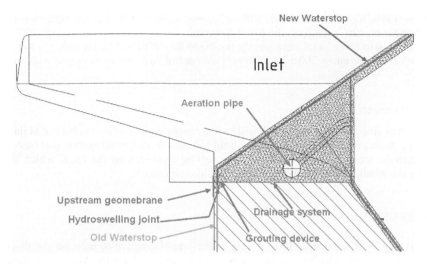

Figure 6. Cross section through an inlet. View of the drainage and waterproofing system.

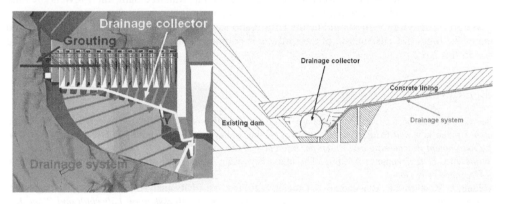

Figure 7. View of the drainage collector system, and detail at the lining-dam junction.

An aeration pipe will be installed through the PKW to ensure atmospheric pressure under the jets, in order to avoid dangerous water pressure fluctuations.

3.4.4 Concrete lining waterproofing and drainage

The concrete lining will be divided into multiple blocks to prevent cracking due to thermal strain, and waterstop joints will be installed at each junction of two slabs. A drainage system will allow reducing the uplift forces and therefore reducing the anchoring of the concrete slabs. In addition, that drainage is very important for the stability of the dam, because it does not raise the dam uplifts.

Foundation will be grouted where excavation is more than 1 m deep, in order to restore watertightness of foundation.

4 NEW FLOOD MANAGEMENT

4.1 Gate opening during floods

Today, the water level is regulated during floods at the Normal Water Level (el. 220 m) by lowering (automatically) the flaps and then, by opening the gates (manually) to a discharge of approx 3500 m³/s. For higher flows, water level will rise up to the Maximum Water Level (el. 221 m) for a discharge of 4000 m³/s.

With the Piano Key Weir, the NWL will be lowered to el. 219.50, but the water level will still be regulated at el. 220. Therefore, the Piano Key Weir will be used for every single flood, with 0 to 50 cm head (0 to 200 m³/s). Consequently, the Piano Key Weir will be the only spillway used for floods with flowrates under 200 m³/s. Current gate control will remain the same with an offset of 200 m³/s.

4.2 Measurements in situ

Due to the new flood management, EDF will have frequent opportunities to observe Malarce PKW functioning during floods with a significant head (50 cm). In situ observations will be confronted to the hydraulic model, velocity and pressures will be measured on the PKW which will allow reviewing the results given by the numerical and physical models.

5 CONCLUSION

The Piano Key Weir has proven to be a very interesting alternative to increase the discharge of existing dams without modifying much the Normal Water Level nor the Maximum Water Level, that is the main reason why EDF has constructed 4 PKW so far: Goulours, St-Marc, L'Etroit and Gloriettes (Vermeulen et al. – 2011). In some cases, such as Malarce dam, the PKW is the only type of free flow spillway compatible with the project constraints.

Malarce Piano Key Weir should be the fifth Piano Key Weir built by Electricité de France, and one of the biggest. Construction of the spillway is planned to last around 9 months and should be over in fall 2012.

REFERENCES

Cicero, G-M., Guene, C., Luck, M., Pinchard, T., Lochu, A. & Brousse, P-H. 2010. Experimental optimization of a piano key weir to increase the spillway of the Malarce dam. *International Association for Hydro-Environment Engineering and Research, Edinburg 2010.*

Lempérière, F. & Ouamane, A. 2003. The piano keys weir: a new cost-effective solutions for spillways. *Hydropower & Dams.*

Pralong, J., Montarros, F., Blancher, B. & Laugier, F. 2011. A sensitivity analysis of Piano Key Weirs geometrical parameters based on 3D numerical modeling. *International Workshop on Labyrinth and Piano Key Weir, Liège. Belgium.*

Truong Chi Hien, Huynh Thanh Son & M. Ho Ta Kanh, M. 2006. Results of some "Piano Keys" Weir hydraulic model tests in Vietnam. Ho Chi Minh.

Vermeulen, J., Faramond, L. & Gille, C. 2011. Lessons learnt from design and construction of EDF first Piano Key Weirs. *International Workshop on Labyrinth and Piano Key Weir, Liège. Belgium.*

Labyrinth and Piano Key Weirs – PKW 2011 – Erpicum et al. (eds)
© 2011 Taylor & Francis Group, London, ISBN 978-0-415-68282-4

Rehabilitation of Sawara Kuddu Hydroelectric Project – Model studies of Piano Key Weir in India

G. Das Singhal
Department of Civil Engineering, COES, University of Petroleum & Eneregy Studies, Dehradun, India
River Engineering Laboratory, WRDM Department, Indian Institute of Technology, Roorkee, India
M3, University of Nottingham, Nottingham, United Kingdom

N. Sharma
River Engineering Laboratory, WRD Management Department, Indian Institute of Technology,
Roorkee, India

ABSTRACT: Labyrinth weirs are normally built to increase the total effective crest length for a given spillway width. They can be used to increase the discharge capacity for a given head or to decrease the head for a given discharge. As their application is sometimes difficult in rehabilitation projects due to inappropriate supporting conditions, a new concept of labyrinth spillways was mooted by Lemperiere et al. with a new shape, called Piano Key Weir. This innovative alternative of labyrinth spillway provides an increase in the stability of the structure which can be placed on the top of most existing or new gravity dams, unlike traditional labyrinth weirs. This paper focuses on the experimental results and physical modelling tests of the evacuation system of Sawara Kuddu Hydro Electric Project 110 MW located in Lower Himalayan mountain ranges of India with the Piano Key Weir and the energy dissipation structure. The physical modelling has been carried out at the River Engineering Laboratory, Water Resources Development and Management, Indian Institute of technology, Roorkee, India in the framework of the rehabilitation project.

1 INTRODUCTION

Weirs are among the oldest and simplest hydraulic structures that have been used for centuries by hydraulic engineers for flow measurement, flow diversion, regulation of flow depth, flood passage, etc. Weirs may be classified according to the shape of opening, the shape of crest, the effect of sides on the issuing nappe and the discharge condition (Borghei et al. 1999, and Falvey 2003). According to the shape of opening, the weirs may be classified as rectangular, triangular and trapezoidal weirs. According to the shape of the crest, the weirs may be classified as sharp crested weir, narrow crested weir, broad crested weir, ogee shaped weir, labyrinth weir, piano key weir, etc.

As projects are reassessed for safety, provision for an increased estimate of the probable maximum flood (PMF) has to be made in many cases. It is therefore necessary to provide more flood storage and/or larger capacity for spillways to pass the PMF safely. If the dam cannot adequately pass the updated flood, the structure requires modification by increasing the flood storage space, and/or increasing the spillway capacity. An innovative and effective way of increasing the spillway capacity is to use a Labyrinth weir. The concept of the Labyrinth weir is to vary the plan shape of the crest to increase the effective crest length (Lempérière and Jun, 2005). This increases the discharge per unit width of the spillway for a given operating head.

Labyrinth spillways are built to provide a longer total effective crest length for a given overall spillway width. This type of structure is used to increase the discharge capacity for a given head or to decrease the head for a given discharge. Such behaviour is useful for limiting the height of the dam or raising the normal water level.

An extensive investigation of labyrinth weirs was performed by Taylor (1968) presenting a capacity ratio by comparison with sharp-crested linear weirs of same width. As a follow up, Hay and Taylor (1970) propose a design procedure for estimating the discharge over triangular and trapezoidal labyrinth weirs. Their work, also presented in Sinniger and Hager (1989), shows a trapezoidal layout of labyrinth weir as being optimum.

Darvas (1971) established a family of curves for designing labyrinth weirs, which are based on experimental results of model studies for the weirs Wonorora and Avon in Australia. Magalhães and Lorena (1989) present curves of discharge coefficients C_d for different ratios "hydraulic head over height of walls" H/P whose values are systematically lower than those obtained by Darvas (1971).

Tullis et al. (1995) propose discharge coefficient values for labyrinth angles between 6° and 35°. The given equations are valid for an apex width between t and 2t, and $H/P < 0.9$ (see Figure 1). The experiments were performed in a structure with a 90° rounded crest shape at the upstream face of radius $R = P/12$. As a result, the authors recommend a design procedure for labyrinth spillways.

In dam rehabilitation projects, the labyrinth spillway solution is sometimes difficult to apply due to the fact that its shape is not appropriate to be installed at the crest of standard gravity dams (Lempérière and Ouamane, 2003).

Over the last few years, a new concept of labyrinth spillways, developed at Hydrocoop France in collaboration with the Laboratory of Hydraulic Developments and Environment of the University of Briska, in Algeria and the National Laboratory of Hydraulic and Environment of Electricité de France (EDF-LNHE Chatou) has been studied. This new shape of spillway, called PK-Weir (piano key weir), is an innovative alternative of labyrinth weirs.

Piano Key Weir (PKW) derives its name from its plan-form shape which looks like the key of the piano. It provides a much longer crest length than the conventional ogee spillways and thereby significantly increases the discharging capacity for same reservoir levels. Furthermore, the construction of PKW will be simpler than that of labyrinth spillway.

This paper presents investigation related to application of typical Piano Key Weir for Sawra Kuddu Hydro Electric Project. Sawra Kuddu HEP with an installed capacity of 111 MW is located on Pabbar River in Himachal Pradesh. Laugier (2007) has studied other form of Piano Key Weir for Goulours Dam in France. He conducted the model test in flume with geometrical similar scale based on Froude law. Laugier has reported that Piano Key Weir is used for rehabilitation project in Goulours dam. In Sawra Kuddu HEP, the flow diversion structure consists of four under-sluices bays on the left and three on the right bank, each of 8.0 m width with 1.5 m thick intermediate piers. A 138 m long Piano Key Weir is proposed in between the two sets of under-sluices. The design discharge of the project is 6,880 m³/s. The Piano Key Weir is designed to handle 3,900 m³/s and the balance discharge 2,980 m³/s passes through under-sluices. This paper focuses on the experimental results and optimization procedure of the evacuation system of Sawra Kuddu HEP with the Piano Key Weir. The physical modeling has been carried out at the laboratory of River engineering at the Water Resources Development and Management department (WRD&M), IITR, Roorkee, India. A comprehensive investigation based on physical model studies on a flume has been undertaken to evolve the best suitable Piano Key Weir elements to assess the behaviour of the Sawra Kuddu HEP.

2 EXPERIMENTAL SET-UP

The experimental set-up consists of the following.

 (i) V-notch: It was used for the measurement of discharge through the PKW. 90° V-notch was used for the discharge measurement.
(ii) Point gauge: It was used to measure the head corresponding to different flows in flume as well as in V-notch.

Water conductor infrastructure consists of the following components.

2.1 Constant head tank

An overflowing tank was installed at the upstream head-end to ensure the supply of steady discharge into the experimental flume.

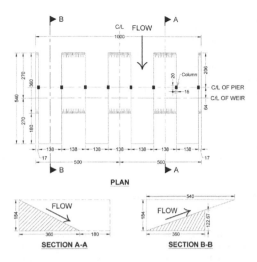

Figure 1. Plan and section of model C_1M_6 for physical model study (dimensions in mm).

2.2 Flume

Rectangular flume of size 100 cm × 110 cm was used for passage of water from upstream head to the PKW installed at the downstream end.

2.3 V-notch

V-notch is used to measure the discharge through PKW. The formula used for discharge of V-notch is

$$Q_{PK} = \frac{8}{15}C_d\sqrt{2g}\tan(\theta/2)H_e^{5/2} \quad Q_{PK} = \frac{8}{15}C_d\sqrt{2g}\tan(\theta/2)H_e^{5/2} \tag{1}$$

where,

$$H_e = H + K_h \tag{2}$$

There, H_e = the effective depth of water above vertex at the upstream of V-notch, $K_h = 0.0008$ m for 90° V-notch, C_d = coefficient of discharge = 0.58 and θ = angle of the V-notch.

2.4 Piano Key Weir Models

Six selected models of PKW have been used in the experimentation. Model of Piano Key Weir is shown in photo no. 1 in dry condition and in photo no. 2 in running condition.

Sawra Kuddu HEP with an installed capacity of 111 MW is under construction on the Pabbar river in Himachal Pradesh state of India, which is located in the Western Himalayan region. The flow diversion structure comprises two under sluices on both sides having three and two bays on left and right banks respectively, each of 8 m width with 1.5 m thick intermediate piers. A 138 m long spillway of Piano Key Weir (PKW) is proposed in between the two sets of under sluices to handle a sudden cloud-burst flood which is an occasional hydrological feature of the region.

One meter wide flume having transparent perpex walls was installed at River Engineering Lab of WRD&M Department of IIT-Roorkee in India. Six transparent acrylic PKW models built to geometrically similar scale of 1:50 were tested in the flume during experimentation. These represented a gross water-way of 50 m including the piers of foot bridge. Plan and section of general model are shown in Fig 1. The maximum prototype discharge adopted for the model studies was 2,500 m³/s. The discharge scale as per Froude Law worked out to 1/17678. The maximum flume discharge was found to be 51.2 litres/sec. The studies were aimed mainly for assessing the enhancement of discharge passing capacity and sediment transport efficiency of various PKW models in comparison to conventional sharp crested weir.

243

Table 1. Model Dimensions.

Model No.	Height of Model (p) (cm)	W_i (cm)	W_o (cm)	$W_i + W_o$ (cm)	L/W
C_1M_1	18.4	14.80	26.28	41.08	3.74
C_1M_2	18.4	26.30	26.30	52.6	2.96
C_1M_3	18.4	19.72	19.72	39.44	3.74
$C1M4$	18.4	13.14	13.14	26.28	5.10
C_1M_5	18.4	16.00	11.60	27.6	4.91
C_1M_6	18.4	13.80	13.80	27.6	4.91

Table 2. Piano Key Weir dimensions for prototype.

Model No.	Height of Model (p) (m)	W_i (cm)	W_o (cm)	$W_i + W_o$ (cm)	L/W (cm)	No. of Element
C_1M_6	9.20	6.90	6.90	13.82	4.91	10

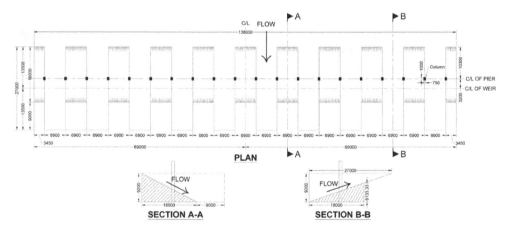

Figure 2. Plan and section of model C_1M_6 for prototype (dimensions in mm).

The size and shape of PKW models are as indicated below in Fig. 1 and their model dimensions in Table 1.

For the last elements on the side of the flume, the width of W_i or W_o is divided by 2.

A comprehensive investigation has been done based on physical model studies on a flume to evolve best suitable Piano Key Weir elements to assess the behaviour of the Sawra Kuddu Barrage system. Comprehensive model constructed at outdoor lab of WRD&M, IIT, Roorkee has reproduced the actual topography of the valley including part of reservoir, designed Piano Key Weir and downstream side of weir. The model extends approximately 400 m upstream of the weir and 150 m downstream of the weir. Here, the Piano Key Weir is installed with full width of 2.76 m in geometric similar model i.e. 138.0 m in Prototype dimension.

The dimensions of Piano Key Weir for prototype are as indicated below in Table 2. The plan and sectional view of Piano Key Weir for prototype with dimension are shown in Fig. 2.

3 DATA ANALYSIS

Graphical plots have been developed by processing the experimental data collected for the six models of PKW with different sizes and shapes for varying flow conditions. The methodology for calculations made for discharge, h/p and L/W using the experimental data is indicated below.

(i) Calculation of discharge through rectangular sharp crested weir is done using the formula,

$$Q_L = \frac{2}{3}C_d\sqrt{2g}Wh^{3/2}$$

(3)

where, Q_L = discharge through rectangular sharp crested weir, h = the head over the crest, C_d = coefficient of discharge and W = the width of channel. In eq. (3)

$$C_d = \left[0.605 + \frac{0.08h}{p} + \frac{0.001}{h}\right]$$

(4)

where, p = height of crest

Formula used for calculation of PKW discharge Q_{PK} through the V-notch is given in the Eq. (1)

Difference of PKW discharge and rectangular sharp crested weir discharge (ΔQ) is obtained as

$$\Delta Q = Q_{PK} - Q_L$$

(5)

where, Q_{PK} = Discharge through PKW

(ii) Ratio of PKW discharge and rectangular sharp crested weir discharge (r) is

$$r = \left(\frac{Q_{PK}}{Q_L}\right)$$

(6)

(iii) Calculation of h/p ratio is done using h as the head over the crest (at one meter u/s of the PKW) and 'p' as the height of PKW.

(iv) Calculation of length magnification ratio, L/W, where L is the total developed length of the PKW along the overflowing crest and W is the total width of PKW.

4 ANALYSIS OF LAB-BASED MODEL EXPERIMENTS

This is done in two steps. Six different configurations of models are tested for their performance in lab based experiments. The test results in the form of discharge passing capacity as net absolute discharge increment is shown in Fig. 3, in which the ordinate 'ΔQ' represents the difference between discharge passing over a Piano Key Weir and sharp crested weir for same h/p. The net absolute value of discharge increment for different models is in the range of 5.0 to 30.0 l/s.

The test results for discharge passing capacity is shown in Fig. 4 where in the ordinate 'r' represents the ratio of discharge passing over a Piano Key Weir and sharp crested weir. Fig. 4 shows that discharge passing over a Piano Key Weir is 1.54 to 4 times higher than the sharp crested weir. From Figs. 3 and 4, it is found that lab based model C_1M_6 performs best in terms of r. For field scale testing, this model is used for construction.

5 ANALYSIS OF COMPREHENSIVE MODEL EXPERIMENTS RESULTS

From laboratory physical model studies, model C_1M_6 is preferred shape of Piano Key Weir. So this model C_1M_6 of Piano Key Weir is used for comprehensive model experiments. Full length of Piano Key Weir is used in comprehensive model study for better analyses of weir with reservoir area in upstream and downstream of weir.

The test results in the form of discharge passing capacity as net absolute discharge increment is shown in Fig. 5, where in the ordinate 'ΔQ' represents the difference between discharge passing

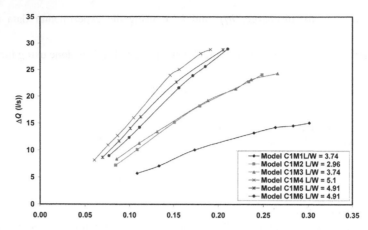

Figure 3.   Plot between ΔQ and h/p for all six models.

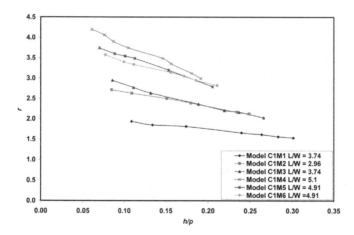

Figure 4.   Plot between r and h/p for all six models.

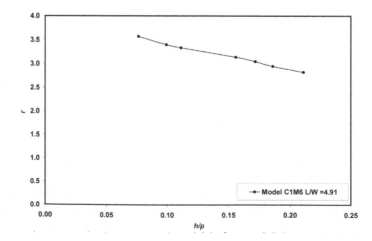

Figure 5.   Plot between ΔQ and h/p for model C_1M_6.

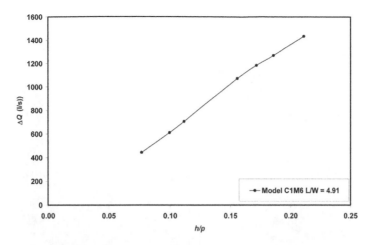

Figure 6. Plot between r and h/p for model C_1M_6.

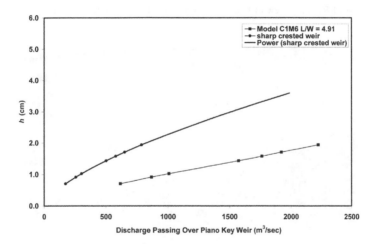

Figure 7. Comparison of Piano Key Weir by to sharp crested weir with head over the crest for model C_1M_6.

over a Piano Key Weir and sharp crested weir for same h/p. The net absolute value of discharge increment for model C_1M_6 lies in the range of 450 to 1,550 m³/sec of prototype discharge.

The test results for discharge passing capacity is shown in Fig. 6 in which the ordinate 'r' represents the ratio of discharge passing over a Piano Key Weir and sharp crested weir. Fig. 6 shows that discharge passing over a Piano Key Weir is 2.65 to 4.00 times higher than sharp crested weir.

Fig. 7 represents the saving of head over the crest of Piano Key Weir against sharp crested weir. This graph shows that saving of head over the crest is 0.80 m (i.e. 58.6%) in Piano Key Weir against sharp crested weir for lower range of discharge (i.e. 500 m³/sec) and is 2.00 m (i.e. 47.6%) in Piano Key Weir against sharp crested weir for higher range of discharge (i.e. 2,500 m³/sec). The running view of the models is shown in Plate no. 1 from downstream side of weir and in plate no. 2 from upstream side of weir.

The maximum water level (MWL) for design flood of 5,240 m³/s was found at El 1,423.12 from model study. The rating curve for the discharge passing over Piano Key Weir is depicted in adjoining Fig. 10. Reservoir level for 4,000 m³/s passes over Piano Key Weir is 1,422 m.

Figure 8.   Plate No. 1 Running view of Piano Key Weir from d/s with under sluice gate.

Figure 9.   Plate No. 2 Running view of Piano Key Weir from u/s with under sluice gate.

6   CONCLUSIONS

Design flood of the Sawara Kuddu HEP is 6,880 m³/s and requires high spilling capacity through weir in limited space. Piano Key Weir is designed with geometrically similar scale factor of 1:50. Six different geometries of Piano Key Weirs have been investigated in the lab. Among them, model C_1M_6 was found to be the most efficient with regard to the weir capacity. Study of model C_1M_6 indicated that the Piano Key Weir gives about 2.62 to 4.20 times higher discharge than sharp crested weir for corresponding head. The best evolved shape of Piano Key Weir from laboratory model study has been used for comprehensive model study. Comprehensive model study shows very interesting result that saving of head over the crest in Piano Key Weir lies in the range of 45 to 58% of sharp crested weir.

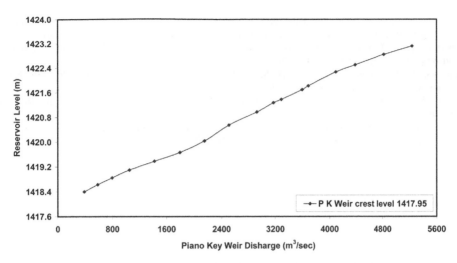

Figure 10.   Discharge passing over Piano Key Weir and corresponding reservoir level.

LIST OF SYMBOLS

B = Upstream-downstream length of the PKW
C_d = Discharge coefficient of rectangular sharp crested weir; and V-notch
g = Gravitational acceleration;
H = Depth of water above vertex at the upstream of V-notch;
h = Head over the crest (at one meter u/s of the PKW);
H_e = Effective depth of water above vertex at the upstream of V-notch;
K_h = 0.0008 m for 90° V-notch;
L = the total developed length of the PKW along the overflowing crest;
p = Height of PKW;
Q_L = Discharge through rectangular sharp crested weir;
Q_{PK} = Discharge through PKW;
W = the total width of PKW;
W_i & W_o = Inlet & Outlet key width;
Z = Height of crest;
ΔQ = Difference of PKW discharge and Rectangular sharp crested weir discharge
θ = Angle of the V-notch

REFERENCES

Blanc, P. & Lemperiere, F. 2001. Labyrinth spillways have a promising future. Hydropower & Dams – Vol. 4
Borghei S.M., Jalili, M.R. & Ghodsian, M. 1999. Discharge Coefficient for Sharp Crested Weir in Subcritical Flow. American Society of Civil Engineering. Journal of Hydraulic Engineering 125(10): 1051–1056.
Darvas, L. A. 1971. Performance and Design of Labyrinth Weirs. American Society of Civil Engineering, Journal of Hydraulic Engineering 97(80): 1246–1251.
Falvey, H.T. 2003. Hydraulic Design of Labyrinth Weir: ACSE Press USA.
Hay, N., & Taylor, G., 1970. Performance and Design of Labyrinth Weirs. American Society of Civil Engineering. Journal of Hydraulic Engineering 96(11): 2237–2357.
Laugier, F. 2007. Design and Construction of the First Piano Key Weir (PKW) Spillway at the Goulours Dam. Hydropower & Dams. 10(5).
Leite Ribeiro M., Boillat, J.-L., Kantoush, S., Albalat, C., Laugier, F. & Lochu, A. 2007. Rehabilitation of St-Marc dam: Model studies for the spillways. Hydro 2007 New approaches for a new era; Granada, 15–17 October 2007. Spain.

Lempérière, F. & Jun, G. 2005. Low Cost Increase of Dams Storage and Flood Mitigation: The Piano Keys Weir", Q. 53 R. 2.06 International Commission on Irrigation and Drainage Nineteenth Congress Beijing.

Lempérière, F. & Ouamane, A. 2003. The Piano Keys weir: a new cost-effective solution for spillways. Hydropower & Dams Vol No 5: 144–149.

Magalhaes, A. P., & Lorena, M. 1989. Hydraulic Design of Labyrinth Weir. Report No. 736, National Laboratory of Civil Engineering, Lisbon, Portugal.

Sinniger, R. O. & Hager, W. H. 1989. Constructions Hydrauliques. Presses Polytechniques Romandes, Lausanne – Switzerland.

Tullis, J. P., Nosratollah, A., & Waldron, D. 1995. Design of labyrinth spillways. American Society of Civil Engineering. Journal of Hydraulic Engineering 121(3): 247–255.

Labyrinth and Piano Key Weirs – PKW 2011 – Erpicum et al. (eds)
© 2011 Taylor & Francis Group, London, ISBN 978-0-415-68282-4

A dam equipped with labyrinth spillway in the Sultanate of Oman

L. Bazerque, P. Agresti & C. Guilbaud
Sogreah Consultants, Grenoble, France

S. Al Harty
Director of Dams Department Ministry of Regional Municipalities and Water Resources, Sultanate of Oman

Theodor Strobl
Technischen Universität München, Munich, Germany

ABSTRACT: Within the context of a project to protect the city of Muscat from flash floods on the wadi Aday, Sogreah Consultants, on behalf of the Ministry of Regional Municipalities and Water Resources, designed an ambitious development scheme based on the construction of seven dams. The main dam in this scheme is equipped with a labyrinth spillway with a flow capacity of 5850 m³/s. The design of this complex structure was initially based on technical literature before being refined through a physical modelling procedure. In this respect, Sogreah carried out a programme of 1:40 scale model tests in its hydraulics laboratory in order to check and finalise the structural design.

1 PRESENT SITUATION

The wadi Aday is one of the main wadis crossing Muscat, the capital of the Sultanate of Oman. With a catchment area extending over 320 km² (Fig. 1), the wadi is fed essentially by the Al Amerat cirque. This cirque represents one of the main urban development areas of the capital; numerous plots of land have been built on and there is very little land left for urban growth.

The cirque is an extensive plain surrounded by low-altitude but steep-sided mountains (from 1 to 10%). This has led to the formation of a network of wadis flowing towards a single narrow gorge (100 to 200 m wide). This 8.5 km long gorge (Fig. 1) crosses the final mountainous rampart before

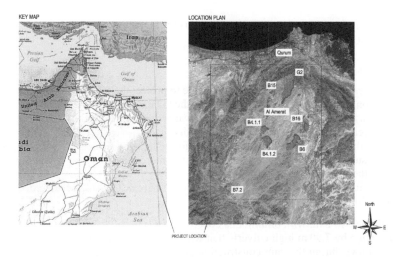

Figure 1.   Catchment area, main reach and dam locations.

flowing into the coastal plain where Muscat has developed. Finally, the wadi discharges into the Oman Sea through a small mangrove swamp, that has been preserved in the middle of the city's main shopping area.

During flash floods, the area downstream of the gorge is regularly inundated as a result of restrictions to the low water bed by buildings, the presence of numerous road embankments sections and the insufficient capacity of the man-made structures built to channel the flow. However, until recently, these inundations have always been fairly limited without any major human or financial impact.

All that changed in June 2007 when the tropical cyclone "Gonu" hit the northern part of the Sultanate, and Muscat in particular. The torrential rainfall from the cyclone continued non-stop for 24 hours and generated floods that led to fifty deaths, destroyed major infrastructure and inundated many shopping areas and homes. Damage was estimated at USD3.9 billion.

In June 2010, a second tropical cyclone, called Phet, hit the Sultanate of Oman. Although this cyclone did not have the intensity of Gonu on the capital, it still caused the death of six people and damage running into tens of millions of dollars.

Following the catastrophic flooding caused by cyclone Gonu, the Omani government decided to upgrade an ongoing project for aquifer recharge dams in the wadi Aday catchment area into a project for flood protection dams. The aim of the revamped project is to protect the downstream urban area from any flooding of the wadi Aday up to a flood of 500-year return period (which was the estimated return period of the rainfall experienced in Muscat during the Gonu cyclone).

2 THE PLANNED PROTECTION PROJECT

Sogreah Consultants, which was awarded the original project study, therefore drew up a flood protection scheme covering the entire catchment area. This scheme is based on the construction of seven dams (Fig. 1): six of these dams are located in the plain of the cirque and one dam, known as G2, is to be built in the gorge. Once completed, this series of seven dams will enable flood flows up to the 500-year flood to be fully controlled and the effects of the 1000-year flood to be significantly attenuated.

The large number of dams, coupled with the considerable urban pressure, make this project highly complex. At the time this article was written, the first of the seven dams (the B15 dam on figure 1) was under construction and is scheduled to be completed in autumn 2011. The six other dams are at the final design stage. Construction of the G2 dam should commence by mid-2011.

When they are all completed, the seven dams will represent a total storage volume of the order of 90 Mm^3 of which 45 Mm^3 will be provided by dams B15 and G2.

3 THE G2 DAM, CORNERSTONE OF THE PROTECTION OF MUSCAT FROM WADI ADAY FLOODS

3.1 *Presentation*

The G2 dam represents the cornerstone of the scheme to protect the city of Muscat from wadi Aday floods. The reason for this is that, of all seven dams, the G2 dam is located the furthest downstream. It is to be built in the gorge which receives all the water draining from the Al Amerat cirque, and will be just upstream of the main Muscat shopping district.

It differs from the other dams in the overall protection scheme by its size, its hydrology, the nature of the soils and subsoil and by its interference with the motorway currently being constructed on the left bank of the wadi.

Given the environmental and technical restrictions, Sogreah Consultants decided to design a dam with the following main characteristics:

− a 35 m high rockfill dam with bituminous concrete core,
− two 5.50 m wide by 7.90 m high culverts that will be used initially for the temporary diversion of the wadi flows during the dam construction period and subsequently equipped with 3.00 m wide by 3.50 m gates for flood damping purposes,

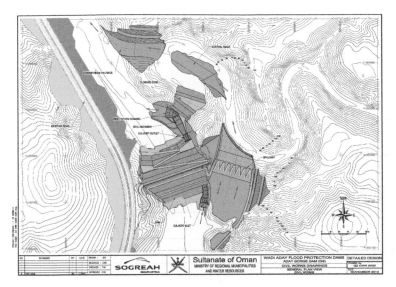

Figure 2. Plan view of the dam + labyrinth spillway + tributary wadi.

– a labyrinth spillway designed to pass the PMF, to be built on the right bank and discharging into a tributary wadi,
– a water discharge channel excavated in the mountain and designed to release the water as far as possible downstream from the foot of the dam.

In view of the considerable thickness of the alluvial cover (>35 m) and the poor quality abutment rock, a rockfill dam was designed with bituminous concrete core. Given that a rockfill dam is likely to be subject to settlement effects that are incompatible with a surface spillway destined to convey high specific discharges, it was decided to build the spillway in the right-bank abutment (Fig. 2) and make use of a secondary tributary as a downstream outlet channel.

3.2 The labyrinth spillway

On one hand, a high flow capacity of 5850 m³/s is required to guarantee the integrity of the dam, but on the other hand, land use restrictions make it essential to have a maximum storage level behind the dam, i.e., at the inlet to the gorge, of 65.50 m above sea level.

The conventional solution would involve building a straight spillway of some considerable length to pass a high flow rate but with a limited head. However, given the geological and topographical constraints of the gorge, any excavation works on the right bank would give rise to a massive volume of waste materials. It is therefore important to restrict the excavation works to the smallest possible volume to avoid penalising the economic rate of return of the project and to minimise the impact of earthworks.

Faced with these two conflicting requirements, Sogreah Consultants devised a labyrinth spillway built on a platform excavated on the right bank and discharging into a tributary of the wadi Aday.

Labyrinth spillways are not commonly found and the design theory behind their construction has not been completely developed. Dimensional design criteria have been obtained by different authors during physical model experiments; these criteria are helpful for designing a labyrinth spillway but are not sufficient for guaranteeing the proper operation of the structure in the future. These criteria are expressed on the basis of the characteristic parameters that define a labyrinth structure (Fig. 3) and usually defined in the technical literature.

From the available literature it would seem that the main criteria to be respected are as follows:

– Headwater Ratio less than 0.50 according to Lux (1989) and less than 0.90 according to Tullis (1996),
– Vertical Aspect Ratio (W/P) greater than 2 (Taylor (1968)),

253

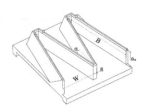

Figure 3. Diagram of a typical labyrinth spillway.

Table 1. Characteristics of the spillway in configuration 1.

	n	Ho	P	Geometry of the cycle						Design parameters			
				a	W	m	L	α	B	Ld	Ho/P	W/P	Ld/B
		m	m	m	m		m	°	m	m			
Configuration 1	7	4.5	9.0	1.0	18.0	4.0	72.0	11.8	33.0	14.8	0.50	2.00	0.44

Table 2. Characteristics of the spillway in configuration 2.

	n	Ho	P	Geometry of the cycle						Design parameters			
				a	W	m	L	α	B	Ld	Ho/P	W/P	Ld/B
		m	m	m	m		m	°	m	m			
Configuration 2	5	4.5	9.0	1.5	26.0	4.0	104.0	11.8	49.0	14.9	0.50	2.89	0.30

– Magnification Ratio ($m = L/W$) between 2 and 10, L being the developed length of one cycle
– Interference Length Ratio (L_d/B) less than 0.30, L_d being the length of interference.

In a preliminary design stage, two spillways were devised. Physical scale model tests were run in the hydraulics laboratory of Sogreah Consultants in order to choose between the two preliminary designs and to test various adaptations.

3.2.1 *Spillway 1*
The first spillway was designed by taking inspiration from and adapting the existing spillway of the Ute dam in the United States; this spillway was designed for a project flood of more than $15,500\ m^3/s$ with a discharge head of less than 6 m, i.e., one of the biggest spillways of its kind in the world. The general dimensions of the cycles were therefore reviewed and the cycle number was halved.

The standard spillway 1 designed for the G2 dam was therefore devised with the following characteristics.

Note that in this configuration the criterion relating to the Interference Length is not respected as it is well over the limit of 0.30.

3.2.2 *Spillway 2*
The second spillway was designed in order to respect the upper limit values of the various criteria referred to in the literature.

The standard spillway 2 designed for the G2 dam was therefore devised with the following characteristics.

Note that the criterion relating to the Interference Length is respected.

Figure 4.   Photographs of the test flume and spillway model.

3.3   Finalisation of the design and scale model study

Verification of spillway operation and adjustment of its geometry for all possible hypotheses go beyond theoretical calculation capacities, notably because of the complex nature of the hydraulic phenomena that come into play in flow conditions at the top of labyrinth spillways.

The physical scale model, on the other hand, can reproduce all the phenomena involved with strict reliability in the particular flow conditions prevailing over complex-shaped spillways.

The two spillways referred to above were tested on 1:40 scale models built and run in the hydraulics laboratory of Sogreah Consultants. The 1:40 scale was chosen in order to respect the Froude similitude, preserve the flow turbulence and eliminate surface tension effects. Given the relatively large scale of the model only a few cycles were represented in a rectangular test flume 1.35 m wide and 15 m long.

For this project, in view of the fact that flow approach conditions, especially the approach angle, do not have any significant impact on spillway discharge capacity, it may be considered that the total spillway flow rate can be extrapolated from the flow rate measured on just a few cycles.

3.3.1   Test conditions

All the comparative tests were the subject of a simulated discharge range from 40 l/s to 320 l/s at model scale, with about one point every 20 l/s, this being equivalent to a real life flow rate ranging from 1000 to about 7000 m³/s for the entire spillway.

For each discharge value, the upstream water level was measured by means of a precision point water level gauge. These values were then converted into head values.

The conditions of upstream and downstream flow, nappe aeration, submersion from downstream, downstream water heights were also analysed but are not discussed in this article.

3.3.2   Comparative tests of the two solutions proposed

These tests were carried out in order to obtain head/discharge curves for both spillway configurations.

According to figure 5, it is clear that both spillways are capable of passing a discharge of 5850 m³/s under a head of 65.50 m amsl.

For an equivalent head, configuration 2, which is totally compliant with the theoretical criteria, can pass a higher discharge than that of the spillway in configuration 1. This difference becomes all the greater as the total overspill discharge increases. However, it is also worth noting that the discharge capacity is well beyond the characteristic dimensional design point (Q = 5850 m³/s; H = 65.50 m).

Moreover, in view of the greater cycle depth, giving rise to a longer slab and side walls, the quantity of concrete required to build configuration 2, for the same total length, is 25% greater than that required for configuration 1.

The spillway in configuration 1 is therefore capable of passing the PMF and is less costly to build. The final choice is therefore for a spillway corresponding to configuration 1 which, although it does not satisfy all the theoretical criteria, is capable of guaranteeing the safety of the dam with a better cost-effectiveness ratio.

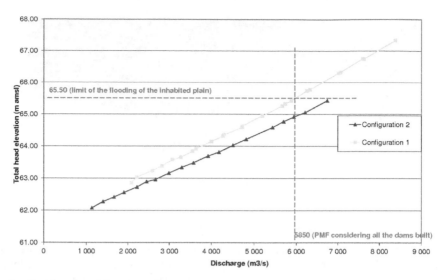

Figure 5.   Rating curve of the two configurations.

3.3.3   *Tests to improve the solution*

Additional tests were carried out to check civil works modifications to the selected solution. These led to finalisation of the definitive solution. The proposed improvements tested included:

– installation of half-rounded upstream cycle apexes instead of flat apexes; the aim of this modification was to improve flow guidance conditions between the upstream cycles, thereby enhancing the discharge flow coefficient.
– reduction of the cycle height (P) from 9 m to 7 m; the purpose of this modification was to study the effect of reducing the cycle height. This point was of particular interest because, in the event of favourable results, the modified solution would have enabled the excavation depth to be reduced by 2 m.
– provision of 3 m steps over half the cycle crest height; this modification was based on proposals made by Lempérière et Ouamane (2007) for the creation of a step made in the cycle apexes to reduce the quantity of steel reinforcement required.
– optimisation of the slab downstream of the spillway; the first tests were carried out with a sloping slab downstream of the spillways which, by causing torrential flow conditions, thus guaranteed no submersion downstream. In order to simplify the geometry and reduce the volume of excavation, new tests were carried out considering a horizontal slab over a distance of about 60 m before a fall into the tributary.

The findings of these tests were as follows:

– the improvement in flow inlet guidance on the upstream face of the cycles does not lead to any change in the head/discharge relationship; for an equivalent head, the difference in discharge is less than 1%. This improvement was nonetheless validated for qualitative considerations regarding the flow and also for its capacity to reduce log-jam blockages.
– The general reduction in cycle height considerably reduces the spillway flow capacity; the target of passing the Probable Maximum Flood can no longer be attained; this modification invalidates the project.
– The arrangement of localised steps did not produce the expected results and led to a significant deterioration in the head/discharge rating curve; the target of passing the Probable Maximum Flood can no longer be attained; this modification invalidates the project.
– The additional tests with a horizontal slab show that for this project, the slab slope does not have any effect on spillway submersion because the head is such that the formation of the hydraulic jump is pushed well beyond the slab. Torrential flow conditions are therefore established until

256

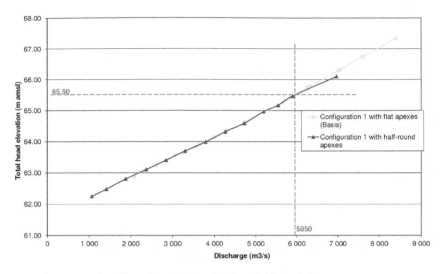

Figure 6. Rating curve of configuration 1 with and without half-rounded apexes.

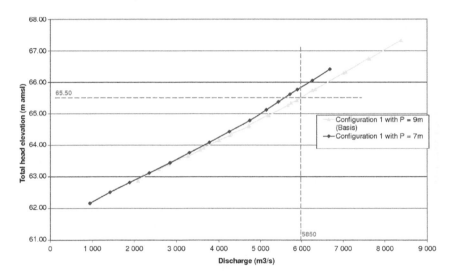

Figure 7. Rating curve of configuration 1 with a reduction in P from 9 m to 7 m.

the fall into the tributary, regardless of the slab slope. This information is important because it allows a substantial reduction in the volume of excavations required.

The additional tests had two results: first a consolidation of the structural design, and secondly, substantial savings in the volume of future excavations to be implemented at the time of the works.

The solution ultimately adopted corresponds to the basic configuration 1 equipped with rounded apexes.

3.3.4 Comparison with theoretical curves

The curves obtained from the results of physical model tests of configurations 1 and 2 were compared with the curves plotted from theoretical values proposed by different authors: Hay & Taylor (1970), Tullis et al. (1995) and Darvas (1971).

For configuration 1, the various theories – that are consistent with each other – give a discharge value overestimated by about +1% compared to the physical model results (Fig. 9).

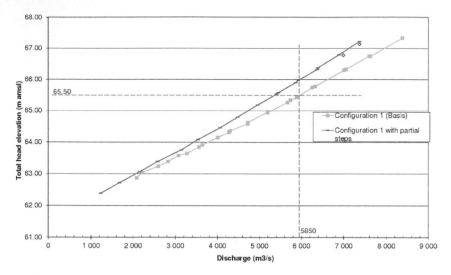

Figure 8.   Rating curve of configuration 1 with and without the partial step.

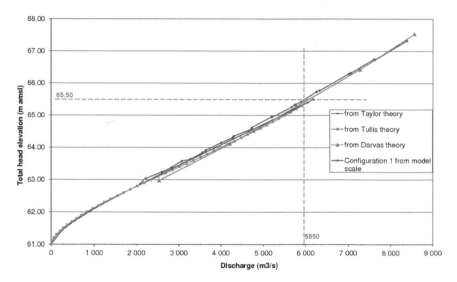

Figure 9.   Rating curve of configuration 1 compared with the theoretical curves.

In configuration 2, the various theories – that are consistent with each other – give a discharge value underestimated by about −7% compared to the physical model results (Fig. 10).

The physical model tested thus proves to be consistent with the predesign curves plotted by the main authors in the literature. Even so, an uncertainty of a few percent still remains. The difference may be perceived as being minimal but its importance becomes obvious when the excavation volume is linked to the additional concrete cost generated by this difference. This clearly shows that it is essential to test this type of structure on a physical scale model.

4   CONCLUSIONS

Use of a physical model in this case made a lot of sense in as much that the tests allowed Sogreah Consultants to opt for a design that did not completely satisfy all the theoretical criteria but allows for simplifications in the construction process and a reduction in the volume of concrete to be

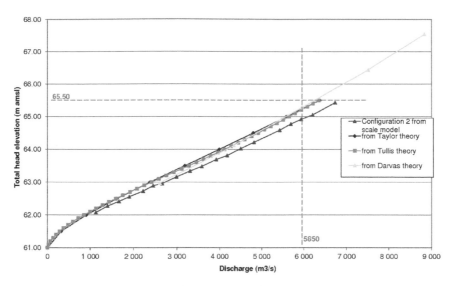

Figure 10. Rating curve of configuration 2 compared with the theoretical curves.

placed. Along the same lines, thanks to the additional tests carried out, it was possible to make substantial savings on excavation works required downstream of the spillway.

This article substantiates the predesign theories available in the literature. However, as their name indicates, they are only predesigns and physical modelling is an essential tool for the consulting engineer to refine his hydraulic design while at the same time making allowance for the structural cost criterion.

To complete the design of this barrage, Sogreah Consultants subsequently built a 1:80 moveable-bed physical model which was used to study potential flow problems in the tributary wadi used as an outflow channel, energy dissipation downstream of the spillway, the flow outlet into the main wadi and the risk of erosion.

Special acknowledgement: Our thoughts go to Antoine Vidal who actively participated in the project but who was forced to curtail his participation for health reasons.

REFERENCES

Cassidy, J.J., Gardner, C.A. & Peacock, R.T., 1985. Boardman labyrinth-crest spillway. *Journal of Hydraulic Engineering, Vol. 111, No. 3.*
Darvas, L.A., 1971. Performance and designof labyrinth weirs. *Journal of Hydraulic Engineering, Vol. 97, No. 8.*
Falvey, H.T., 2003. Hydraulic design of labyrinth spillways. *ASCE Press.*
Houston, K.L. 1982. Hydraulic Model of Ute dam labyrinth spillway. *USBR Report.*
Houston, K.L. 1983. Hydraulic model of Hyrum dam labyrinth spillway. *USBR Report.*
Khatsuria, R.M., 2005. Hydraulics of spillway and energy dissipators. *Marcel Dekker Edition.*
Lux, F. & Hinchcliff, D.L., 1985. Design and construction of labyrinth spillways. *ICOLD Report, Lausanne Congress.*
Magahales, A.P., 1985. Labyrinth weir spillways. *ICOLD Report, Lausanne Congress.*
Ouamane, M. & Lempérière, F., 2007. Amélioration du rendement des évacuateurs de crue en labyrinthe. *Colloque International sur l'Eau et l'Environnement.*
Savage, B., Frizell, K. & Crowder, J., 2004. Brain versus Brawn: the changing world of hydraulic model studies. *USBR Report.*
Tullis, J.P., Amanian, N. & Waldron, D., 1995. Design of labyrinth spillways. *Journal of Hydraulic Engineering, Vol. 121, No. 3.*

Labyrinth and Piano Key Weirs – PKW 2011 – Erpicum et al. (eds)
© 2011 Taylor & Francis Group, London, ISBN 978-0-415-68282-4

Labyrinth fusegate applications on free overflow spillways – Overview of recent projects

M. Le Blanc, U. Spinazzola & H. Kocahan
Hydroplus, Rueil-Malmaison, France

ABSTRACT: Fusegate system have been designed and installed by Hydroplus in more than 55 dams all over the world. These installations involve various types of dams, various type and size of Fusegates ranging from 0.60 m high to over 9 m high, and various purposes including increasing reservoir storage capacity or improving flood discharge potential. The Fusegate System offers a versatile solution, where the project requirements can be met by customizing the geometry of the gates or optimizing the construction methodology. The following topics are detailed in this paper: optimization of labyrinth crested Fusegates, progressive release of flows and varying stability margin and construction methodology.

1 INTRODUCTION

Fusegate system have been designed and installed by Hydroplus in more than 55 dams all over the world. These installations involve various types of dams, various type and size of Fusegates ranging from 0.60 m high to over 9 m high, and various purposes including increasing reservoir storage capacity or improving flood discharge potential. The Fusegate System offers a versatile solution, where the project requirements can be met by customizing the geometry of the gates or optimizing the construction methodology.

The following topics have been covered in this article:

- Optimization of labyrinth crested Fusegates: It is possible to optimize the shape of the Fusegate for the common floods, and depending on the project, all floods with return periods of less than 100 or 1000 years can be passed without any Fusegate tipping.
  That way, the labyrinth Fusegates are optimized for medium range flows. The discharge coefficient obtained by the Fusegates crest depends on the dimension of the gates and modification of the standards shape allows to improve their characteristics. For example, at Urra dam application in Colombia, the Fusegates are designed to have a very high discharge coefficient, where the discharge capacity has reached up to + 25% of the standard fusegates parameters.
- Progressive release of flows and varying stability margins: The spillway equipped with Fusegates allows for a progressive release of floods during a peak flood event and ensure that the spillway can pass the design inflow unencumbered. Unitary discharge flows can be increased up to more than 100 cumecs/m of the spillway depending on the project. Stability analysis of each fusegate allows to define operation of the fusegate including in extreme load cases.
- Construction methodology: various solutions have been developed in order to reduce the construction works on site. For example, an innovative method has been developed for Little Para Dam in Australia that involves the use of composite materials for the construction of 6.5 m high Fusegates.

2 OPTIMIZATION OF FUSEGATE'S LABYRINTH CREST

A spillway equipped with Fusegates operates as a fixed labyrinth weir on a free overflow spillway before any tipping occurs. This configuration corresponds to the normal routing of all the floods

Table 1. Type of application of Hydroplus Fusegate system realized up to date.

Application	Number of projects
Increasing existing spillway discharge capacity	30
Increasing reservoir storage capacity of the dam	12
Increasing both the reservoir storage and spillway discharge capacities	8
Optimization of design for a new spillway on the existing or on a new dam	3
Other applications	2

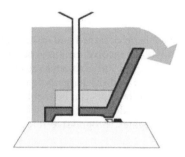

Figure 1. Overspilling over the crest of a Fusegate under normal conditions.

until the reservoir level is high enough to tip a Fusegate. For floods ranging between the first tipping flood (usually at least a 100 year flood) and the design flood (usually the PMF), the Fusegates then start tipping consecutively in order to have progressive release of the flows.

Design of Fusegates with high overspilling head (before tipping) depends on:

- site requirements: normal pool and maximum water levels, dimension of the spillway and attenuation offered by the reservoir
- environmental and economical project considerations
- Fusegates stability criteria

The lower the overspilling head for a determined flow is, the higher the storage capacity of the reservoir will be.

Various standard shapes for labyrinth Fusegates have been developed by Hydroplus in order to optimize the design in a given project. The main criteria in the optimization of the design involve the selection of the required discharge coefficient. These standard shapes have been generated by using trapezoidal shapes with extended length of labyrinth crest reaching as much as 3 times the width of the fusegate ($Ld = 3*W$).

For the design of Urra project, a specific shape of Fusegate was developed to increase the discharge coefficient. This configuration involves an extended length of crest up to $Ld = 3.7*W$ for 2 m high Fusegates. Model test were also performed to validate the discharge capacity of the subject Fusegates and they further highlighted the importance of the upstream cantilever section for the design and the minimum contribution of the lateral wall angles of the labyrinth section. The proposed design also offers advantages for the construction works as it minimizes the number of inlet wells, which at the end reduces the impact of floating debris.

For this project involving 2 m high Fusegates, a linear relationship between the flow discharge and the overspilling nape of more than 0.30 m has been established with the following formula:

$$Qs = 4.64 \times h - 0.52 \qquad (1)$$

where $Qs =$ unitary discharge flow for one meter [m³/s/m]; and $h =$ elevation of the overspilling nape on the crest of the Fusegate [m].

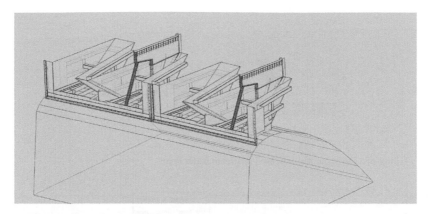

Figure 2.   Uptream view of Fusegates installed on the ogee crest of the spillway: optimized labyrinth shape for 100 years return period flood routing.

Figure 3.   Fusegates in operation during rare flood event in December 2010.

3   DISCHARGE CHARACTERISTICS AND STABILITY OF FUSEGATES

3.1   *Flood routing of Fusegated spillways*

During a major flood event, each Fusegate will operate independently. Tilting of a Fusegate can only occur when water gets into the inlet well steadily. Each Fusegate's tilting level is adjusted by setting the inlet wells at different elevations and by using different amounts of concrete ballast. The combination of both parameters ensures that each Fusegate has the required stability while preventing the Fusegates to tip at the same time.

As the tilting of each Fusegate occurs for different upstream water elevations, the release of flows through the spillway is progressive, and the flood attenuation capacity of the reservoir is maintained.

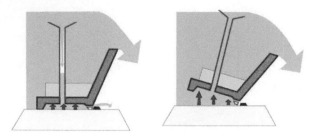

Figure 4.   Tilting of a Fusegate for increased upstream water level.

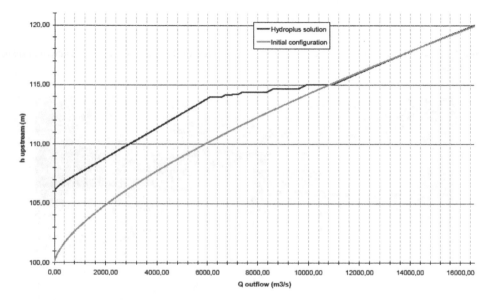

Figure 5.   Discharge rating curve of a 100 m wide spillway equipped with 8 Fusegates.

After a Fusegate tips, this section of the spillway operates as a free overflow spillway with unitary flow discharge described by the following formula:

$$Qs = Cd \times \sqrt{2 * g} \times h_1^{1,5} \qquad (2)$$

where Qs = unitary discharge flow for one meter [m³/s/m]; and h = elevation of the overspilling nape on the crest of the spillway [m] ($h_1 = h + h_{Fusegate}$ at a tipping event).

Depending on site conditions, the unitary discharge capacity can then reach to 100 m³/s/ml.

The discharge rating curve shown in Figure 4 illustrates the increasing flow discharge and the progressive release of flows through a Fusegated spillway. For a complete assessment, the operation of the reservoir, its attenuation capacity and the project hydrographs should also be taken into account. The following curve illustrates the discharge capacity of 6.5 m high Fusegates having 100% overspilling head (before first Fusegate tipping).

3.2   *Adjusting the stability of Fusegates*

Stability of a Fusegate is based on various balancing forces (such as weight of the Fusegate and the vertical component of the upstream water pressure in the bucket, etc.) and overturning forces (such as horizontal component of the upstream water pressure on the Fusegates body and the uplift pressure in the chamber, etc.). Several articles have been published on these topics, which are available at Hydroplus's website.

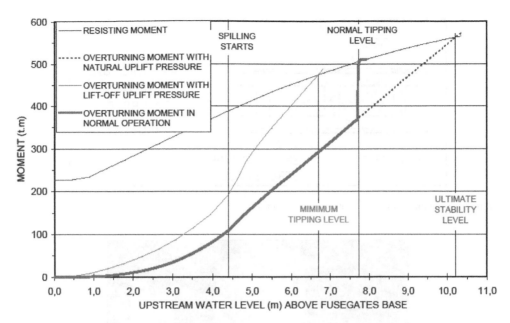

Figure 6. Stability levels for 4.40 m high Fusegates.

Table 2. Material used for construction of Hydroplus fusegates.

Type of material	No. of projects	Fusegate min height	Fusegate max height
Pre-cast concrete	8	1.1 m	2.7 m
Concrete cast-in-situ	16	1.8 m	7.6 m (9.74 m)
Combination of steel and concrete	7	1.5 m	6.5 m
Steel	24	0.6 m	6.5 m

The stability analysis of Fusegates will also help determining the extreme load cases, such as the ultimate stability level. The ultimate stability level is defined as the upstream water level of the reservoir for which the Fusegate will tilt without any uplift pressure in the chamber is generated. As long as the dry freeboard permits, it can be set below the dam crest level to prevent the risk of the dam from being overtopped during a PMF event. The ultimate stability level analysis is included in the scope of each project depending on the geometry of the Fusegate and the configuration of the spillway.

It is possible to ensure safe operation of the Fusegate System without reducing the discharge capacity for any loading cases such as floating debris.

The following figure illustrates minimum and ultimate stability levels obtained through model studies by Hydroplus.

4 CONSTRUCTION AND INSTALLATION OF FUSEGATES

Various construction and installation methods have been developed over the years to meet specific project requirements.

Installation of the Fusegates is generally performed on broad crested spillways. Toe abutments are constructed either as a part of the new spillway or installed afterwards by anchoring them into the sill. They transfer the loads exerted by the Fusegates to the spillway.

In the case of major flood events, the spillway will operate as a free overflow spillway (after all Fusegates tip) and the toe abutment will have an insignificant contraction effect.

Table 2 shows the distribution of Fusegate applications in terms of the material used.

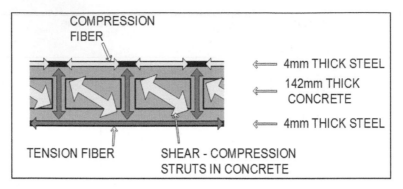

Figure 7. Principle of the composite wall construction.

Figure 8. Little Para Dam Hydroplus project: Steel elements in workshop.

In order to shorten the installation duration on site, Hydroplus have developed an innovative construction solution for a recent project in Australia. Apart from cost savings, this solution also allows for reducing significantly the thickness of the Fusegate lateral walls, while increasing the structural performance.

The 6.5 m high Fusegates have been constructed with a composite structure involving inox and concrete.

The final structure is shown in the following figure. Structural calculations have been done in accordance with the Australian construction standards and were reviewed by an independent structural engineer.

The external structures with the double purpose of coffer and reinforcement have been constructed in workshop with internal reinforcement and reservation. The steel elements have been assembled by welding on site with adapted rebars. Following the placement of the rebars in the assembled permanent steel panels, a high viscosity and high performance low shrinkage grout has been poured.

Metallic parts of the Fusegates are inox type LDX 2101 that offers many benefits as compared to the common Gr 316L inox such as:

- LDX 2101 has similar corrosion resistance to Gr 316L stainless
- Improved Yield strength of LDX 2101 = 450 MPa cw 250 MPa for 316
- LDX is less expensive than 316
- LDX uses approximately 70% of recycled material

Figure 9. Little Para Dam Hydroplus project: Assembling the elements on the final location.

Figure 10. Little Para Dam Hydroplus project: Completed Fusegate installation.

The highlights of the installation are shown in the above photos.

For a 52 m wide spillway to be equipped with five 6.5 m high Fusegates, this solution helped to reduce the construction works on site to only 8 weeks.

It is also important to emphasize that the Fusegate application has minimized the construction footprint, thus the environmental impacts.

5   CONCLUSION

Fusegate system offers large range of application on existing and new dam. Existing standard for fusegate offers a good balance between dimensions of the structure, discharge capacity and minimum preparation works on the spillway, but optimization can ever been done for improving the geometry and the construction works. Those optimizations have to take into account the specific requirement of dam site, the stability of the fusegates and the economical parameters.

By improving the discharge capacity of the labyrinth crest of the fusegates, the fusegate can offer better flood routing for common flood. The case of Urra illustrates the possible optimization of a fusegate and the complexity of the structural shape.

By improving the fabrication process, it is possible to reduce installation works. Those solutions offer new perspectives for construction of large fusegates.

REFERENCES

Lemperiere F., Bessiere C., (1992). "HYDROPLUS submersible Fusegates for surface spillways." *Modification of dams to accommodate major floods,* USCOLD, Fort Worth, Dallas, Texas

H. Falvey, P. Treille "Hydraulic and Design of Fusegates " *Journal of Hydraulic Engineering*, July 1995, pages 512/518

Aït Alla A. "The Role of Fusegates in Dam Safety". *Hydropower & Dams*. 3(6), 1996

S. Chevalier, J. Rayssiguier "Optimization of existing dams", *ICOLD Montreal* 2003

Hasan T. Kocahan "Hydraulics and design of fusegates", *USCOE Corps of Engineers Infrastructure Systems Conference* 2003

M. Barcouda, O. Cazaillet, P. Cochet, B.A. Jones, S. Lacroix, C. Odeyer, F. Laugier, J.P. Vigny, "Cost effective increase in storage and safety of most dams using Fusegates or P.K. Weirs", *ICOLD Barcelone* 2006

Silva-Monteil E., Piedrahita de Leon R., Le Blanc M., "Maximizing the benefits of Urra dam, Colombia", *Hydropower and dams*, Issue three, 2009

Le Blanc M., Silva-Monteil E., Solano A. Piedrahita de Leon R. "Recrecimiento del vertedero de la presa de Urra: Optimizar la capacidad de embalse preservando el medio ambiente en Colombia", *II International Congress on Dam Maintenance and Rehabilitation* Zaragoza 2010

Nomenclature, data base,
future developments

Labyrinth and Piano Key Weirs – PKW 2011 – Erpicum et al. (eds)
© *2011 Taylor & Francis Group, London, ISBN 978-0-415-68282-4*

A naming convention for the Piano Key Weirs geometrical parameters

J. Pralong, J. Vermeulen, B. Blancher & F. Laugier
EDF – Hydro Engineering Center, France

S. Erpicum, O. Machiels & M. Pirotton
Hydrology, Applied Hydrodynamics and Hydraulic Constructions (HACH), University of Liège, Belgium

J.-L. Boillat, M. Leite Ribeiro & A.J. Schleiss
Laboratory of Hydraulic Constructions (LCH), Ecole Polytechnique Fédérale de Lausanne, Switzerland

ABSTRACT: Flood management is more than ever an issue for dam designers and engineering consulting firms in charge of rehabilitation works. Piano Key Weirs are a new cost-effective type of spillway designed to improve dams discharge capacity. These structures are particularly attractive: they can easily be built on existing structures and enable very high discharge capacities. Therefore, Piano Key Weirs are nowadays studied worldwide. Piano Key Weir description involves a lot of geometrical parameters (more than 30), which designations are not already universally defined. A naming convention is required to enhance exchanges and cooperation between the numerous developers. A naming convention has been developed at *EDF – Hydro Engineering Center* in cooperation with the *Laboratory of Hydraulic Constructions (LCH), Ecole Polytechnique Fédérale de Lausanne* and the *Laboratory of Hydrology, Applied Hydrodynamics and Hydraulic Constructions (HACH), University of Liège*. This paper describes the proposed naming convention and gives definitions and notations of the various geometrical parameters. This work represents a first attempt which should be updated with the contribution of stakeholders involved in this topic.

1 INTRODUCTION

In the frame of rehabilitation programs and new dams design, Piano Key Weirs represent a new advantageous alternative for spillways. PKW provide very high discharge capacity and can easily be built on the crest of existing dams. Moreover, they represent a cost-effective option for designers. This explains the worldwide interest for PKW development and implementation.

The scientific interest about PKW requires a unicity in terminology. Indeed, PKW geometry involves a lot of parameters making its configuration description uneasy.

A naming convention needs to be applied by the different stakeholders involved in PKW development. A workgroup gathering *EDF – Hydro Engineering Center* in cooperation with the *Laboratory of Hydraulic Constructions (LCH), Ecole Polytechnique Fédérale de Lausanne* and the *Laboratory of Hydrology, Applied Hydrodynamics and Hydraulic Constructions (HACH), University of Liège* developed a specific nomenclature. The naming convention aims to propose a uniform description to designers while keeping the number of parameters to a reasonable amount. The idea is to describe the general shape of PKW with a fixed number of geometrical inputs.

2 PRINCIPLES OF THE NAMING CONVENTION

The naming convention only cares about the structure of the PKW in itself. It does not consider any annex structures that could be required in some cases (bridge pier, anchorage, etc.). This nomenclature does not describe the lateral boundaries of the PKW neither as these parts are likely to differ from an implementation to another. However, it includes notations for physical parameters such as head or depths.

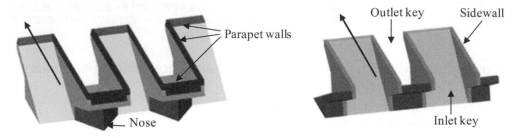

Figure 1.   Components considered in the convention (left) and constitutive basic elements (right).

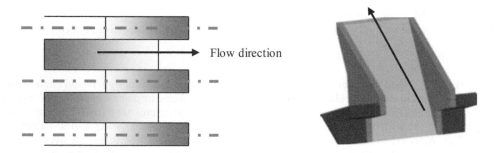

Figure 2.   Plan view of the segmentation principle (left) and 3D-view of a PKW unit (right).

The structure of the PKW can be described as the gathering of different elements. The components considered in this nomenclature are the basic structure, the parapet walls and the noses (Fig. 1 left). The inlet keys, outlet keys and sidewalls compose the basic structure (Fig. 1 right) while noses and parapet walls are both optional in the PKW design. They are however considered in order to define as much configurations as possible.

In order to keep the nomenclature as simple as possible, the number of geometrical parameters was limited to fundamental characteristics. The choice has been made to assign an index to the notations referring to the component or to the subcomponent (inlet key, outlet key, etc.) they are part of. Some of the parameters are overlapping but it has appeared that they all have advantages depending on circumstances. The number of parameters is thus not a minimum, but still remains reasonable.

The proposed description of the PKW structure relies on the assumption that PKW have a regular shape that can be divided into similar representative "PKW units" (Fig. 2 left). The unit represents the smallest extent of a complete structure and is composed of an entire inlet key with a sidewall and half an outlet key on both sides (Fig. 2 right).

This PKW unit is interesting as the whole structure can be reconstituted from it by juxtaposition. It is sufficient to describe most of the geometrical parameters of PKW. Only a few global dimensions need to be added to the unit description to fully define a PKW configuration.

3   FUNDAMENTAL PARAMETERS

All the parameters described in this part are related to the global structure of the PKW. They are the most commonly used and reveal sufficient to provide an overall description of the weir. The parameters can be divided into categories depending on whether they represent lengths, widths, heights or thicknesses. In accordance with the nomenclature used for Labyrinth weirs (Falvey, 2003) and considering different previous publications about Labyrinth and PK-Weirs (US Dpt. of Interior 1987, Tullis 1995, Lempérière & Ouamane 2003, Leite Ribeiro et al. 2009) the following rules have been applied: B is used for lengths; W for widths; P for heights; L for the developed lengths; S for slopes and T for thicknesses.

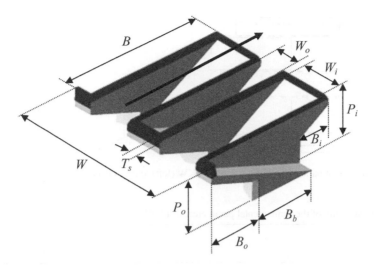

Figure 3.   Fundamental parameters on an entire PKW – 3D-view.

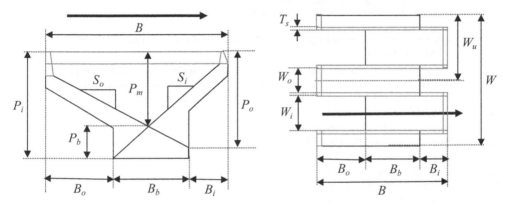

Figure 4.   Fundamental parameters on an entire PKW – plan view (left) and cross section (right).

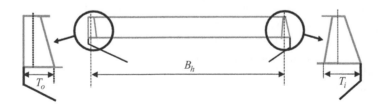

Figure 5.   Detail cross sections of the PKW crests.

Indexes are applied to define the location of the geometrical parameter: for instance, index i refers to inlet key; index o stands for outlet key and index b refers to parameters linked to the base of the structure.

Based on these rules, 23 parameters have been defined to describe the PKW global structure general shape (Fig. 3 to Fig. 6 and Tab. 1). Parapet walls are taken into account as they influence the height of the weir, but are not described in this part.

The last five parameters of Table 1 are not required for the PKW description as they can be deduced from the previous ones. However, they are interesting for comparisons between configurations as they give a first idea of the dimensions and geometric aspect of the weir.

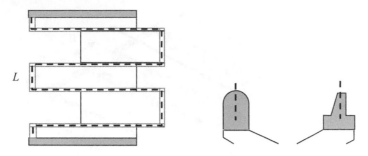

Figure 6.   Plan view of parameter L over an entire PKW (left) and typical crest cross sections (right).

Table 1.   Nomenclature of the fundamental parameters of PKW geometry.

Parameter symbol	Meaning
B	Upstream-downstream length of the PKW $B = B_b + B_i + B_o$
B_o	Upstream (outlet key) overhang crest length
B_i	Downstream (inlet key) overhang crest length
B_b	Base length
B_h	Sidewall overflowing crest length measured from the outlet key crest axis to the inlet key crest axis (Fig. 5)
P_i	Height of the inlet entrance measured from the PKW crest (including possible parapet walls)
P_o	Height of the outlet entrance measured from the PKW crest (including possible parapet walls)
P_b	Height of the apron level at inlet key and outlet key intersection
P_m	Difference between P_i and P_b
S_i	Slope of the inlet key apron (length over height)
S_o	Slope of the outlet key apron (length over height)
W	Total width of the PKW
W_u	Width of a PKW unit
W_i	Inlet key width (sidewall to sidewall)
W_o	Outlet key width (sidewall to sidewall)
T_s	Sidewall thickness
T_i	Horizontal crest thickness at inlet key extremity (measured at the basis of possible parapet walls)
T_o	Horizontal crest thickness at outlet key extremity (measured at the basis of possible parapet walls)
L	Total developed length along the overflowing crest axis
L_u	Developed length of the PKW unit along the overflowing crest axis $L_u = W_i + W_o + 2B_h + 2T_s$
N_u	Number of PKW units constituting the structure
n	Developed length ratio of the PKW: $n = \frac{L}{W}$
n_u	Developed length ratio of a PKW unit: $n_u = \frac{L_u}{W_u}$

*Some physical parameters also need to be added to this naming convention. The proposed se*lection has been made to limit their number to a minimum (Fig. 7 and Tab. 2).

4   OPTIONAL DETAIL PARAMETERS

The fundamental parameters above enable to describe the global configuration of the PKW, but are not sufficient to define nor the geometry of optional features nor the structure of the overhangs. Additional parameters are then necessary to complete a full description of the structure. They are presented hereafter, based on the current knowledge about PKW optimization.

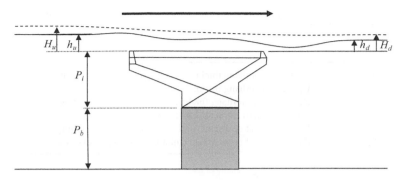

Figure 7. Cross section of the physical parameters nomenclature.

Table 2. Physical parameters nomenclature.

Parameter symbol	Meaning
P_d	Dam height
H_u	Total head over crest upstream from weir
h_u	Upstream flow depth over crest
H_d	Total head over crest downstream from weir (can be negative)
H_d	Downstream flow depth over crest (can be negative)
Q	Flow discharge
q_{sW}	Specific discharge referred to total width of the PKW $q_{sW} = \frac{Q}{W}$
C_{dW}	Discharge coefficient related to PKW total width as $Q = C_{dW}.W.\sqrt{2g}.H^{3/2}$
q_{sL}	Specific discharge referred to developed length of the PKW $q_{sL} = \frac{Q}{L}$
C_{dL}	Discharge coefficient related to PKW developed length as $Q = C_{dL}.L.\sqrt{2g}.H^{3/2}$

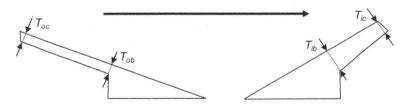

Figure 8. Overhangs base and extremities thicknesses – upstream (left) and downstream (right).

Figure 9. Overhangs extremity batter angle – upstream (left) and downstream (right) cross sections.

4.1 Overhangs structural description

There is a need for engineers to be able to assess concrete volumes involved in a PKW structure. This requires a description of overhangs thicknesses. Moreover overhangs mass balance is crucial for PKW stability. The information provided by the description should enable a first assessment of these two items. The proposed parameters concern the thicknesses of the overhangs and the shape at their extremities (Fig. 8, Fig. 9 and Tab. 3). Index b still refers to the base of the PKW and index c stands for the crest of the keys.

Table 3. Nomenclature of overhangs thicknesses and batter angle parameters.

Parameter symbol	Meaning
T_{ob}	Upstream overhang thickness at upstream base extremity
T_{oc}	Upstream overhang thickness at upstream extremity
T_{ib}	Downstream overhang thickness at downstream base extremity
T_{ic}	Downstream overhang thickness at downstream extremity
Γ_o	Batter angle at upstream (outlet key) overhang extremity
Γ_i	Batter angle at downstream (inlet key) overhang extremity

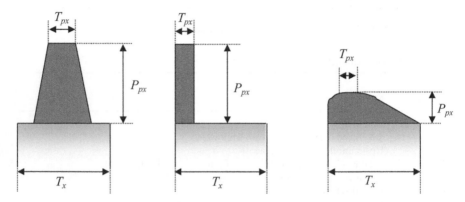

Figure 10. Parapet walls parameters in cross section.

Table 4. Nomenclature of parapet walls parameters.

Parameter symbol	Meaning
P_{px}	Height of the parapet wall on x-component
T_{px}	Crest thickness of the parapet wall on x-component

4.2 *Parapet walls*

Parapet walls have been developed while optimizing the hydraulic efficiency of PKW. They consist of vertical extensions that can be placed over the crest of the PKW. Three components of parapet wall can be distinguished: the inlet key one, the outlet key one, and the sidewall one. The two first enable to modify the longitudinal cross-sections of the keys while the latter especially aims in modifying the crest profile of the sidewall. The three components are independent, and do not need to have the same height. In this nomenclature, index p refers to parapet walls.

The naming convention describes each of the three features following the same parameterization. Then, index i refers to the downstream parapet wall of inlet key, index o to the upstream parapet wall of outlet key, and index s to sidewall parapet wall. Hence, for $x \in [i, o, s]$ (Fig. 10 and Tab. 4).

As shown on Figure 10, there is an infinite number of possible shapes for parapet walls. The two proposed parameters are obviously not enough to fully describe the geometry of the parapet walls. This nomenclature cannot take into account such a variety of shapes. The naming convention gives the global dimensions of the parapet walls and recommends to the authors to provide a detailed scheme to define their geometries.

4.3 *Noses and side wall angle*

Noses are features placed under the upstream overhangs in order to increase the discharge capacity of PKW by improving the flow pattern at inlet keys entrance. Their shape can vary from triangular

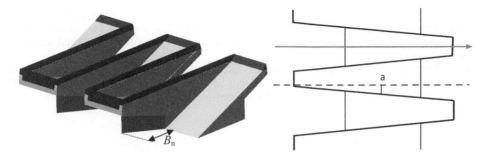

Figure 11.   Nose parameter – 3D-view (left) and sidewall angle – plan view (right).

Table 5.   Nomenclature of nose parameter and sidewall angle.

Parameter symbol	Meaning
B_n	Length of the nose
α	Sidewall angle

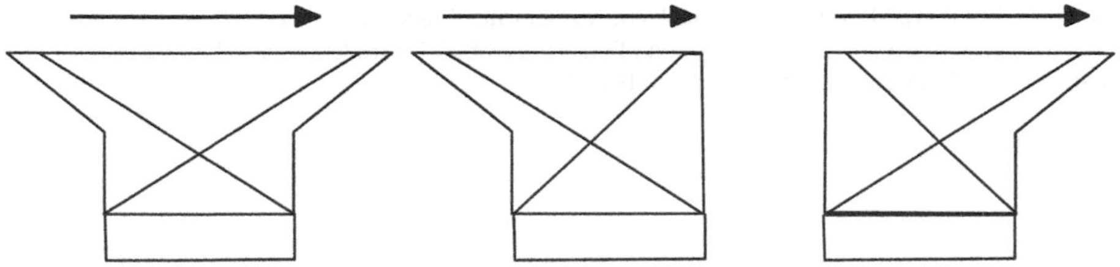

Figure 12.   Typical shapes of PKW: Type A (left); Type B (middle); Type C (right).

profile to rounded one. In this naming convention, their description only requires one parameter (Fig. 11 left and Tab. 5), but the nose shape needs to be mentioned. In the case when other parameters would be required to define the nose shape, they would be characterised with index n.

Labyrinth design is familiar with the notion of sidewall angle α. This parameter can be transposed to PKW structures. Indeed, it is likely to improve the discharge capacities, especially for high upstream heads. The notation used by Falvey (2003) has been adopted for PKW (Fig. 11 right and Tab. 5).

5   REFERENCE LAYOUTS

The global cross section layout enables to distinguish several categories of PKW. A basic classification has been proposed by Lempérière a few years ago with respect to the arrangement of the upstream and downstream overhangs (Fig. 12):
   Type A – with symmetrical upstream and downstream overhangs
   Type B – with only upstream overhang
   Type C – with only downstream overhang

6   CONCLUSIONS

The developed naming convention enables to describe globally a PKW using 22 parameters. Additional parameters allow engineers and developers to assess concrete volume and mass balance and to integrate optional features to their design.

Recent PKW optimization investigations have focused on noses and parapet walls, and this is the reason why specific parameters are proposed for these optional features. Due to the variety of designs, this nomenclature recommends to join schemes to the descriptions of special features or new design options.

Future developments will probably introduce other particularities and require completing the parametric description. The naming convention will have to evolve with PKW development, but already accounts for a very large number of PKW configurations.

ACKNOWLEDGEMENTS

Special thanks are extended to F. Lempérière, M. Ho Ta Kanh, G. Degoutte and J. P. Tullis for their helpful contributions to the redaction of this naming convention.

REFERENCES

Falvey, T. H. (2003). Hydraulic design of labyrinth weirs. *ASCE Press*.

Lempérière, F., Ouamane, A. (2003). The PK Weir: a new cost-effective solution for spillways. *Hydropower & Dams*, Issue five.

Tullis, J. P. (1995). Design of Labyrinth Spillways. *Journal of Hydraulic Engineering*.

United States Department of Interior, (1987). Design of Small Dams. *A water resources technical publication*, Third edition.

Leite Ribeiro, M., Bieri, M., Boillat, J-L., Schleiss, A., Delorme, F. and Laugier, F. (2009). "Hydraulic capacity improvement of existing spillways – Design of piano key weirs". 23rd Congress of Large Dams. Question 90, Response 43. 25–29 May 2009. Brasilia, Brazil.

Labyrinth and Piano Key Weirs – PKW 2011 – Erpicum et al. (eds)
© *2011 Taylor & Francis Group, London, ISBN 978-0-415-68282-4*

Creation of a PKW Database – Discussion

J.-L. Boillat & M. Leite Ribeiro
*Laboratory of Hydraulic Constructions (LCH), Ecole Polytechnique Fédérale de Lausanne
(EPFL), Switzerland*

J. Pralong
Electricité de France (EDF), Centre d'Ingénierie Hydraulique, France

S. Erpicum & P. Archambeau
*Laboratory of Hydrology, Applied Hydrodynamics and Hydraulic Constructions (HACH), ArGEnCo
Department, Liège University, Belgium*

ABSTRACT: Piano Key Weir description involves a lot of geometrical parameters which designations are now well defined. A naming convention has been developed to enhance exchanges and cooperation between the numerous developers. This nomenclature was presented and discussed during the workshop and the updated version is included in the present proceedings (Pralong et al., pp. 277–284). In order to make easier the access to experimental or field data from different sources, a Database is required. This document relates the proposal which was presented and discussed during the workshop. The carrying out of this Database is expected in a near future.

1 INTRODUCTION

The naming convention for PKW presented in the proceedings (Pralong et al., pp. 277–284) was developed and approved in order to enhance exchanges and cooperation between the numerous developers and for the designers of PKWs. This nomenclature offers a unicity in terminology for the hydraulic and geometrical related parameters.

The next step to valorize the existing information consists in the development of a Database to gather the results of studies performed all around the world. The idea for a Database lies on the existing need to share knowledge for PKW design and for PKW understanding.

2 DATABASE OVERVIEW

A Database for PKW is simplified by the fact that the geometry can be globally described through a reasonable amount of parameters. Likewise from a hydraulic point of view, essentially discharge capacity curves are required.

The database layout presented during the workshop is divided into two Excel sheets. The first, so-called "*Parameters*" aims to describe the PKW structures and the framework of the projects, the second, so-called "*Measurements*" contains the rating curves providing the relation between the overflowing discharge and the upstream total hydraulic head.

The *Parameters* sheet is organized as summarized in Table 1 and the *Measurements* sheet content is presented in Table 2. The Database enables quick comparisons thanks to Filters on geometrical parameters values as well as on upstream heads and discharge capacity. It also allows identifying the sources of Data with contact addresses and publication references.

Table 1. Content summary of the Excel *Parameters* sheet.

Column	Description
[°]	Name of the series, It contains the provenance of the data (E = experimental, P = prototype and N = numerical), the source and year
Description	Framework of the project
Model scale	Geometrical scale of the study
Geometry	Columns dedicated to the geometrical parameters
Main dimensionless parameters	Columns automatically filled from the corresponding geometrical parameters
Approach conditions	Description of the approach flow, either "channel" or "reservoir"
General comments	Important remarks about measurements or type of the structure
Source	Source of data
Website	Internet address of the Institute or Company owner of the data
Contact	Contact person at the Institute or Company

Table 2. Content summary of the Excel *Measurements* sheet.

Column	Description
[°]	Same name mentioned in the sheet "*Parameters*"
H_u [m]	Total upstream hydraulic head ($h_u + V^2/2g$). If only the upstream flow height (h_u) is available, it will be mentioned in "*Parameters*" sheet
Q_{PKW} [m³/s]	Discharge overflowing the PKW for a given total hydraulic head
H_u/P_i	Ratio between upstream hydraulic head and height of the inlet key. This column is automatically filled
q_{sW} [m³/s.m]	Specific discharge related to PKW width (W). This column is automatically filled

3 TERMS AND CONDITIONS FOR USING THE PKW DATABASE

The use of the Database will be submitted to rules concerning the *Introduction of Data*, the *Use of Data*, the *Responsibility* and some additional agreements. It has still to be defined if the access to the Database will be open or restricted to the Data providers.

4 DATABASE ACCOMODATIONS

The Database will be hosted at the University of Liège where an administrator will be in charge of the logistic aspects.

5 CONCLUSIONS AND PERSPECTIVES

A working group, constituted of members of the author's institutions, will develop a detailed project which will then be submitted to all potential Data providers. After validation, the carrying out of this Database is expected in a near future.

REFERENCE

Pralong, J., Vermeulen, J., Blancher, B., Laugier, F., Erpicum, S., Machiels, O., Pirotton, M., Boillat. J-L., Leite Ribeiro M. and Schleiss, A.J. 2011. Proposal of a naming convention for the Piano Key Weir geometrical parameters. *International Workshop on Labyrinth and Piano Key weirs*, Liège, Belgium.

Labyrinth and Piano Key Weirs – PKW 2011 – Erpicum et al. (eds)
© 2011 Taylor & Francis Group, London, ISBN 978-0-415-68282-4

Development of a new concept of Piano Key Weir spillway to increase low head hydraulic efficiency: Fractal PKW

F. Laugier, J. Pralong, B. Blancher & F. Montarros
EDF – Hydro Engineering Center

ABSTRACT: Piano Keys Weir (PKW) spillways are a technical and economical optimization of traditional labyrinth spillways and an interesting opportunity to increase the discharge capacity of many dams. The concept was first introduced by Lemperiere and Ouamane in 2003. Compared to traditional labyrinth spillways, their main advantage is that they can be installed at the top of most existing concrete dams (especially gravity dams). A new concept of PKW has been developed aiming at increasing PKW low head efficiency. It is called fractal PKW. It consists of adding a smaller PKW on the crest of a bigger PKW spillway. The concept was first tested through 3D numerical simulations with Flow3D® software. Several different configurations were tested and converged to an improved design. Finally, this configuration was tested on a physical model, which confirms the results given by the 3D simulation. It appears that the fractal PKW is very efficient for low upstream head smaller than 1 m, especially lower than 0.50 m. Above this point, the flow over the fractal PKW tends to saturate and this option is not very interesting any more. As the discharge capacity is important for heads of few centimeters, fractal PKW are an interesting option for reservoir level control to replace gates or other mechanical components. Finally, fractal PKW configuration seems to be a promising option as civil engineering additional cost is limited by using small prefabricated steel units anchored on a traditional PKW unit with thickened walls.

1 INTRODUCTION

PKW labyrinth spillways are innovative solutions which allow getting discharge capacity from 4 to 5 times higher than usual free flow Creager spillway one. Their main feature, thanks to their optimized design, as compared to traditional vertical wall labyrinths spillways, is that they can be installed on the crest of main concrete dams, especially gravity ones.

In France and for EDF dams, water head above free flow crest is generally low and is typically worth 0.5 to 1.5 m. Aiming at improving discharge capacity for smaller water heads, the concept of "fractalizing" the PKW crest is introduced. It consists of creating a labyrinth of labyrinth spillway in order to drastically increase the overall total crest developed length.

It actually appears that the total crest developed length, is a major factor of PKW discharge capacity for moderate water heads. Fractalizing PKW might thus appear to be a way to achieve the aforementioned goal.

The issue raised by this study is to know whether expected improvements are worth developing this concept further more.

This study is based on both numerical and physical models. 3D numerical simulations were first carried out to test many different configurations within limited time and cost development.

Finally a configuration was chosen and physically tested in Liège university hydraulic laboratory.

2 FRACTALIZATION PRINCIPLE

Fractalization principle consists of superimposing a smaller PKW unit on the crest of a usual PKW spillway. This concept is called "fractal PKW".

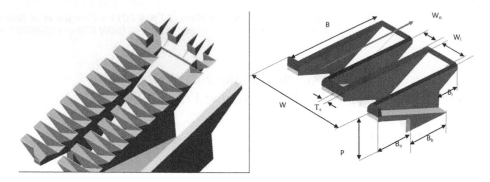

Figure 1.   a) CAD view of fractal PKW concept – b) PKW main parameters.

Naming convention corresponds to the one proposed by Pralong et al. (2011) whose mains parameters are reminded here below.

Geometrical parameters of fractal PKW will be marked with "f" symbol. However, their geometry might not be entirely defined by usual PKW parameters. In order to simplify their description, a plan view will be presented for each case.

3   REFERENCE CONFIGURATION C_0

Small fractal PKW are installed on the crest of a reference configuration C_0. This configuration will be used to quantify benefits of each tested fractal PKW configuration. The reference configuration was chosen to keep constant as many geometrical parameters as possible. It is a slightly modified configuration of Model A PKW proposed by Lempérière et al. – (2003). Modifications include wall thickness and a 1 m high parapet wall to allow a fair comparison with a 1 m high fractal PKW.

$W_i = 2.4$ m $W_o = 2.4$ m $B_o = 3$ m $B_i = 3$ m $B = 12$ m
$P = 4 + 1$ (parapet) $T_s = 0.5$ m $n = 5.14$

In configurations 1 to 8, fractal walls are supposed to be extremely thin (steel plates) compared to concrete structure of the PKW support.

4   3D NUMERICAL SIMULATION

A *FLOW-3D®* model has been developed at EDF-CIH in 2008, calibrated on experimental data from Lempérière et al. – (2003) before being validated on data collecting from physical model studies. The model has demonstrated a good correlation with the experimental results.

For each fractal configuration, two water heads are tested: 0.5 m and 1 m (corresponding to $H/P = 0{,}1$ and $0{,}2$)

Overall results are given after the description of all configurations.

4.1   *Fractal PKW configuration C_1*

This first configuration was chosen because its geometrical features are straightforward and easy to design. Main fractal geometrical parameters are as follows:

$W_{if} = 0.56$ m $W_{of} = 0.56$ m $B_{if} = 1$ m $B_{of} = 1$ m $B_f = 2.5$ m
$P_f = 1$ m $n_f = 5.2$ $n_{total} = n \times nf \sim 25$

4.2   *Fractal PKW configuration C_2*

Configuration C_2 is also quite simple. It consists in installing basic crenels on the crest of the PKW. Upstream and downstream fractal overhangs are removed to decrease flow disturbances in main

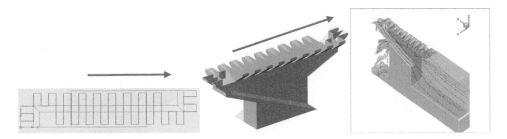

Figure 2.   Fractal PKW – Configuration C_1 plan and 3D view – 3D view of numerical flow patterns.

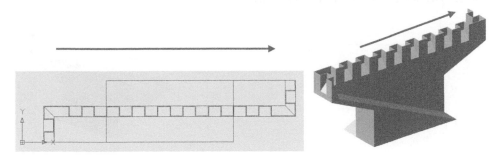

Figure 3.   Fractal PKW – Configuration C_2 plan and 3D view.

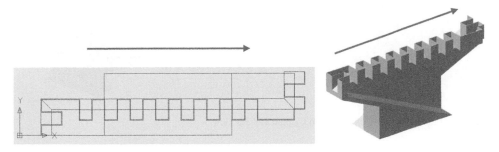

Figure 4.   Fractal PKW – Configuration C_3 plan and 3D view.

PKW inlets and outlets. Overall total developed length is rather smaller than configuration C_1. Main fractal geometrical parameters are as follows:

$W_{if} = 0.56\,m$ $W_{of} = 0.56\,m$ $B_{if} = 0\,m$ $B_{of} = 0\,m$ $B_f = 0.5\,m$

$P_f = 1\,m$ $n_f = 1.8$ $n_{total} = n \times nf \sim 9$

4.3   *Fractal PKW configuration C_3*

Configuration C_3 is something of a compromise between configurations C_1 and C_2. Upstream overhangs are removed and downstream are half reduced.

Main fractal geometrical parameters are as follows:

$W_{if} = 0.56\,m$ $W_{of} = 0.56\,m$ $B_{if} = 0\,m$ $B_{of} = 0.5\,m$ $B_f = 1\,m$

$P_f = 1\,m$ $n_f = 2.7$ $n_{total} = n \times nf \sim 14$

4.4   *Fractal PKW configuration C_4*

Configuration C_4 aims at increasing the length of hydraulically efficient crest length because previous configurations showed an important fraction of fractal crest length whose specific discharge was very low. Overall developed length is reduced.

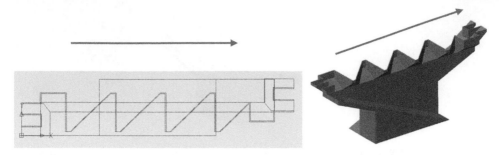

Figure 5.   Fractal PKW – Configuration C4 plan and 3D view.

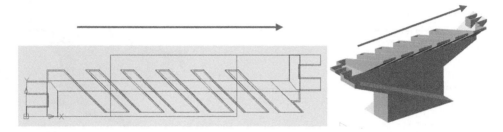

Figure 6.   Fractal PKW – Configuration C_5 plan and 3D view.

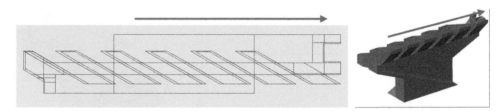

Figure 7.   Fractal PKW – Configuration C_6 plan and 3D view.

Main fractal geometrical parameters are as follows:
$B_{if} = 0.5$ m $B_{of} = 1$ m $B_f = 2$ m
$P_f = 1$ m $n_f = 2.1$ $n_{total} = n \times nf \sim 10$

4.5   *Fractal PKW configuration C_5*

Configuration C_5 corresponds to configuration C_1 whose fractal units are oriented towards a 45° angle to better fit with the main direction of the incoming flow. Outlet overhangs are also reduced from 1 to 0.6 m to decrease flow obstruction in PKW inlets.
Main fractal geometrical parameters are as follows:
$W_{if} = 1$ m $W_{of} = 0.6$ m $B_{if} = 0.5$ m $B_{of} = 1$ m $B_f = 2$ m
$P_f = 1$ m $n_f = 4.5$ $n_{total} = n \times nf \sim 23$

4.6   *Fractal PKW configuration C_6*

Configuration C_6 approximately corresponds to configuration C_5 with an angle of 30° rather than 45°. The outlet width of supporting PKW is also reduced from 2.4 m to 1.6 m ($n = 6,2$)
Main fractal geometrical parameters are as follows:
$W_{if} = 1$ m $W_{of} = 0.6$ m $B_{if} = 0.5$ m $B_{of} = 0.4$ m $B_f = 1.4$ m
$P_f = 1.2$ m $n_f = 4.5$ $n_{total} = n \times nf \sim 28$

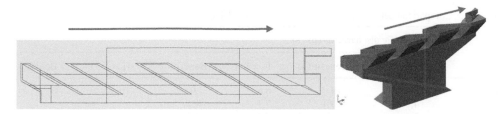

Figure 8.   Fractal PKW – Configuration C_7 plan and 3D view – Cross section.

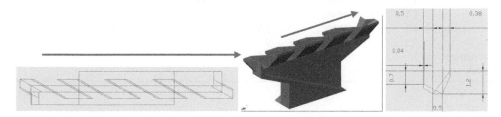

Figure 9.   Fractal PKW – Configuration C_8 plan and 3D view.

4.7   Fractal PKW configuration C_7

Configuration C_7 differs from configuration C_6 by reducing the number of fractal units on PKW crest. Thus fractal PKW are significantly wider for this configuration.

Main fractal geometrical parameters are as follows:

$W_{if} = 1.6\,m$ $W_{of} = 1.2\,m$ $B_{if} = 0.5\,m$ $B_{of} = 0.4\,m$ $B_f = 1.4\,m$
$P_f = 1.2\,m$ $n_f = 2{,}7$ $n_{total} = n \times nf \sim 17$

4.8   Fractal PKW configuration C_8

Configuration C_8 follows the same concept as configuration C_7 with several minor modifications affecting the main PKW support (inlet width reduced from 2.4 to 1.8 m, wall thickness reduced from 0.5 to 0.4 m $=> n = 6{,}7$) and fractal PKW units (reduction of upstream overhang length).

Main fractal geometrical parameters are as follows:

$W_{if} = 1.6\,m$ $W_{of} = 1\,m$ $B_{if} = 0.4\,m$ $B_{of} = 0.4\,m$ $B_f = 1.2\,m$
$P_f = 1.2\,m$ $n_f = 2.7$ $n_{total} = n \times nf \sim 18$

4.9   Summary of numerical simulations

Results of numerical simulations are given in Table 1.

Numerical simulations show that the fractal concept really improves PKW discharge capacity. Expected benefits can be more than 100% of the initial basic PKW for low heads such as 0.5 m. This benefit is approximately divided by 2 for 1 m head and by 4 for 1.5 m head. This result seems logic as flow in small fractal PKW tend to quickly saturate when upstream water head increase.

From all tested configurations, configurations C_4 to C_8 seem to be more or less equivalent. Configuration C_5 and C_7 are probably slightly less expensive to build and we would therefore recommend it for the moment.

5   PHYSICAL MODEL – VALIDATION OF 3D NUMERICAL MODEL

A physical model of a fractal PKW was tested in 2010 in the hydraulic laboratory of Liège university (Belgium). The fractal part of the model was physically built by the mean of stereolithographic technics which allow forming very thin and complex structures.

The physical model corresponds to a new fractal configuration C_9, equivalent to a project optimization. 3D simulations were performed to compare numerical and physical results in order to validate the use of 3D numerical simulations for very complex hydraulic structures.

Table 1. Comparison between physical and numerical model specific discharges.

Config.	Specific discharge qs (m^2/s)					
	H = 0,5 m	Benefit/Co	H = 1 m	Benefit/Co	H = 1,5 m	Benefit/Co
C_0	2,6	0	6,6	0	10, 6	0
C_1	4,7	80%	7,1	8%		
C_2	3,7	40%	7,1	9%		
C_3	4,2	59%	7,3	11%		
C_4	4,7	79%	7,7	18%		
C_5	5,4	108%	9,5	45%	12,0	14%
C_6	5,7	118%	9,6	47%	13,2	25%
C_7	5,5	110%	9,7	48%	13,5	28%
Lempèriére Type A	3,6		7,7		12,5	

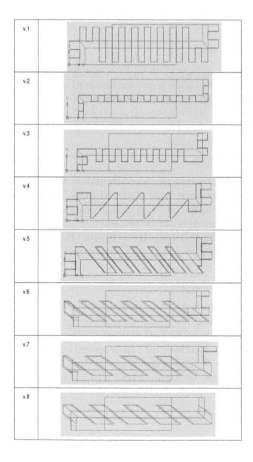

Results are given in figure 12. They show a very good correlation between numerical and physical results (10% error bar in figure 12) in spite of extreme complexity of flow patterns.

These results validate the use of 3D numerical simulations for fractal PKW. They also validate configuration C_1 to C_8 results and the choice of the best configuration.

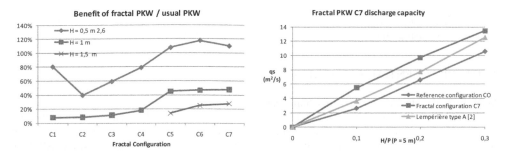

Figure 10. a) Benefit of fractal PKW – b) Fractal PKW C_7 discharge capacity.

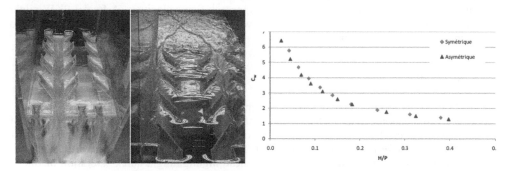

Figure 11. a) Physical model of fractal PKW – b) Discharge capacity.

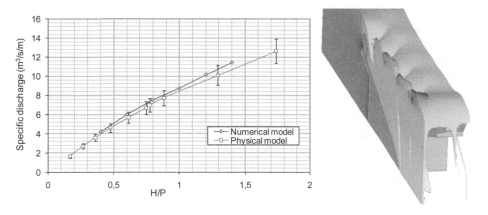

Figure 12. a) Comparison numerical / physical simulation – b) View of 3D simulation.

6 ECONOMICAL ISSUES

From an economical point of view, fractal PKW concept seems to be interesting as it is expected to design fractal units with steel prefabricated units. These units could be anchored on a usual PKW spillway; whose walls might be slightly enlarged for structural reasons.

Prefabrication should allow getting competitive costs of production. Unit pieces will be typically 1 m high and 2 m long. Thickness should be designed and optimized taking into account vibration issues. For water heads smaller than 1 m, corresponding loads are minor and the steel thickness should be less than 5 mm.

Installation on site is a task that could be achieved within one or two days by anchoring individual pieces weighting less than 50 kg and which can be easily handled by basic means.

Finally, the overcost coming from the steel fractal units might be minor compared to the overall PKW project, maybe 5 to 10%. This overcost has to be compared with the discharge capacity benefit which can be more than 100% for appropriate water heads. Fractal PKW cost might also be reduced by using other composite material such as glass fiber.

7 OTHER POTENTIAL APPLICATION

Apart from the classical use as a discharge capacity tool, fractal PKW being very efficient at low heads could also be used to control water levels of river, channels of lakes at a given head. It would prevent the use of gates or mechanical machinery which implies maintenance issues and significant construction costs. The use of fractal PK weirs in wastewater flocculation tanks, where extremely low tank velocities allow flocculated particulates to settle out of solution, might represent another potential application. Labyrinth weirs are commonly used as the outlet flow control structure; with the long weir length being used to maintain low approach flow velocities. A fractal PKW might be more economical to construct than a labyrinth weir for that application. It might be required to install boom in case of risk of floating debris.

8 CONCLUSIONS

Fractal PKW are still a concept under development.

Numerical simulations, based on 8 different configurations, show that there is a real potential of benefit in terms of discharge capacity compared to usual PKW spillways. This benefit is approximately worth 100% for low heads such as 0.5 m. The benefit rapidly decreases for higher heads, being around 50% for 1 m head and 25% for 1,5 head highlighting a quick saturation of fractal PKW for higher heads.

Simulations show many relevant points:

– Simple crenels on the PKW crest are not efficient.
– 30° angle seems to be the a good compromise as far as the orientation of fractal PKW according to main PKW flow axis is concerned.
– Increasing the developed length (fractal ratio) is not efficient beyond a given value for a given head.

In addition the construction cost of the fractal PKW is probably minor compared to overall civil engineering cost of the main PKW unit. Being prefabricated in steel units or composite materials, they can be installed on the top crest of an usual concrete PKW whose walls would have been slightly enlarged (typically 40 to 50 cm thick walls).

Fractal PKW can be used as spillways. Nevertheless it seems that they could also be used as tools to control water level of rivers, channels or lakes.

Finally, the results given in this study are only partial and aim at providing preliminary advice on this new type of structure. In terms of discharge capacity per linear meter of developed free flow crest, it appears that fractal PKW might still be far from an optimized shape and that significant improvement might be expected from farther development.

REFERENCES

Lempérière, F. & Ouamane, A. 2003. The piano keys weir: a new cost-effective solutions for spillways. *Hydropower & Dams* (5): 144–149.
Laugier, F., Blancher, B., Guyot, G., Valette, E., Pralong, J., Breysse, M., 2009. Assesment of numerical flow model for standard and complex cases of free flow spillway discharge capacity, *Hydro 2009*, Lyon "Progress, potential, plans"
Pralong, J., Blancher, B., Laugier, F., Machiels, O., Erpicum, S., Pirotton, M., Leite Ribeiro, M., Boillat. J-L., & Schleiss, A.J. 2011. Proposal of a naming convention for the Piano Key Weir geometrical parameters. *International Workshop on Labyrinth and Piano Key weirs*, Liège, Belgium.
Pralong J., Montarros F., Blancher B. & Laugier. F. 2011. A sensitivity analysis of Piano Key Weirs geometrical parameters based on 3D numerical modeling *International Workshop on Labyrinth and Piano Key weirs*, Liège, Belgium.

Labyrinth and Piano Key Weirs – PKW 2011 – Erpicum et al. (eds)
© 2011 Taylor & Francis Group, London, ISBN 978-0-415-68282-4

General comments on labyrinths and Piano Key Weirs – The future

F. Lempérière & J.-P. Vigny
Hydrocoop, France

ABSTRACT: The first part of this article deals with comments on costs and economy. The future possible utilizations of the PKW are then listed. Lastly, the fields of enhancement are described for research and developments tests to come.

1 COMMENTS ON COSTS AND ECONOMY

1.1 *Costs and quantities*

Until now, PKW have been built to upgrade the discharge capacity of existing spillway. The adjunction of a PKW on existing dams includes annex works such as the construction of a heavy downstream concrete apron. And unit costs are much higher than for new dams. The cost ratios following are then to be used with care and each project deserves specific cost study.

- A PKW with a Pm of 2 m discharging 12 m^3/s/m for a head H of 2 m (instead of 6 m^3/s/m for a Creager sill) needs 3 m^3/m of reinforced concrete and 3 m^3/m of ordinary concrete, when the quantity of ordinary concrete for a corresponding Creager sill upon a gravity dam is in the range of 8 m^3/m
- A PKW with a Pm of 6 m discharging 63 m^3/s/m for a head H of 6 m (instead of 32 m^3/s/m for a Creager sill) needs 15 m^3/m of reinforced concrete and 32 m^3/m of ordinary concrete, when the quantity of ordinary concrete for a corresponding Creager sill upon a gravity dam is in the range of 70 m^3/m.

Compared to a Creager sill, a PKW requires about 0.5 m^3 of reinforced concrete and saves 1 m^3 of ordinary concrete to earn 1 m^3/s of extra discharge. On the other hand, a PKW requires 3 to 5 m^3 of reinforced concrete and saves 5 to 10 m^3 of ordinary concrete to earn 1 m of upstream head along 1 m of spillway.

1.2 *Economical optimization*

To evaluate at what extent it is economically reasonable to increase the performance, it is interesting to calculate (Fig. 1) the ratio "k" between the discharge coefficient of the PKW and the discharge coefficient of a Creager sill for an equal upstream water head and the corresponding cost increase.

Such economical limit may be different depending of the sought goal: to increase the discharge (which means to reduce the spillway length when keeping the same water level) or to increase the storage (which means to increase the normal water level when keeping the maximum water head and the same spillway length). For instance, when the ratio "k" is increased from 2 to 4, the discharge capacity is increased from 1 to 3. when the nape depth, which varies in proportion with $1/k^{2/3}$, decreases from about 60% of the nape depth of a Creager sill to 40% of it. Similarly, the saving on the length of the sill is 50% for $k = 2$ and 75% for $k = 4$, corresponding to an increase in savings of only 50%. In both cases, the savings (and benefits) increase by half when the saving in discharge is multiplied by 3.

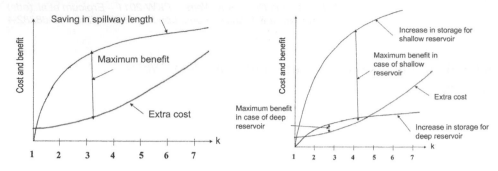

Figure 1.   Economical optimizations with constant water level (left) and constant spillway length (right).

To sum up:

- The increase in discharge is proportional to k − 1.
- The extra cost for a PKW versus a Creager is quickly increasing with the developed length ratio of the PKW (n). This extra cost is very roughly proportional to $(k-1)^{1.5}$ or $(k-1)^2$ plus a fix amount (studies, etc).

The figure 1 roughly indicate how vary these various parameters versus "k". Despite of the imprecision, it appears for a new dam that, most often:

- The benefit increases rapidly for k > 1.5 and is already important for k = 2 or 2.5 and will be close to the maximum for k = 3.
- A "k" value above 5 is usually not economically justified
- For a same saving in the nape depth, the increase in storage being more important for a shallow reservoir than for a deep one, the optimal "k" ratio will vary from about 2.5 to 4 or 5 with the depth of the reservoir.

2   POSSIBLE UTILIZATIONS

The main possible uses of Labyrinths and PKW are:

- Free flow spillways for new dams
- Additional free flow spillways to a gated spillway
- Upgrading of existing free flow spillways
- Run of river dams
- Morning glory spillways
- Other utilizations

For each target the possible utilization of following solutions is commented below:

- PKW A (upstream and downstream overhangs)
- PKW B (upstream overhang)
- PKW C (downstream overhang)
- PKW E (no overhang)

and sometimes compared with fuse devices (Fuse gates or Concrete fuse plugs).

2.1   Free flow spillways for new dams

The 4 models of PKW can be implemented by the dam contractor using the same equipment.

The model B (upstream overhang) may be the most interesting for large spillways but attention has to be paid to the floating debris. For embankment dams, it may be cost effective to use a small number of high elements and to reduce the width of the spillway.

The model C should be used in case of high hazard of floating debris.

Table 1. Dams and PKW descriptions.

P_m (m)	Q (m³/s)			Spillway width (m)		
	$N_u = 2$	$N_u = 5$	$N_u = 10$	$N_u = 2$	$N_u = 5$	$N_u = 10$
2	70	175	350	7.2	18	36
4	380	950	1900	14.4	36	72
6	1050	2575	5250	21.6	54	108
8	2200	5500	11000	28.8	72	144

To avoid overhangs when there are no experienced contractors, it is possible to modify the cross section of the spillway to obtain a labyrinth with vertical walls (model E).

For usual values of upstream head H between 0.5 Pm and Pm, the total discharge of one PKW element 1.8P_m wide is close to 1.8P_m × 4.5H × $\sqrt{P_m}$ that is to say 6$P_m^{2.5}$ for H = 0.75P_m.

For N_u elements, approximate values of total discharge and spillway length are as follow (Table 1):

It means that a flow of 100 m³/s may be discharged by 3 elements ($P_m = 2$ m) along 10 m. A flow of 1000 m³/s may be discharged by 2 to 5 elements ($P_m = 4$ to 6 m) along 20 to 35 m. 5000 m³/s may be discharged by 5 to 10 elements ($P_m = 6$ to 8 m) along 70 to 100 m.

PKW will be usually much more cost efficient than traditional Creager Weirs: For a cost of reinforced concrete of 500 to 1000 €/m³ and a cost of ordinary concrete of 100 to 200 €/m³, the cost per m of spillway for raising the storage by 1 m is 1 200 to 2 400 €/m of spillway i.e. 20 000 to 200 000 € for most spillways.

It will thus be possible, keeping the same reservoir level, to reduce the dam height and this will be cost effective for quite all dams, except for low concrete dams or for very small discharges for which concrete fuseplugs may be a safe solution at low cost. Another possibility is to reduce by over half the length of the spillway of an embankment dam.

Another use of PKW is to keep the same dam height and to increase the storage: The cost of extra storage will be usually in the range of 0.1 to 0.5 €/m³, which is much lower than the usual cost of new reservoirs.

For discharges up to 5 000 m³/s, PKW solution is probably the most cost effective. For larger discharges, fusegates, which require less reinforced concrete than PKW may be more cost effective than Creager Weirs. Tilting should be chosen for floods of yearly probability under 1/100.

For discharges over 10 000 m³/s, the best solution may be to associate a gated spillway for the 100 years flood and PKW or fusegates for the extra discharge.

In any case, PKW or fuse devices favour the choice of free flow spillways even for large discharges.

2.2 Additional free flow spillway to a gated spillway

This will apply essentially for large discharges. The nape depth will be usually part or whole of the freeboard. Using a free board of 3 to 5 m with rather high PKW A or B will allow a specific flow of 30 to 50 m³/s/m. This solution will operate also in case of gates jamming. It is much better than a Creager Weir limited to 10 or 20 m³/s/m.

Fusegates may be also an alternative especially for very high specific discharges (as example a solution with a 10 m high fusegate and a 4 m freeboard discharges after tilting up to 100 m³/s/m, similar to the specific flow of a large gated spillway).

2.3 Upgrading existing free flow spillways

There are 50 000 dams higher than 15 m in the world. Over 30 000 have free flow spillways; their usual capacity is between 100 and 1000 m³/s but sometimes over 10 000 m³/s. There are also 100 000 dams lower than 15 m storing over 100 000 m³ each (20 000 storing over 1 million m³ each) with free flow spillways usually in the range of 100 m³/s or few hundred m³/s. Most are in Asia. Dozens of thousands spillways may deserve extra capacity or extra storage.

Labyrinths with vertical walls may hardly be used and PKW A may be easier to build than PKW B. They may increase the specific discharge in m³/s by about 2 $P_m^{1.5}$ or the storage level by 0.4 or 0.5 P_m and be very cost effective.

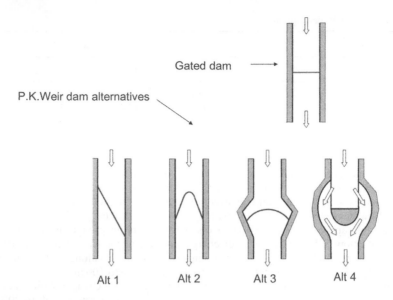

Gated dam

P.K.Weir dam alternatives

Alt 1   Alt 2   Alt 3   Alt 4

Figure 2.   Various possible layouts.

The cost will be over the cost for new dams because it includes various adjustments of the existing spillway and because unit costs are higher. The optimization requires specific studies.

Fusegates for large discharges and simple concrete fuse plugs for small ones may be alternatives but PKW keep the advantage to avoid tilting.

2.4   Run of river dams

2.4.1   Traditional solution and P.K. Weir alternative

Run of river dams may be chosen for large rivers in flat areas for navigation, irrigation or hydropower. The width of the river is usually 50 to 300 m and exceptional floods 1 000 to 15 000 m^3/s.

The reservoir level is often close to the natural level of extreme floods; i.e. a few meters over the banks. The head over the usual downstream level is in the range of 4 to 10 m.

The traditional solution is a gated dam across the river with often a specific flow of 20 to 40 m^3/s/m for extreme floods. It is associated with upstream dykes along the banks levelled at about 2 m over the reservoir level.

A cost effective alternative may be a free flow PKW dam with a length 2 or 3 fold the length of a gated dam, with specific flows for extreme floods 10 to 15 m^3/s/m (Fig. 2). If using PKW with $P_m = 4$ or 5 m and L/W of 5 or 6, the nappe depth is limited to 1 or 1.5 m for extreme floods. Model B may be the most attractive and should be optimized for these specific dams.

The cost per m of such PKW dam may be in the range of 20% of the cost per m of a gated dam (concrete volume much smaller, no gates, shorten construction period limited to one or two dry seasons). Even with a length 2 or 3 fold the length of a gated dam, PKW may thus reduce by half the cost of many run of river dams.

As shown on the figures here after, many alternatives are possible for the dam lay out for a same increased dam length: choice will be according to local conditions and construction methods.

2.4.2   Performance of drowned PKW

For run of river dams, PKW may be used with a maximum downstream level of the river higher than the top of the weir:

We call H_u and H_d the upstream and the downstream water levels over the PKW (Fig. 3). For a same specific flow, as long as the downstream level remains lower than or equal to the top of the weir, the upstream level is H_o (for $H_d \leq 0$, then $H_u = H_o$).

H_u increases when the downstream level is rising over the top of the weir by H_d, but this increase $H_u - H_o$ is much less than H_d.

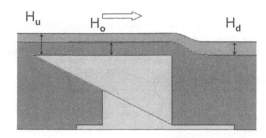

Results of model tests:

H_d/H_o	H_u/H_o	$\dfrac{H_u-H_d}{H_o}$
0	1	1
0.2	1.05	0.85
0.5	1.15	0.65
1	1.4	0.4
1.5	1.75	0.25

Figure 3.   Performance of drowned PKW.

Figure 4.   Morning glory spillways equipped with fusegates.

2.4.3   *Comparison between PKW and gated dam*

With a same reservoir level and a PKW dam twice longer than a gated dam:

- If the reservoir level is the natural level of the extreme flood, the upstream level will be about the same for both solutions.
- If the reservoir level is 1 m lower than the natural level of the extreme flood, the upstream level will be higher for the gated dam than for the PKW dam (about 1 m over the downstream level against 0.40 m).

Using PKW for most or all discharge of many run of rivers dams may well be very cost effective and modify totally the traditional design of such dams.

2.4.4   *Morning glory spillways*

Using PKW on existing or new Morning glory spillways may increase the reservoir level and thus the storage capacity.

The discharge is anyway limited by the capacity of the shaft but this capacity may possibly be increased by a better hydraulic efficiency of the shaft inlet linked with the PKW design.

The design of corresponding labyrinths will be rather different from PKW usual designs. Promising specific model tests have been made in EDF-LNHE laboratory (Chatou).

Two morning glory spillways equipped with fusegates have already been implemented (Fig. 4).

2.5   *Other utilizations*

The principle of labyrinths and PKW may apply also to small specific flows such as for irrigation canals or sewage tunnels. Low elements in the range of 1 m or under 1 m may be used, probably in standardized prefabricated concrete or steel elements. The costs and weighs may be low.

3 SOME POSSIBLE FUTURE STUDIES AND MODEL TESTS

PKW are cost efficient spillways and existing results of hydraulic tests are promising. However it is probably possible to optimize the models already studied and to adapt them to the very different dams data and targets. Other designs may be studied, especially for specific targets. Physical or Mathematical models may be used.

A first target may be the optimization of 4 models already studied:

- A model with 2 overhangs (A)
- A model with 1 upstream overhang (B)
- A model with 1 downstream overhang (C)
- A model with no overhang (E)

Many studies can be made by hydraulic or mathematical tests on the impact of modifying one parameter such as the width or the length or the longitudinal section, or the general spillway lay out and entrance, the thickness of walls, the impact of floating debris.

For models B, C and E the discharge could be more precised in a wide fume with at least 5 elements.

3.1 *New solutions*

A model with an upstream overhang longer than the downstream one may be of interest. Other shapes may be studied, especially for specific targets such as Morning Glory spillways. For run of river dams, the shape of model B may be significantly modified.

3.2 *New studies*

The downstream impact, such as the association with a stepped dam or the need of toe protection, deserve further studies such as already made in Vietnam. Impact of floating debris should be compared for various solutions.

Labyrinth and Piano Key Weirs – PKW 2011 – Erpicum et al. (eds)
© 2011 Taylor & Francis Group, London, ISBN 978-0-415-68282-4

Research axes and conclusions

S. Erpicum
Laboratory of Engineering Hydraulics – HACH, University of Liège, Belgium

J.-L. Boillat
Laboratory of Hydraulic Constructions (LCH), Ecole Polytechnique Fédérale de Lausanne (EPFL), Switzerland

ABSTRACT: The aim of this short paper is to summarize main outcomes of the workshop, in order to guide future developments of research on Piano Key Weirs (PKW) towards most effective and integrated directions. The main research topics highlighted in this text came out from the 34 high-quality presentations given during the conference, as well as from the fruitful discussions in which the participants were actively involved. Many more truly innovative and promising ideas can be found throughout the other papers gathered in this proceedings book.

1 INTRODUCTION

As a closure of the 1st International Workshop on Labyrinth and Piano Key Weirs, an integrative lecture has been given with the aim of highlighting most promising future research directions in order to further develop focused and coordinated work towards deeper understanding of flow characteristics on PKW and more innovations in their design. This paper aims at summarizing key issues addressed in this presentation. The topics identified hereafter came out as a result of the high level of the 34 technical presentations given during the conference, as well as from the interactive discussions in which workshop participants took a very active part. Far from exhaustive, this summary needs to be complemented by the many promising paths set in the technical papers presented in this book. In addition, some of the topics listed below have already been analyzed partly by different researchers, as detailed in their respective papers.

2 RESEARCH – PHYSICAL-NUMERICAL MODELING

Regarding research projects, much work remains to be undertaken concerning the comparison between the labyrinth weir and the Piano Key weir, using both physical and a numerical modeling approaches. The same applies for parametric studies of the PKW, which remain a challenging step towards a better support for finding the optimal design in each specific application.

In addition, more fundamental experimental research is needed to enhance our present understanding of the detailed flow conditions on a PKW, especially the different flow phases depending on the upstream head as well as the spatial patterns of pressure and velocity fields.

The related problems of floating debris, sediment deposition, sediment release and ice also need to be more deeply addressed, such as the restitution conditions, the energy dissipation and the effect of the downstream water level.

Scale effects in experimental studies constitute surely one of the key challenge of the coming years, as well as the advancement of numerical modeling approaches towards truly predictive capacities. Sharing feedback on the application of both physical and numerical modeling techniques may become a crucial asset in achieving breakthroughs as regards the abovementioned research topics.

Finally, defining a single widely-recognized protocol for testing PKW as well as the use of a reference model for calibration of experimental facility, would definitely contribute to maximizing the impact of the significant amount of experimental tests carried out in laboratories worldwide.

3 PROJECTS – PHYSICAL-NUMERICAL MODELING

In terms of projects studies, the key topic for progress remains the integration of PKW on dams, based on comprehensive evaluations accounting for abutment characteristics, real-life environment conditions (bridge piers, reservoir bathymetry...) and energy dissipation downstream. Structural aspects also need to be more deeply addressed, especially regarding the use of other materials than reinforced concrete.

4 REAL STRUCTURES/PROJECTS

A number of real PKW are now in operation and several more, among which very large projects are due to be completed within the next few years. Detailed monitoring of these structures and feedback about their behavior is crucial to support validation and improvement of current experimental research as well as numerical models. This is particularly important for scale effects assessment or respond to floating debris. Such feedback from the field is thus strongly needed to guide future research.

5 HOW TO BE EFFICIENT?

A workgroup on PKWs, bringing together researcher and engineers from Hydrocoop, EDF-CIH, EDF-R&D, LCH-EPFL, ULg-HACH and University of Biskra, has been active for two years through informal one day technical meetings organized every six months. This workgroup is open to everybody interested in sharing data and knowledge about PKW. Such meetings will continue to be regularly organized in the future.

It might also be interesting to set up a common research project on the topic of PKW, at least at the scale of the European community (EU funded project), to mobilize the means necessary to undertake a large coordinated research project.

Finally, in view of the success of the first workshop, it is quite sure that you will be invited to take part in a second edition in 2013!

Author index

Agresti, P. 251
Al Harty, S. 251
Anderson, R.M. 75
Archambeau, P. 59, 105, 151, 199, 279

Barcouda, M. 81
Bazerque, L. 251
Belaabed, F. 89
Ben Saïd, M. 67
Bieri, M. 123
Blancher, B. 133, 141, 159, 271, 281
Boillat, J.-L. 35, 113, 123, 183, 271, 279, 295
Boutet, J.-M. 233

Chi Hien, T. 191
Cicero, G.M. 81, 167, 207, 233
Crookston, B.M. 25

Das Singhal, G. 241
Daux, C. 105
Delorme, F. 123
Dewals, B. 59, 105, 151, 199
Dugué, V. 35

Erpicum, S. 43, 59, 105, 151, 199, 271, 279, 295

Faramond, L. 123, 215
Federspiel, M. 123

Gille, C. 215
Guilbaud, C. 251

Hachem, F. 35
Ho Ta Khanh, M. 191, 225
Houdant, B. 123

Kocahan, H. 261

Laugier, F. 35, 43, 113, 133, 141, 159, 183, 215, 271, 281
Le Blanc, M. 261
Le Doucen, O. 183
Leite Ribeiro, M. 113, 183, 271, 279
Lempérière, F. 17, 289
Lopes, R. 97
Luck, M. 81, 167

Machiels, O. 59, 105, 151, 199, 271
Matos, J. 97
Melo, J.F. 97
Menon, J.M. 167
Montarros, F. 133, 141, 281

Nagel, V. 35, 43
Noui, A. 175

Ouamane, A. 17, 51, 67, 89, 175

Pinchard, T. 167, 233
Pirotton, M. 59, 105, 151, 199, 271
Pralong, J. 133, 159, 271, 279, 281

Roca, J.-P. 35

Schleiss, A.J. 3, 113, 183, 271
Sharma, N. 241
Spinazzola, U. 261
Strobl, T. 251
Sy Quat, D. 225

Thanh Hai, N. 191
Tullis, B.P. 25, 75

Vermeulen, J. 215, 271
Vettori, E. 81
Vigny, J.-P. 17, 289

Xuan Thuy, D. 225

Printed and bound by CPI Group (UK) Ltd, Croydon, CR0 4YY

21/10/2024

01777095-0005